THE
THIRD
CHIMPANZEE

THE THIRD CHIMPANZEE

The Evolution and Future
OF THE
Human Animal

JARED DIAMOND

HarperPerennial
A Division of HarperCollinsPublishers

Designed by Ruth Kolbert

The Library of Congress has catalogued the hardcover edition as follows:
Diamond, Jared M.
    The third chimpanzee : the evolution and future of the human animal / Jared Diamond. — 1st ed.
    p.   cm.
    Includes bibliographical references and index.
    ISBN 0-06-018307-1 (cloth)
    1. Human evolution.   2. Social evolution.   3. Man—Influence on nature.   I. Title.
GN281.D53   1992
573.2—dc20   91-50455

ISBN 0-06-098403-1 (pbk.)

98 99 RRD-H 20 19

Dedicated to my sons
Max and Joshua,
to help them understand
where we came from
and where we may be heading

Theme

How the human species changed, within a short time,
from just another species of big mammal
to a world conqueror;
and how we acquired the capacity
to reverse all that progress overnight

CONTENTS

Prologue

IT'S OBVIOUS THAT humans are unlike all animals. It's also obvious that
we're a species of big mammal, down to the minutest details of our
anatomy and our molecules. That contradiction is the most fascinating feature of the human species. It's familiar, but we still have
difficulty grasping how it came to be and what it means.

On the one hand, between us and all other species lies a seemingly
unbridgeable gulf that we acknowledge by defining a category called
"animals." It implies that we consider centipedes, chimpanzees, and
clams to share decisive features with each other but not with us, and
to lack features restricted to us. Among those unique characteristics
of ours, we talk, write, and build complex machines. We depend on
tools, not on our bare hands, to make a living. Most of us wear clothes
and enjoy art, and many of us believe in a religion. We are distributed
over the whole Earth, command much of its energy and production,
and are beginning to expand into the ocean depths and into Space.
We're also unique in darker behaviors, including genocide, delight in
torture, addiction to toxic drugs, and extermination of other species
by the thousands. While a few animal species have one or two of

these behaviors in rudimentary form (like tool use), we still far eclipse animals in even those respects.

Thus, for practical and legal purposes, humans aren't considered animals. When Darwin proposed in 1859 that we had evolved from apes, it's no wonder that most people initially regarded his theory as absurd and continued to insist that we had been separately created by God. Many people, including a quarter of all American college graduates, still hold to that belief today.

But, on the other hand, we obviously are animals, with the usual animal body parts, molecules, and genes. It's even clear what particular type of animal we are. Externally, we're so similar to chimpanzees that eighteenth-century anatomists who believed in divine creation could already recognize our affinities. Just imagine taking some normal people, stripping off their clothes, taking away all their other possessions, depriving them of the power of speech, and reducing them to grunting, without changing their anatomy at all. Put them in a cage in the zoo next to the chimp cages, and let the rest of us clothed and talking people visit the zoo. Those speechless caged people would be seen for what we all really are: chimps that have little hair and walk upright. A zoologist from Outer Space would immediately classify us as just a third species of chimpanzee, along with the pygmy chimp of Zaire and the common chimp of the rest of tropical Africa.

Molecular genetic studies of the last half-dozen years have shown that we continue to share over 98 percent of our genetic program with the other two chimps. The overall genetic distance between us and chimps is even smaller than the distance between such closely related bird species as red-eyed and white-eyed vireos. Thus, we still carry most of our old biological baggage with us. Since Darwin's time, fossilized bones of hundreds of creatures variously intermediate between apes and modern humans have been discovered, making it impossible for a reasonable person to deny the overwhelming evidence. What once seemed absurd—our evolution from apes—actually happened.

Yet the discoveries of many missing links have only made the problem more fascinating, without fully solving it. The few bits of new baggage we acquired—the 2 percent difference between our genes and those of chimps—must have been responsible for all of our seemingly unique properties. We underwent some small changes with big consequences rather quickly and recently in our evolutionary history. In fact, as recently as a hundred thousand years ago that zoologist from Outer Space would have viewed us as just one more species of big mammal. Granted, we had a couple of curious behaviors, notably our control of fire and our dependence on tools. But those behaviors would have seemed no more curious to the extraterrestrial visitor than would the behaviors of beavers and bowerbirds. Somehow, within a few tens of thousands of years—a period that is almost infinitely long when measured against one person's memory but is only a tiny fraction of our species' separate history—we had begun to demonstrate the qualities that make us unique and fragile.

What were those few key ingredients that made us human? Since our unique properties appeared so recently and involved so few changes, those properties or at least their precursors must already be present in animals. What are those animal precursors of art and language, of genocide and drug abuse?

OUR UNIQUE QUALITIES have been responsible for our present biological success as a species. No other large animal is native to all the continents, or breeds in all habitats from deserts and the Arctic to tropical rainforests. No large wild animal rivals us in numbers. But among our unique qualities are two that now jeopardize our existence: our propensities to kill each other and to destroy our environment. Of course, both propensities occur in other species: lions and many other animals kill their own kind, while elephants and others damage their environment. However, these propensities are much more threatening in us than in other animals because of our technological power and exploding numbers.

There's nothing new about prophecies to the effect that the world's end is near if we don't repent. What's new is that this prophecy is now likely to come true, for two obvious reasons. First, nuclear weapons give us the means to wipe ourselves out quickly; no humans possessed this means before. Second, we already appropriate about 40 percent of the Earth's net productivity (i.e., the net energy captured from sunlight). With the world's human population now doubling every forty-one years, we soon shall reach the biological limit to growth, at which point we shall have to start fighting each other in dead earnest for a share of the world's fixed pie of resources. In addition, given the present rate at which we are exterminating species, most of the world's species will become extinct or endangered within the next century, but we depend on many species for our own life support.

Why rehearse these familiar depressing facts? And why try to trace the animal origins of our destructive qualities? If they really are part of our evolutionary heritage, that seems to say that they are genetically fixed and hence unchangeable.

In fact, our situation is not hopeless. Perhaps the urge to murder strangers or sexual rivals is innate in us. But that still hasn't prevented human societies from attempting to thwart those instincts, and from succeeding in sparing most people the fate of being murdered. Even taking two world wars into account, proportionately far fewer people have died violent deaths in twentieth-century industrialized states than in Stone Age tribal societies. Many modern populations enjoy longer life spans than did humans of the past. Environmentalists don't always lose in battles with developers and destroyers. Even some genetic infirmities, such as phenylketonuria and juvenile-onset diabetes, can now be mitigated or cured.

My purpose in rehearsing our situation is to help us avoid repeating our mistakes—to use knowledge of our past and our propensities in order to change our behavior. That's the hope behind the dedication of this book. My twin sons were born in 1987 and will reach my present age in the year 2041. What we are doing now is shaping their world.

It is not the goal of this book to propose specific solutions to our predicament, because the solutions we should adopt are already clear in broad outline. Some of those solutions include halting population growth, limiting or eliminating nuclear weapons, developing peaceful means for solving international disputes, reducing our impact on the environment, and preserving species and natural habitats. Many excellent books make detailed proposals for how to carry out these policies. Some of these policies are being implemented in some cases now; we "just" need to implement them consistently. If we all became convinced today that they were essential, we would already know enough to start carrying them out tomorrow.

Instead, what is lacking is the necessary political will. Through this book I seek to foster that will, by tracing our history as a species. Our problems have deep roots tracing back to our animal ancestry. They have been growing for a long time with our growing power and numbers, and are now steeply accelerating. We can convince ourselves of the inevitable outcome of our current shortsighted practices just by examining the many past societies that destroyed themselves by destroying their own resource base, despite having less potent means of self-destruction than ours. Political historians justify the study of individual states and rulers by the resulting opportunity to learn from the past. That justification applies even more to the study of our history as a species, because the lessons of that study are simpler and clearer.

A VOLUME that ranges over such a broad canvas as this one has to be selective. Every reader is bound to find some absolutely crucial favorite subjects omitted, some other subjects pursued in inordinate detail. So that you won't feel you were misled, I'll lay out at the start my own particular interests, and where they come from.

My father is a physician, my mother a musician with a gift for languages. Whenever I was asked as a child about my career plans, my response was that I wanted to be a doctor like my father. By my last year in college, that goal had become gently transformed into the related goal of medical research. And so I trained in physiology, the area in which I now teach and do research at the University of California Medical School in Los Angeles.

However, at the age of seven I had also become interested in bird-watching, and I had been fortunate to go to a school that let me delve into languages and history. After I got my Ph.D., the prospect of devoting the rest of my life to the single professional interest of physiology began to look increasingly oppressive. At that point, a happy constellation of events and people gave me the chance to spend a summer in the highlands of New Guinea. Ostensibly, the purpose of my trip was to measure nesting success of New Guinea birds, a project that collapsed dismally within a few weeks when I found myself unable to locate even a single bird nest in the jungle. Yet the real purpose of the trip succeeded completely: to indulge my thirst for adventure and bird-watching in one of the wildest remaining parts of the world. What I saw then of New Guinea's fabulous birds, including its bowerbirds and birds of paradise, led me to develop a parallel second career, in bird ecology, evolution, and biogeography. Since then, I've returned to New Guinea and neighboring Pacific islands a dozen times to pursue my bird research.

But I found it hard to work in New Guinea amid the accelerating destruction of the birds and forests that I loved, without getting involved in conservation biology. So I began to combine my academic research with practical work as a consultant for governments, by applying what I knew about animal distributions to designing national park systems and surveying proposed national parks. It was also hard to work in New Guinea, where languages replace each other every twenty miles, and where learning bird names in each local language proved to be the key to tapping New Guineans' encyclopedic knowledge of their birds, without returning to my earlier interest in languages. Most of all, it was hard to study the evolution and extinction of bird species without wanting to understand the evolution and possible extinction of *Homo sapiens,* by far the most interesting species of all. That interest, too, was especially hard to ignore in New Guinea, with its enormous human diversity.

Those are the paths by which I came to be interested in the particular aspects of humans that are emphasized in this book. Numerous excellent books by anthropologists and archaeologists already discuss human evolution in terms of tools and bones, which this book can therefore summarize briefly. However, those other volumes devote much less space to my particular interests of the human life cycle, human geography, human impacts on the environment, and humans as animals. Those subjects are as central to human evolution as are the more traditional subjects involving tools and bones.

I believe that what may at first seem to be a plethora of examples drawn from New Guinea is appropriate. Granted, New Guinea is just one island, located in a particular part of the world (the tropical Pacific), and hardly providing a random cross-section of modern humanity. But New Guinea harbors a much bigger slice of humanity than you would at first guess from its area. About a thousand of the world's approximately five thousand languages are spoken only in New Guinea. Much of the cultural diversity that survives in the modern world is contained within New Guinea. All highland peoples in New Guinea's mountainous interior were Stone Age farmers until very recently, while many lowland groups were nomadic hunter-gatherers and fishermen practicing somewhat casual agriculture. Local xenophobia was extreme, cultural diversity correspondingly so, and travel outside of one's tribal territory would have been suicidal. Many of the New Guineans who have worked with me are deadly expert hunters who lived out their childhood in the days of stone tools and xenophobia. Thus, New Guinea is as good a model as we have left today of what much of the rest of the human world was once like.

THE STORY of our rise and fall divides itself into five natural parts. In the first part I'll follow us from several million years ago until just before agriculture's appearance ten thousand years ago. These two chapters deal with the evidence of bones, tools, and genes—the evidence that is preserved in the archaeological and biochemical record, and that gives us our most direct information about how we have changed. Fossilized bones and tools can often be dated, permitting us to deduce in addition just when we changed. We'll examine the basis of the conclusion that we're still 98 percent chimps in our genes, and we'll try to figure out what difference of 2 percent was responsible for our great leap forward.

The second part deals with changes in the human life cycle, which were as essential to the development of language and art as were the skeletal changes discussed in Part One. It's restating the obvious to mention that we feed our children after the age of weaning, instead of leaving them to find food on their own; that most adult men and women associate in couples; that most fathers as well as mothers care for their children; that many people live long enough to experience grandchildren; and that women undergo menopause. To us, these traits are the norm, but by the standards of our closest animal relatives they are bizarre. They constitute major changes from our ancestral condition, though they don't fossilize and so we don't know when they arose. For that reason they receive much briefer treatment in books on human paleontology than do our changes in brain size and pelvis. But they were crucial to our uniquely human cultural development, and merit equal attention.

With Parts One and Two thus having surveyed the biological underpinnings of our cultural flowering, Part Three proceeds to consider the cultural traits that we consider as distinguishing us from animals. Those that come first to mind are the ones of which we are proudest: language, art, technology, and agriculture, the hallmarks of our rise. Yet our distinguishing cultural traits also include black marks on our record, such as abuse of toxic chemicals. While one can debate whether all these hallmarks rank as uniquely human, they at least constitute huge advances on animal precursors. But animal precursors there must have been, since these traits flowered only recently on an evolutionary time scale. What were those precursors? Was their flowering inevitable in the history of life on Earth? For example, so inevitable that we suspect there to be many other planets out in Space, inhabited by creatures as advanced as we are?

Besides chemical abuse, our black traits include two so serious that they may lead to our fall. Part Four considers the first of these: our propensity for xenophobic killing of other human groups. This trait has direct animal precursors—namely, the contests between competing individuals and groups that, in many species besides our own, may be resolved by murder. We've merely used our technological prowess to improve our killing power. In Part Four we'll consider the xenophobia and extreme isolation that marked the human condition before the rise of political states began to make us more homogeneous culturally. We'll see how technology, culture, and geography affected the outcome of two of the most familiar historical sets of contests between human groups. We'll then survey the worldwide recorded history of xenophobic mass murder. This is a painful subject, but here above all is an example of how our refusal to face up to our history condemns us to repeat past mistakes on a more dangerous scale.

The other black trait that now threatens our survival is our accelerating assault on our environment. This behavior too has its direct animal precursors. Animal populations that for one reason or another escaped control by predators and parasites have in some cases also escaped their own internal controls on their numbers, multiplied until they damaged their resource base, and occasionally eaten their way into extinction. Such a risk applies with special force to humans, because predation on us is now negligible, no habitat is beyond our influence, and our power to kill individual animals and destroy habitats is unprecedented.

Unfortunately, many people still cling to the Rousseauian fantasy that this behavior did not appear in us until the Industrial Revolution, before which we lived in harmony with Nature. If that were true, we would have nothing to learn from the past except how virtuous we once were and how evil we have now become. Part Five seeks to dismantle this fantasy by facing up to our long history of environmental mismanagement. In Part Five as in Part Four, the emphasis is on recognizing that our present situation is not novel, except in degree. The experiment of trying to manage a human society while mismanaging its environment has already been run many times, and the outcome is there for us to learn from.

This book concludes with an epilogue that traces our rise from animal status. It also traces the acceleration in our means to bring about our fall. I wouldn't have written this book if I had thought that the risk was remote, but I also wouldn't have written it if I had considered us doomed. Lest any readers get so discouraged by our track record and present predicament that they overlook this message, I point out the hopeful signs and the ways in which we can learn from the past.

PART ONE

JUST ANOTHER
SPECIES OF
BIG MAMMAL

The clues about when, why, and in what ways we ceased to be just another species of big mammal come from three types of evidence. Part One considers some of the traditional evidence from archaeology, which studies fossil bones and preserved tools, plus newer evidence from molecular biology.

One basic question concerns just how extensive the genetic differences between us and chimps are. That is, do our genes differ by 10, 50, or 99 percent from chimpanzee genes? Merely eyeballing humans and chimps or counting up visible traits wouldn't be any help, because many genetic changes have no visible effects at all, while other changes have sweeping effects. For example, the visible differences between breeds of dogs such as Great Danes and Pekinese are far greater than those between chimps and us. Yet all dog breeds are interfertile, breed with each other (insofar as it's mechanically feasible) when given the opportunity, and belong to the same species. To a naïve observer, a glance at Great Danes and Pekinese would have suggested them to be genetically much further apart than chimps are from humans. Those visible differences among dog breeds in size, proportions, and hair color depend on relatively few genes, which have negligible consequences for reproductive biology.

How, then, can we estimate our genetic distance from chimps? This problem has been solved only within the past half-dozen years, by molecular biologists. The answer is not just intellectually surprising but may also have some practical ethical implications for how we treat chimps. We'll see that gene differences between us and chimps, although large compared to those among living human populations or among breeds of dogs, are still small compared to differences among many other familiar pairs of related species. Evidently, changes in only a small percentage of the chimpanzee genetic program had enormous consequences for our behavior. It has also proved possible to work out a calibration between genetic distance and elapsed time, and thereby to get an approximate answer to the question of when we and chimps split apart from our common ancestor. That turns out to be somewhere around seven million years ago, give or take a few million years.

While these molecular biological results yield overall measures of genetic distance and elapsed time, they tell us nothing about how specifically we differ from chimps, and when those specific differences appeared. Hence we'll go on to consider what more can be learned from bones and tools left by creatures variously intermediate between our apelike ancestor and modern humans. The changes in bones constitute the traditional subject matter of physical anthropology. Especially important were our increase in brain size, skeletal changes associated with walking upright, and decreases in skull thickness, tooth size, and jaw muscles.

Our large brain was surely prerequisite for the development of human language and innovativeness. One might therefore expect the fossil record to show a close parallel between increased brain size and sophistication of tools. In fact, the parallel is not at all close. This proves to be the greatest surprise and puzzle of human evolution. Stone tools remained very crude for hundreds of thousands of years after we had undergone most of our expansion of brain size. As recently as forty thousand years ago, Neanderthals had brains even larger than those of modern humans, yet their tools show no signs of innovativeness and art. Neanderthals were still just another species of big mammal. Even for tens of thousands of years after some other human populations had achieved virtually modern skeletal anatomy, their tools remained as boring as those of Neanderthals.

These paradoxes sharpen the conclusion drawn from the evidence of molecular biology. Within that modest percentage of difference between our genes and chimpanzee genes, there must have been an even smaller percentage not involved in the shapes of our bones but responsible for the distinctively human attributes of innovation, art, and complex tools. At least in Europe, those attributes appeared unexpectedly suddenly, at the time of the replacement of Neanderthals by Cro-Magnons. That is the time when we finally ceased to be just another species of big mammal. At the end of Part One I'll speculate about what those few changes were that triggered our steep rise to human status.

CHAPTER I

A Tale of Three Chimps

T HE NEXT TIME YOU VISIT A ZOO, MAKE A POINT OF WALKING PAST the ape cages. Imagine that the apes had lost most of their hair, and imagine a cage nearby holding some unfortunate people who had no clothes and couldn't speak but were otherwise normal. Now try guessing how similar those apes are to us in their genes. For instance, would you guess that a chimpanzee shares 10 percent, 50 percent, or 99 percent of its genetic program with humans?

Then ask yourself why those apes are on exhibit in cages, and why other apes are being used for medical experiments, while it's not permissible to do either of those things to humans. Suppose it turned out that chimp genes were 99.9 percent identical to our genes, and that the important differences between humans and chimps were due to just a few genes. Would you still think it's okay to put chimps in cages and to experiment on them? Consider those unfortunate mentally defective people who have much less capacity to solve problems, to care for themselves, to communicate, to engage in social relationships, and to feel pain than do apes. What is the logic that forbids medical experiments on those people, but not on apes?

You might answer that apes are "animals," while humans are humans, and that's enough. An ethical code for treating humans shouldn't be extended to an "animal," no matter how similar its genes are to ours, and no matter what its capacity for social relationships or feeling pain. That's an arbitrary but at least self-consistent answer that can't be lightly dismissed. In that case, learning more about our ancestral relationships won't have any ethical consequences, but it will still satisfy our intellectual curiosity to understand where we come from. Every human society has felt a deep need to make sense of its origins, and has answered that need with its own story of the Creation. The Tale of Three Chimps is the Creation Story of our time.

FOR CENTURIES it's been clear approximately where we fit into the animal kingdom. We are obviously mammals, the group of animals characterized by having hair, nursing their young, and other features. Among mammals we are obviously primates, the group of mammals including monkeys and apes. We share with other primates numerous traits lacking in most other mammals, such as flat fingernails and toenails rather than claws, hands for gripping, a thumb that can be opposed to the other four fingers, and a penis that hangs free rather than being attached to the abdomen. Already by the second century A.D., the Greek physician Galen deduced our approximate place in nature correctly when he dissected various animals and found that a monkey was "most similar to man in viscera, muscles, arteries, veins, nerves, and in the form of bones."

It's also easy to place us within the primates, among which we are obviously more similar to apes (the gibbons, orangutan, gorilla, and chimpanzees) than to monkeys. To name only one of the most visible signs, monkeys sport tails, which we lack along with apes. It's also clear that gibbons, with their small size and very long arms, are the most distinctive apes, and that orangutans, chimpanzees, gorillas, and humans are all more closely related to each other than any is to gibbons. But to go further with our relationships proves unexpectedly difficult. It has provoked an intense scientific debate, which revolves around three questions:

What is the detailed family tree of relationships among humans, the living apes, and extinct ancestral apes? For example, which of the living apes is our closest relative?

When did we and that closest living relative, whichever ape it is, last share a common ancestor?

What fraction of our genetic program do we share with that closest living relative?

At first, it would seem natural to assume that comparative anatomy had already solved the first of those three questions. We look especially like chimpanzees and gorillas, but differ from them in obvious features like our larger brains, upright posture, and much less body hair, as well as in many subtler points. However, on closer examination these anatomical facts aren't decisive. Depending on what anatomical characters one considers most important and how one interprets them, biologists differ as to whether we are most closely related to the orangutan (the minority view), with chimps and gorillas having branched off our family tree before we split off from orangutans, or whether we are instead closest to chimps and gorillas (the majority view), with the ancestors of orangutans having gone their separate way earlier.

Within the majority, most biologists have thought that gorillas and chimps are more like each other than either is like us, implying that we branched off before the gorillas and chimps diverged from each other. This conclusion reflects the commonsense view that chimps and gorillas can be lumped in a category termed "apes," while we're something different. However, it's also conceivable that we look distinct only because chimps and gorillas haven't changed much since we shared a common ancestor with them, while we were changing greatly in a few important and highly visible features like upright posture and brain size. In that case, humans might be most similar to gorillas, or humans might be most similar to chimps, or humans and gorillas and chimps might be roughly equidistant from each other in overall genetic makeup.

Thus, anatomists have continued to argue about the first question, the details of our family tree. Whichever tree one prefers, anatomical studies by themselves tell us nothing about the second and third questions, our time of divergence and genetic distance from apes. Perhaps, however, fossil evidence might in principle solve the questions of the correct ancestral tree and of dating, though not the question of genetic distance. That is, if we had abundant fossils, we might hope to find a series of dated protohuman fossils and another series of dated protochimp fossils converging on a common ancestor around ten million years ago, converging in turn on a series of protogorilla fossils twelve million years ago. Unfortunately, that hope for insight from the fossil record has also been frustrated, because almost no ape fossils of any sort have been found for the crucial relevant period between five and fourteen million years ago in Africa.

THE SOLUTION to these questions about our origins came from an unexpected direction: molecular biology as applied to bird taxonomy. About thirty years ago, molecular biologists began to realize that the chemicals of which plants and animals are composed might provide "clocks" by which to measure genetic distances and to date times of evolutionary divergence. The idea is as follows. Suppose there is some class of molecules that occurs in all species, and whose particular structure in each species is genetically determined. Suppose further that that structure changes slowly over the course of millions of years because of genetic mutations, and that the rate of change is the same in all species. Two species derived from a common ancestor would start off with identical forms of the molecule, which they inherited from that ancestor. But mutations would then occur independently and produce structural changes between the molecules of the two species. Thus, the two species' versions of the molecule would gradually diverge in structure. If we knew how many structural changes occur on the average every million years, we could then use the present difference in the molecule's structure between any two related animal species as a clock, to calculate how much time had passed since the species shared a common ancestor.

For instance, suppose one knew from fossil evidence that lions and tigers diverged five million years ago. Suppose the molecule in lions was 99 percent identical in structure to the corresponding molecule in tigers and differed only by 1 percent. If one then took a pair of species of unknown fossil history and found that the molecule differed by 3 percent between those two species, the molecular clock would say that they had diverged three times five million, or fifteen million, years ago.

Neat as this scheme sounds on paper, testing whether it succeeds in practice has cost biologists much effort. Four things had to be done before molecular clocks could be applied: scientists had to find the best molecule; find a quick way of measuring changes in its structure; prove that the clock runs steady (i.e., that the molecule's structure really does evolve at the same rate among all species that one is studying); and measure what that rate is.

Molecular biologists solved the first two of these problems by around 1970. The best molecule proved to be deoxyribonucleic acid (abbreviated DNA), the famous substance whose structure James Watson and Francis Crick showed to consist of a double helix, thereby revolutionizing the study of genetics. DNA is made up of two complementary and extremely long chains, each made up of four types of small molecules whose sequence within the chain carries all the genetic information transmitted from parents to offspring. A quick method of measuring changes in DNA structure is to mix the DNA from two species, then measure by how many degrees of temperature the melting point of the mixed (hybrid) DNA is reduced below the melting point of pure DNA from a single species. The method is generally referred to as DNA hybridization. As it turns out, a melting point lowered by one degree centigrade (abbreviated: $\Delta T = 1° C$) means that the DNAs of the two species differ by roughly 1 percent.

IN THE 1970s most molecular biologists and most taxonomists had little interest in each other's work. Among the few taxonomists who appreciated the potential power of the new DNA hybridization technique was Charles Sibley, an ornithologist then serving as Professor of Ornithology and Director at Yale's Peabody Museum of Natural History. Bird taxonomy is a difficult field because of the severe anatomical constraints imposed by flight. There are only so many ways to design a bird capable, say, of catching insects in midair, with the result that birds of similar habits tend to have very similar anatomies, whatever their ancestry. For example, American vultures look and behave much like Old World vultures, but biologists have come to realize that the former are related to storks, the latter to hawks, and that their resemblances result from their common life-style. Frustrated by the shortcomings of traditional methods for deciphering bird relationships, Sibley and Jon Ahlquist turned in 1973 to the DNA clock, in the most massive application to date of the methods

of molecular biology to taxonomy. Not until 1980 were Sibley and
Ahlquist ready to begin publishing their results, which eventually
came to encompass applying the DNA clock to about seventeen
hundred bird species—nearly one-fifth of all living birds.

While Sibley's and Ahlquist's achievement was a monumental one,
it initially caused much controversy because so few other scientists
possessed the mixture of expertises required to understand it. Here
are typical reactions I heard from my scientist friends:

"I'm sick of hearing about that stuff. I no longer pay attention to
anything those guys write." (An anatomist.)

"Their methods are okay, but why would anyone want to do
something so boring as all that bird taxonomy?" (A molecular biologist.)

"Interesting, but their conclusions need a lot of testing by other
methods before we can believe them." (An evolutionary biologist.)

"Their results are the Revealed Truth, and you better believe it."
(A geneticist.)

My own assessment is that the last view will prove to be the most
nearly correct one. The principles on which the DNA clock rests are
unassailable; the methods used by Sibley and Ahlquist are state-of-
the-art; and the internal consistency of their genetic distance measures from over eighteen thousand hybrid pairs of bird DNA testifies
to the validity of their results.

Just as Darwin had the good sense to marshal his evidence for
variation in barnacles before discussing the explosive subject of human variation, Sibley and Ahlquist similarly stuck to birds for most
of the first decade of their work with the DNA clock. Not until 1984
did they publish their first conclusions from applying the same DNA
methods to human origins, and they refined their conclusions in later
papers. Their study was based on DNA from humans and from all
of our closest relatives: the common chimpanzee, pygmy chimpanzee,
gorilla, orangutan, two species of gibbons, and seven species of Old
World monkeys. Figure 1 summarizes the results.

As any anatomist would have predicted, the biggest genetic difference, expressed in a big DNA melting-point lowering, is between
monkey DNA and the DNA of humans or of any ape. This simply
puts a number on what everybody has agreed ever since apes first
became known to science: that humans and apes are more closely
related to each other than either are to monkeys. The actual number

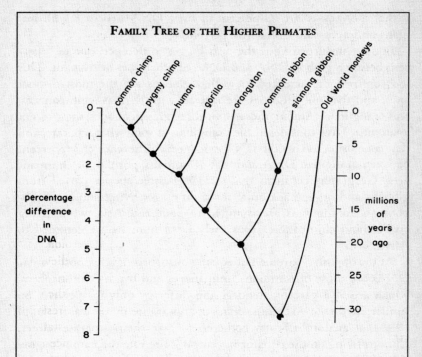

Figure 1. Trace back each pair of modern higher primates to the black dot connecting them. The numbers to the left then give the percentage difference between the DNAs of those modern primates, while the numbers to the right give the estimated number of millions of years ago since they last shared a common ancestor. For example, the common and pygmy chimps differ in about 0.7 percent of their DNA and diverged around three million years ago; we differ in 1.6 percent of our DNA from either chimp and diverged from their common ancestor around seven million years ago; and gorillas differ in about 2.3 percent of their DNA from us or chimps and diverged from the common ancestor leading to us and the two chimps around ten million years ago.

is that monkeys share 93 percent of their DNA structure with humans and apes, and differ in 7 percent.

Equally unsurprising is the next-biggest difference, one of 5 percent between gibbon DNA and DNA of other apes or humans. This too confirms the accepted view that gibbons are the most distinct apes, and that our affinities are instead with gorillas, chimpanzees, and orangutans. Among those latter three groups of apes, most recent anatomists have considered the orangutan as somewhat separate, and that conclusion too fits the DNA evidence: a difference of 3.6 percent between orangutan DNA and that of humans, gorillas, or chimpanzees. Geography confirms that the latter three species parted from gibbons and orangutans quite some time ago: living and fossil gibbons and orangutans are confined to Southeast Asia, while living gorillas and chimpanzees plus early fossil humans are confined to Africa.

At the opposite extreme but equally unsurprising, the most similar DNAs are those of common chimpanzees and pygmy chimpanzees, which are 99.3 percent identical and differ by only 0.7 percent. So similar are these two chimp species in appearance that it was not until 1929 that anatomists even bothered to give them separate names. Chimps living on the equator in central Zaire rate the name "pygmy chimps" because they average slightly smaller (and have more slender builds and longer legs) than the widespread "common chimps" ranging across Africa just north of the equator. However, with the increased knowledge of chimp behavior acquired in recent years, it has become clear that the modest anatomical differences between pygmy and common chimps mask considerable differences in reproductive biology. Unlike common chimps but like us, pygmy chimps assume a wide variety of positions for copulation, including face to face; copulation can be initiated by either sex, not just by the male; females are sexually receptive for much of the month, not just for a briefer period in midmonth; and there are strong bonds among females or between males and females, not just among males. Evidently, those few genes (0.7 percent) differing between pygmy and common chimps have big consequences for sexual physiology and roles. That same theme—a small percentage of gene differences having big consequences—will recur later in this and the next chapter in regard to the gene differences between humans and chimps.

In all these cases that I've discussed so far, anatomical evidence of relationships was already convincing, and the DNA-based conclusions confirmed what the anatomists had already concluded. But DNA was also able to resolve the problem at which anatomy had failed—the relationships between humans, gorillas, and chimpanzees. As Figure 1 shows, humans differ from either common chimps or pygmy chimps in about 1.6 percent of their (our) DNA, and share 98.4 percent. Gorillas differ somewhat more, by about 2.3 percent, from us or from either of the chimps.

Let's pause to let some of the implications of these momentous numbers sink in:

The gorilla must have branched off from our family tree slightly before we separated from the common and pygmy chimpanzees. The chimpanzees, not the gorilla, are our closest relatives. Put another way, the chimpanzees' closest relative is not the gorilla but humans. Traditional taxonomy has reinforced our anthropocentric tendencies by claiming to see a fundamental dichotomy between mighty man, standing alone on high, and the lowly apes all together in the abyss of bestiality. Now, future taxonomists may see things from the chimpanzees' perspective: a weak dichotomy between slightly higher apes (the *three* chimpanzees, including the "human chimpanzee") and slightly lower apes (gorillas, orangutans, gibbons). The traditional distinction between "apes" (defined as chimps, gorillas, etc.) and humans misrepresents the facts.

The genetic distance (1.6 percent) separating us from pygmy or common chimps is barely double that separating pygmy from common chimps (0.7 percent). It's less than that between two species of gibbons (2.2 percent), or between such closely related North American bird species as red-eyed vireos and white-eyed vireos (2.9 percent). The remaining 98.4 percent of our DNA is just normal chimp DNA. For example, our principal hemoglobin, the oxygen-carrying protein that gives blood its red color, is identical in all of its 287 units with chimp hemoglobin. In this respect as in most others, we are just a third species of chimpanzee, and what's good enough for common and pygmy chimps is good enough for us. Our important visible distinctions from the other chimps—our upright posture, large brains, ability to speak, sparse body hair, and peculiar sexual lives—must be concentrated in a mere 1.6 percent of our genetic program.

If genetic distances between species accumulated at a uniform rate with time, they would function as a smoothly ticking clock. All that would be required to convert genetic distance into absolute time since the last common ancestor would be a calibration, furnished by a pair of species for which we know *both* the genetic distance *and* the time of divergence as dated independently by fossils. In fact, two independent calibrations are available for higher primates. On the one hand, monkeys diverged from apes between twenty-five and thirty million years ago according to fossil evidence, and now differ in about 7.3 percent of their DNA. On the other hand, orangutans diverged from chimps and gorillas between twelve and sixteen million years ago and now differ in about 3.6 percent of their DNA. Comparing these two examples, a doubling of evolutionary time, as one goes from twelve-to-sixteen to twenty-five-to-thirty million years, leads to a doubling of genetic distance (3.6 to 7.3 percent of DNA). Thus, the DNA clock has ticked relatively steadily among higher primates.

With those calibrations, Sibley and Ahlquist estimated the following time scale for our evolution. Since our own genetic distance from chimps (1.6 percent) is about half the distance of orangutans from chimps (3.6 percent), we must have been going our separate way for about half of the twelve to sixteen million years that orangutans had to accumulate their genetic distinction from chimps. That is, the human and "other chimp" evolutionary lines diverged around six to eight million years ago. By the same reasoning, gorillas parted from the common ancestor of us three chimpanzees around nine million years ago, and the pygmy and common chimps diverged around three million years ago. In contrast, when I took physical anthropology as a college freshman in 1954, the assigned textbooks said that humans diverged from apes fifteen to thirty million years ago. Thus, the DNA clock strongly supports a controversial conclusion also drawn from several other molecular clocks based on amino-acid sequences of proteins, mitochondrial DNA, and globin pseudogene DNA. Each clock indicates that humans have had only a short history as a species distinct from other apes, much shorter than paleontologists used to assume.

WHAT DO THESE RESULTS imply about our position in the animal kingdom? Biologists classify living things in hierarchical categories, each less distinct than the next: subspecies, species, genus, family, superfamily, order, class, and phylum. The *Encyclopedia Britannica* and all the biology texts on my shelf say that humans and apes belong to the same order, called Primates, and the same superfamily, called Hominoidea, but to separate families, called Hominidae and Pongidae. Whether Sibley's and Ahlquist's work changes this classification depends on one's philosophy of taxonomy. Traditional taxonomists group species into higher categories by making somewhat subjective evaluations of how important the differences between species are. Such taxonomists place humans in a separate family because of distinctive functional traits like large brain and bipedal posture, and this classification would remain unaffected by measures of genetic distance.

However, another school of taxonomy, called cladistics, argues that classification should be objective and uniform, based on genetic distance or times of divergence. All taxonomists agree now that red-eyed and white-eyed vireos belong together in the genus *Vireo,* the various species of gibbons in the genus *Hylobates.* Yet the members of these pairs of species are genetically more distant from each other than are humans from the other two chimpanzees, and diverged longer ago. On this basis, then, humans don't constitute a distinct family, or even a distinct genus, but belong in the same genus as common and pygmy chimps. Since our genus name *Homo* was proposed first, it takes priority, by the rules of zoological nomenclature, over the genus name *Pan* coined for the "other" chimps. So there are not one but three species of genus *Homo* on earth today: the common chimpanzee, *Homo troglodytes;* the pygmy chimpanzee, *Homo paniscus;* and the third chimpanzee or human chimpanzee, *Homo sapiens.* Since the gorilla is only slightly more distinct, it has almost equal right to be considered a fourth species of *Homo.*

Even taxonomists espousing cladistics are anthropocentric, and the lumping of humans and chimps into the same genus will undoubtedly be a bitter pill for them to swallow. There is no doubt, however, that whenever chimpanzees learn cladistics, or whenever taxonomists from Outer Space visit Earth to inventory its inhabitants, they will unhesitatingly adopt the new classification.

WHICH PARTICULAR GENES are the ones that differ between humans and chimps? Before we can consider this question, we need first to understand what it is that DNA, our genetic material, does.

Much or most of our DNA has no known function and may just constitute "molecular junk": i.e., DNA molecules that have become duplicated or have lost former functions, and that natural selection hasn't eliminated from us because they do us no harm. Of our DNA that does have known functions, the main functions have to do with the long chains of amino acids called proteins. Certain proteins make up much of our body's structure (such as the proteins keratin of hair or collagen of connective tissue), while other proteins termed enzymes synthesize and break down most of our body's remaining molecules. The sequences of the component small molecules (nucleotide bases) in DNA specify the sequence of amino acids in our proteins. Other parts of our functional DNA regulate protein synthesis.

Those of our observable features that are easiest to understand genetically are ones arising from single proteins and single genes. For instance, our blood's oxygen-carrying protein, which I have already mentioned, hemoglobin, consists of two amino-acid chains, each specified by a single chunk of DNA (a single "gene"). Those two genes have no observable effects except through specifying the structure of hemoglobin, which is confined to our red blood cells. Conversely, hemoglobin's structure is totally specified by those genes. What you eat or how much you exercise may affect how much hemoglobin you make, but not the details of its structure.

That's the simplest situation, but there are also genes influencing many observable traits. For example, the fatal genetic disorder known as Tay-Sachs disease involves many behavioral as well as anatomical anomalies: excessive drooling, rigid posture, yellowish skin, abnormal head growth, and other changes. We know in this case that all these observable effects result somehow from changes in a single enzyme specified by the Tay-Sachs gene, but we don't know exactly how. Since that enzyme occurs in many tissues of our bodies and breaks down a widespread cellular constituent, changes in that one enzyme have wide-ranging and ultimately fatal consequences. Conversely, some traits, such as your height as an adult, are influenced simultaneously by many genes and also by environmental factors (e.g., your nutrition as a child).

While scientists understand well the function of numerous genes that specify known individual proteins, we know much less about the function of genes involved in complexly determined traits, like most behaviors. It would be absurd to think that a human hallmark such as art, language, or aggression depends on a single gene. Behavioral differences among individual humans are obviously subject to enormous environmental influences, and it's very controversial what role genes play in such individual differences. However, for those behaviors that differ consistently between chimps and humans, genetic differences are likely to be involved, even though we can't yet specify the genes responsible. For instance, the ability of humans but not chimps to speak surely depends on differences in genes specifying the anatomy of the voice box and the wiring of the brain. A young chimpanzee brought up in a psychologist's home along with the psychologist's human baby of the same age still continued to look like a chimp and didn't learn to talk or walk erect. But whether an individual human grows up to be fluent in English or Korean is independent of genes and dependent solely on childhood linguistic environment, as proved by the linguistic attainments of Korean infants adopted by English-speaking parents.

With this as background, what can we say about the 1.6 percent of our DNA that differs from chimp DNA? We know that the genes for our principal hemoglobin don't differ, and that certain other genes do exhibit minor differences. In the nine protein chains studied to date in both humans and common chimps, only five out of a total of 1,271 amino acids differ: one amino acid in a muscle protein called myoglobin, one in a minor hemoglobin chain called the delta chain, and three in an enzyme called carbonic anhydrase. But we don't yet know which chunks of our DNA are responsible for the functionally significant differences between humans and chimps to be discussed in Chapters 2 to 7: the differences in brain size, anatomy of the pelvis and voice box and genitalia, amount of body hair, female menstrual cycle, menopause, and other traits. Those important differences certainly don't arise from the five amino-acid differences detected to date. At present, all we can say with confidence is this: much of our DNA is junk; at least some of the 1.6 percent that differs between us and chimps is already known to be junk; and the functionally significant differences must be confined to some as-yet-unidentified small fraction of 1.6 percent.

Within that small differing fraction of our DNA, some differences have bigger consequences for our bodies than do others. To begin with, most amino acids of proteins can be specified by at least two alternative sequences of nucleotide bases in DNA. Changes in nucleotide bases from one such sequence to an alternative one are "silent" mutations: they produce no change in the amino-acid sequences of proteins. Even when a change in one base does cause one amino acid to be replaced by another, some amino acids are very similar to certain others in their chemical properties, or are located in relatively insensitive parts of a protein.

But other parts of a protein are crucial to the protein's function. Replacing an amino acid in such a part with a chemically dissimilar amino acid is likely to produce some detectable effect. For instance, the disease sickle-cell anemia is an often-fatal condition, resulting from a change in our hemoglobin's solubility, resulting in turn from a change in just one of hemoglobin's 287 amino acids, resulting in turn from a change in just one of the three nucleotides specifying that amino acid. That change, however, replaces a negatively charged amino acid with one lacking net charge, thereby changing the electrical charge on the whole hemoglobin molecule.

While we don't know which particular genes or nucleotide bases are the crucial ones accounting for our observed differences from chimps, there are numerous precedents for one or a few genes having big impacts. I just mentioned the many big and visible differences between Tay-Sachs patients and normal people, all somehow arising from a single change in one enzyme. That's an example of differences among individuals of the same species. As for differences between related species, a good example is provided by the cichlid fishes of Africa's Lake Victoria. Cichlids are popular aquarium species, of which about two hundred are confined to that one lake, where they evolved from a single ancestor within perhaps the last 200,000 years. Those two hundred species differ among themselves in their food habits as much as do tigers and cows. Some graze on algae, others catch other fish, and still others variously crush snails, feed on plankton, catch insects, nibble the scales off other fish, or specialize in grabbing fish embryos from brooding mother fish. Yet all those Lake Victoria cichlids differ from each other on the average by only about 0.4 percent of their DNA studied. Thus, it took even fewer genetic mutations to change a snail crusher into a specialized baby killer than it took to produce us from an ape.

*   *   *

Do THE NEW findings about our genetic distance from chimps have any broader implications, besides technical questions of taxonomic names? Probably the most important implications concern how we think about the place of humans and apes in the universe. Names are not just technical details but express and create attitudes. (To convince yourself, try greeting your spouse this evening either as "my darling" or as "you swine," using the same expression and tone of voice). The new findings do not specify how we *should* think about humans and apes. But, just as did Darwin's *Origin of Species,* they will probably influence how we *do* think, and it will probably take us many years to readjust our attitudes. I'll mention just one example of a controversial area that might be affected: our use of apes.

At present we make a fundamental distinction between animals (including apes) and humans, and this distinction guides our ethical code and actions. For instance, as I noted at the start of this chapter, it's considered acceptable to exhibit caged apes in zoos, but it's not acceptable to do the same with humans. I wonder how the public will feel when the identifying label on the chimp cage in the zoo reads "*Homo troglodytes.*" Yet, if it were not for the sympathetic interest in apes that many people gain at zoos, there might be much less public financial support for conservationists' efforts to protect apes in the wild.

I also noted earlier that it's considered acceptable to subject apes, but not humans, without their consent to lethal experiments for purposes of medical research. The motive for doing so is precisely that apes are so similar to us genetically. They can be infected with many of the same diseases as we can, and their bodies respond similarly to the disease organisms. Thus, experiments on apes offer a far better way to devise improved medical treatments for humans than would experiments on any other animals.

This ethical choice poses an even more difficult problem than does caging apes in zoos. After all, we regularly cage millions of human criminals under worse conditions than zoo apes, but there is no socially accepted human analogue of medical research on animals, even though lethal experiments on humans would provide medical scientists with far more valuable information than do lethal experiments on chimps. Yet the human experiments performed by Nazi concentration-camp physicians are widely viewed as one of the most abominable of all the Nazis' abominations. Why is it okay to perform such experimentation on chimps?

Somewhere along the scale from bacteria to humans, we have to decide where killing becomes murder, and where eating becomes cannibalism. Most people draw those lines between humans and all other species. However, quite a few people are vegetarians, unwilling to eat any animal (yet willing to eat plants). And an increasingly vocal minority, belonging to the animal-rights movement, object to medical experiments on animals—or at least on certain animals. That movement is especially exercised about research on cats and dogs and primates, less concerned about mice, and generally silent about insects and bacteria.

If our ethical code makes a purely arbitrary distinction between humans and all other species, then we have a code based on naked selfishness devoid of any higher principle. If our code instead makes distinctions based on our superior intelligence, social relationships, and capacity for feeling pain, then it becomes difficult to defend an all-or-nothing code that draws a line between all humans and all animals. Instead, different ethical constraints should apply to research on different species. Perhaps it's just our naked selfishness, reemerging in a new disguise, that would advocate granting special rights to those animal species genetically closest to us. But an objective case, based on the considerations I just mentioned (intelligence, social relationships, etc.), can be made that chimps and gorillas qualify for preferred ethical consideration over insects and bacteria. If there's any animal species currently used in medical research for which a total ban on medical experimentation can be justified, that species is surely the chimpanzee.

The ethical dilemma posed by animal experiments is compounded for chimps by the fact that they are endangered as a species. In this case, medical research not only kills individuals but threatens to kill the species itself. That's not to say that demands for research have been the sole threat to wild chimp populations; habitat destruction and capture for zoos have also been major threats. But it's enough that research demands have been a significant threat. The ethical dilemma is further compounded by other considerations: that on the average several wild chimps are killed in the process of capturing one alive (often a young animal that was being carried by its mother) and delivering it to a medical research lab; that medical scientists have played little role in the struggle to protect wild chimp populations, despite their obvious self-interest in doing so; and that chimps used for research are often caged under cruel conditions. The first chimp that I saw being used for medical research had been injected with a slow-acting lethal virus and was being kept alone for the several years until it died, in a small indoor cage devoid of play objects, at the U.S. National Institutes of Health.

Breeding chimps in captivity for research use avoids objections based on depleting wild chimp populations. But that still doesn't get around the basic dilemma, any more than enslaving children of U.S.-born blacks after abolition of the African slave trade made black slavery in the nineteenth-century U.S. acceptable. Why is it okay to experiment on *Homo troglodytes,* but not on *Homo sapiens?* Conversely, how should we explain to parents whose children are at risk of dying from diseases now being studied in captive chimps that their children are less important than chimps? Ultimately, we the public, not just scientists, will have to make these terrible choices. All that is certain is that our view of man and apes will determine our decision.

Finally, changes in our attitudes about apes may be crucial in determining whether apes will survive at all in the wild. At present, their populations are threatened especially by destruction of their rainforest habitats in Africa and Asia, and by legal and illegal capture and killing. If present trends continue, the mountain gorilla, orangutan, pileated gibbon, Kloss's gibbon, and possibly some other apes as well will exist only in zoos by the time this year's crop of human babies enters college. It's not enough for us to preach to the governments of Uganda, Zaire, and Indonesia about their moral obligation to protect their wild apes. These are impoverished countries, and national parks are expensive to create and maintain. If we as the third chimpanzee decide that the other two chimpanzees are worth saving, those of us in the richer countries will have to bear most of the expense. From the point of view of the apes themselves, the most important effect of what we've recently learned about the tale of three chimpanzees will be on how we feel about footing that bill.

CHAPTER 2

The Great Leap
Forward

FOR MOST OF THE MANY MILLIONS OF YEARS SINCE OUR LINEAGE diverged from that of apes, we remained little more than glorified chimpanzees in how we made our living. As recently as forty thousand years ago, western Europe was still occupied by Neanderthals, primitive beings for whom art and progress scarcely existed. Then there came an abrupt change, as anatomically modern people appeared in Europe, bringing with them art, musical instruments, lamps, trade, and progress. Within a short time, the Neanderthals were gone.

That Great Leap Forward in Europe was probably the result of a similar leap that had occurred over the course of the preceding few tens of thousands of years in the Near East and Africa. Even a few dozen millennia, though, is a trivial fraction (less than 1 percent) of our long history separate from ape history. Insofar as there was any single point in time when we could be said to have become human, it was at the time of that leap. Only a few more dozen millennia were needed for us to domesticate animals, develop agriculture and metallurgy, and invent writing. It was then but a short further step to those monuments of civilization that distinguish humans from animals across what used to seem an unbridgeable gulf—monuments such as the "Mona Lisa" and *Eroica Symphony,* the Eiffel Tower and Sputnik, Dachau's ovens and the bombing of Dresden.

This chapter will confront the questions posed by our abrupt rise to humanity. What made it possible, and why was it so sudden? What held back the Neanderthals, and what was their fate? Did Neanderthals and modern peoples ever meet? If so, how did they behave toward each other?

Understanding the Great Leap Forward isn't easy, and writing about it isn't easy either. The immediate evidence comes from technical details of preserved bones and stone tools. Archaeologists' reports are full of terms obscure to the rest of us, such as "transverse occipital torus," "receding zygomatic arches," and "Châtelperronian backed knives." What we really want to understand—the way of life and the humanity of our various ancestors—isn't directly preserved but only inferred from those technical details of bones and tools. Much of the evidence is missing, and archaeologists often disagree over the meaning of such evidence as has survived. Since the books and articles listed in Further Readings, pages 373–74, will slake the interest of readers curious to learn more about receding zygomatic arches, I'll emphasize instead the inferences from bones and tools.

To PLACE HUMAN EVOLUTION in a time perspective, recall that life originated on Earth several billion years ago, and that the dinosaurs became extinct around sixty-five million years ago. It was only between six and ten million years ago that our ancestors finally became distinct from the ancestors of chimps and gorillas. Hence human history constitutes only an insignificant portion of the history of life. Science-fiction films that depict cavemen fleeing from dinosaurs are just that, science fiction.

The shared ancestor of humans, chimps, and gorillas lived in Africa, to which chimps and gorillas are still confined, and to which we remained confined for millions of years. Initially, our own ancestors would have been classified as merely another species of ape, but a sequence of three changes launched us in the direction of modern humans. The first of these changes had occurred by around four million years ago, when the structure of fossilized limb bones shows that our ancestors were habitually walking upright on the two hind limbs. In contrast, gorillas and chimps walk upright only occasionally, and usually proceed on all fours. The upright posture freed our ancestors' forelimbs to do other things, among which toolmaking proved the most important.

The second change occurred around three million years ago, when our lineage split into at least two distinct species. As background, reflect that members of two animal species living in the same area must fill different ecological roles and do not normally interbreed with each other. For example, coyotes and wolves are obviously closely related and (until wolves were exterminated in most of the United States) lived in many of the same areas of North America. However, wolves are larger, mainly hunt big mammals like deer and moose, and often live in large packs, whereas coyotes are smaller, mainly hunt small mammals like rabbits and mice, and usually live in pairs or small groups. Coyotes usually mate with coyotes, wolves with wolves. In contrast, every human population living today has interbred with every other human population with which it has had extensive contact. Ecological differences among existing humans are entirely a product of childhood education: it's not the case that some of us are born with sharp teeth and equipped to hunt deer, while others are born with grinding teeth, gather berries, and don't marry the deer hunters. Hence all modern humans belong to the same species.

On perhaps two occasions in the past, though, the human lineage split into separate species, as distinct as wolves and coyotes. The most recent such occasion, which I'll describe later, may have been at the time of the Great Leap Forward. The earlier such occasion was around three million years ago, when our lineage split into two: a man-ape with a robust skull and very big cheek teeth, assumed to eat coarse plant food, and often referred to as *Australopithecus robustus* (meaning "the robust southern ape"); and a man-ape with a more lightly built skull and smaller teeth, assumed to have an omnivorous diet, and known as *Australopithecus africanus* ("the southern ape of Africa"). The latter man-ape evolved into a larger-brained form termed *Homo habilis* ("man the handyman"). However, fossil bones that some paleontologists consider to represent male and female *Homo habilis* differ so much from each other

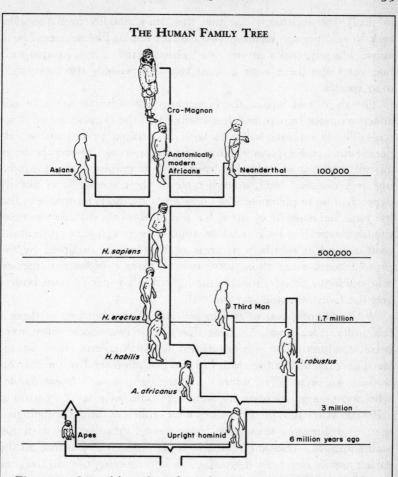

THE HUMAN FAMILY TREE

Cro-Magnon

Asians

Anatomically modern Africans

Neanderthal

100,000

H. sapiens

500,000

H. erectus

Third Man

1.7 million

H. habilis

A. robustus

A. africanus

3 million

Apes

Upright hominid

6 million years ago

Figure 2. Several branches of our family tree have become extinct, including those belonging to the robust australopithecines, Neanderthals, and possibly a poorly understood "Third Man" and an Asian population contemporary with Neanderthals. Some descendants of *Homo habilis* survived to evolve into modern humans. To recognize by different names the changes in fossils representing this line, they are somewhat arbitrarily divided into *Homo habilis,* then *Homo erectus* beginning around 1.7 million years ago, then *Homo sapiens* beginning around 500,000 years ago. *A.* stands for the genus name of *Australopithecus, H.* for *Homo.*

in skull size and tooth size that they may actually imply another fork in our lineage yielding two distinct *habilis*-like species: *Homo habilis* himself, and a mysterious "Third Man." Thus, by two million years ago there were at least two and possibly three protohuman species.

The third and last of the big changes that began to make our ancestors more human and less apelike was the regular use of stone tools. This is a human hallmark with clear animal precedents: woodpecker finches, Egyptian vultures, and sea otters are among the other animal species that evolved independently to employ tools in capturing or processing food, though none of these species is as heavily dependent on implements as we now are. Common chimpanzees also use tools, occasionally of stone, but not in numbers sufficient to litter the landscape. But by around 2½ million years ago very crude stone tools appear in numbers in areas of East Africa occupied by the protohumans. Since there were two or three protohuman species, who made the tools? Probably the light-skulled species, since both it and the tools persisted and evolved.

With only one human species surviving today but two or three a few million years ago, it's clear that one or two species must have become extinct. Who was our ancestor, which species ended up instead as a discard in the trash heap of evolution, and when did this shakedown occur? The winner was the light-skulled *Homo habilis,* who went on to increase in brain size and body size. By around 1,700,000 years ago the differences were sufficient for anthropologists to give our lineage a new name, *Homo erectus,* meaning "the man that walks upright." (*Homo erectus* fossils were discovered before all the earlier fossils I've been discussing, so anthropologists didn't realize that *Homo erectus* wasn't the first protohuman to walk upright). The robust man-ape disappeared somewhat after 1,200,000 years ago, and the "Third Man" (if he ever existed) must have disappeared by then also. As for why *Homo erectus* survived and the robust man-ape didn't, we can only speculate. A plausible guess is that the robust man-ape could no longer compete, since *Homo erectus* ate both meat and plant food, and since tools and a larger brain made *Homo erectus* more efficient at getting even the plant food on which his robust sibling depended. It's also possible that *Homo erectus* gave his sibling a direct push into oblivion, by killing him for meat.

All the developments that I've been discussing so far were played out within the continent of Africa. The shakedown there left *Homo erectus* as the sole protohuman on the African stage. It was only around one million years ago that *Homo erectus* finally expanded his horizons. His stone tools and bones show that he reached the Near East, then the Far East (where he is represented by the famous fossils known as Peking Man and Java Man) and Europe. He continued to evolve in our direction by an increase in brain size and in skull roundness. By around 500,000 years ago, some of our ancestors looked sufficiently like us, and different from earlier *Homo erectus,* that they are classified as our own species (*Homo sapiens,* meaning "the wise man"), though they still had thicker skulls and brow ridges than we do today.

Readers unfamiliar with details of our evolution might be forgiven for assuming that the appearance of *Homo sapiens* constituted the Great Leap Forward. Was our meteoric ascent to *sapiens* status half-a-million years ago the brilliant climax of Earth history, when art and sophisticated technology finally burst upon our previously dull planet? Not at all: the appearance of *Homo sapiens* was a nonevent. Cave paintings, houses, and bows and arrows still lay hundreds of thousands of years off in the future. Stone tools continued to be the crude ones that *Homo erectus* had been making for nearly a million years. The extra brain size of those early *Homo sapiens* had no dramatic effect on our way of life. That whole long tenure of *Homo erectus* and early *Homo sapiens* outside Africa was a period of infinitesimally slow cultural change. In fact, the sole candidate for a major advance was possibly the control of fire, for which caves occupied by Peking Man provide one of the earliest indications in the form of ash, charcoal, and burnt bones. Even that advance—if those cave fires really were man-lit rather than caused by lightning—would belong to *Homo erectus,* not *Homo sapiens.*

The emergence of *Homo sapiens* illustrates the paradox discussed in the previous chapter: that our rise to humanity was not directly proportional to the changes in our genes. Early *Homo sapiens* had progressed much further in anatomy than in cultural attainments along the road up from chimpanzeehood. Some crucial ingredients still had to be added before the Third Chimpanzee could conceive of painting the Sistine Chapel.

*   *   *

How DID OUR ANCESTORS make their living during the 1½ million years that spanned the emergence of *Homo erectus* and *Homo sapiens*?

The only surviving tools from this period are stone tools that can charitably be described as very crude, in comparison with the beautiful polished stone tools made until recently by Polynesians, American Indians, and other modern Stone Age peoples. Early stone tools vary in size and shape, and archaeologists have used those differences to give the tools different names, such as "hand axe," "chopper," and "cleaver." These names conceal the fact that none of those early tools had a sufficiently consistent or distinctive shape to suggest any specific function, as do the obvious needles and spear points left by the much later Cro-Magnons. Wear marks on the tools show that they were variously used to cut meat, bone, hides, wood, and nonwoody parts of plants. But any size or shape of tool seems to have been used to cut any of those things, and the tool names applied by archaeologists may be little more than arbitrary divisions of a continuum of stone forms.

Negative evidence is also significant here. Many advances in tools that appear after the Great Leap Forward were unknown to *Homo erectus* and early *Homo sapiens*. There were no bone tools, no ropes to make nets, and no fishhooks. All the early stone tools may have been held directly in the hand; they show no signs of having been mounted on other materials for increased leverage, as we mount steel axe blades on wooden handles.

What food did our early ancestors get with those crude tools, and how did they get it? At this point, anthropology books usually insert a long chapter entitled something like "Man the Hunter." The point here is that baboons, chimps, and some other primates occasionally prey on small vertebrates, but recently surviving Stone Age people (like Bushmen) did a lot of big-game hunting. So did Cro-Magnons, according to abundant archaeological evidence. There's no doubt that our early ancestors also ate some meat, as shown by marks of their stone tools on animal bones and by wear marks on their stone tools caused by cutting meat. The real question is: how *much* big-game hunting did our early ancestors do? Did big-game hunting skills improve gradually over the past 1½ million years, or was it only after the Great Leap Forward that they made a large contribution to our diet?

Anthropologists routinely reply that we have been successful big-game hunters for a long time. The supposed evidence comes mainly from three archaeological sites occupied around 500,000 years ago: a cave at Zhoukoudian near Beijing, containing bones and tools of *Homo erectus* ("Peking Man") and bones of many animals; and two noncave (open-air) sites at Torralba and Ambrona in Spain, with stone tools plus bones of elephants and other large animals. It's usually assumed that the people who left the tools killed the animals, brought their carcasses to the site, and ate them there. But all three sites also have bones and fecal remains of hyenas, which could equally well have been the hunters. The bones of the Spanish sites in particular look as if they came from a collection of scavenged, water-washed, trampled carcasses such as one can find around African waterholes today, rather than from human hunters' camps.

Thus, while early humans ate some meat, we don't know how much meat they ate, or whether they got the meat by hunting or scavenging. It's not until much later, around 100,000 years ago, that we have good evidence about human hunting skills, and it's clear that humans then were still very *ineffective* big-game hunters. Hence human hunters of 500,000 years ago and earlier must have been even more ineffective.

The mystique of Man the Hunter is now so rooted in us that it's hard to abandon our belief in its long-standing importance. Today, shooting a big animal is regarded as an ultimate expression of macho masculinity. Trapped in this mystique, male anthropologists like to stress the key role of big-game hunting in human evolution. Supposedly, big-game hunting was what induced protohuman males to cooperate with each other, develop language and big brains, join into bands, and share food. Even women were supposedly molded by men's big-game hunting: women suppressed the external signs of monthly ovulation that are so conspicuous in chimps, so as not to drive men into a frenzy of sexual competition and thereby spoil men's cooperation at hunting.

As an example of the purple prose spawned by this men's locker-room mentality, consider the following account of human evolution by Robert Ardrey in his book *African Genesis:* "In some scrawny troop of beleaguered not-yet-men on some scrawny forgotten plain a radian particle from an unknown source fractured a never-to-be-forgotten gene, and a primate carnivore was born. For better or for worse, for tragedy or for triumph, for ultimate glory or ultimate damnation, intelligence made alliance with the way of the killer, and

Cain with his sticks and his stones and his quickly running feet emerged on the high savannah." What pure fantasy!

Western male writers and anthropologists aren't the only men with an exaggerated view of hunting. In New Guinea I've lived with real hunters, men who recently emerged from the Stone Age. Conversations at campfires go on for hours over each species of game animal, its habits, and how best to hunt it. To listen to my New Guinea friends, you would think that they eat fresh kangaroo for dinner every night and do little each day except hunt. In fact, when pressed for details, most New Guinea hunters admit that they have bagged only a few kangaroos in their whole lives.

I still recall my first morning in the New Guinea highlands, when I set out with a group of a dozen men armed with bows and arrows. As we passed a fallen tree, there was suddenly much excited shouting, men surrounded the tree, some spanned their bows, and others pressed forward into the brush pile. Convinced that an enraged boar or kangaroo was about to come out fighting, I looked for a tree that I could climb to a perch of safety. Then I heard triumphant shrieks, and out of the brush pile came two mighty hunters holding aloft their prey: two baby wrens, not quite able to fly, weighing about one-third of an ounce each, and promptly plucked, roasted, and eaten. The rest of that day's catch consisted of a few frogs and many mushrooms.

Studies of most modern hunter-gatherers with far more effective weapons than early *Homo sapiens* show that most of a family's calories comes from plant food gathered by women. Men catch rabbits and other small game never mentioned in the heroic campfire stories. Occasionally the men do bag a large animal, which does indeed contribute significantly to protein intake. But it's only in the Arctic, where little plant food is available, that big-game hunting becomes the dominant food source. And humans didn't reach the Arctic until within the last few dozen millennia.

I would guess that big-game hunting contributed only modestly to our food intake until *after* we had evolved fully modern anatomy and behavior. I doubt the usual view that hunting was the driving force behind our uniquely human brain and societies. For most of our history we were not mighty hunters but skilled chimps, using stone tools to acquire and prepare plant food and small animals. Occasionally, men did bag a large animal, and then retold the story of that rare event incessantly.

*   *   *

IN THE PERIOD just before the Great Leap Forward, at least three distinct human populations occupied different parts of the Old World. These were the last truly primitive humans, supplanted by fully modern people at the time of the Great Leap. Let's consider those among the last primitives whose anatomy is best known and who have become a metaphor for brutish subhumans: the Neanderthals.

Where and when did they live? Their geographic range extended from western Europe, through southern European Russia and the Near East, to Uzbekistan in Central Asia near the border of Afghanistan. (The name "Neanderthal" comes from Germany's Neander Valley [valley = *Thal*, or *Tal*, in German], where one of the first skeletons was discovered). As to the time of their origin, that's a matter of definition, since some old skulls have characteristics anticipating later full-blown Neanderthals. The earliest "full-blown" examples date to around 130,000 years ago, and most specimens postdate 74,000 years ago. While their start is thus arbitrary, their end is abrupt: the last Neanderthals died somewhat after 40,000 years ago.

During the time that Neanderthals flourished, Europe and Asia were in the grip of the last Ice Age. Neanderthals must have been a cold-adapted people—but only within limits. They got no further north than southern Britain, northern Germany, Kiev, and the Caspian Sea. The first penetration of Siberia and the Arctic was left to later, fully modern humans.

Neanderthals' head anatomy was so distinctive that, even if a Neanderthal dressed in a business suit or designer dress were to walk down the streets of New York or London today, everybody else (all the homines sapientes) on the street would be staring in shock. Imagine converting a modern face to soft clay, gripping the middle of the face from the bridge of the nose to the jaws in a vise, pulling the whole midface forward, and letting it harden again. You'll then have some idea of a Neanderthal's appearance. Their eyebrows rested on prominently bulging bony ridges, and their noses and jaws and teeth protruded far forward. Their eyes lay in deep sockets, sunk behind the protruding nose and brow ridges. Their foreheads were low and sloping, unlike our high vertical modern foreheads, and their lower jaws sloped back without chins. Despite these startlingly primitive features, Neanderthals' brain size was nearly 10 percent *greater* than ours!

A dentist who examined a Neanderthal's teeth would have been in for a further shock. In adult Neanderthals the incisors (front teeth) were worn down on the outer-facing surface, in a way found in no modern people. Evidently, this peculiar wear pattern somehow resulted from a use of their teeth as tools, but what exactly was that function? As one possibility, they may have routinely used their teeth as vises to grip objects, like my baby sons, who gripped their milk bottles in their teeth and ran around with their hands free. Alternatively, Neanderthals may have bitten hides with their teeth to make leather, or bitten wood to make wooden tools.

While a Neanderthal in a business suit or dress would attract attention today, one in shorts or a bikini would draw gasps. Neanderthals were more heavily muscled, especially in their shoulders and neck, than all but the most avid modern bodybuilders. Their limb bones, which took the force of those big muscles' contracting, had to be considerably thicker than ours to withstand the stress. Their arms and legs would have looked stubby to us, because the lower leg and forearm were relatively shorter than ours. Even their hands were much more powerful than ours; a Neanderthal's handshake would have been literally bone-crushing. While their average height was only around five feet four inches, their weight would have been at least twenty pounds more than that of a modern person of that height, and this excess was mostly in the form of lean muscle.

One other possible anatomical difference is intriguing, though its reality as well as its interpretation are quite uncertain. A Neanderthal woman's birth canal may have been wider than a modern woman's, permitting her baby to grow inside her to a bigger size before birth. If so, a Neanderthal pregnancy might have lasted one year, instead of our nine months.

Besides their bones, our other main source of information about Neanderthals is their stone tools. As I described for earlier human tools, Neanderthal tools may have been simple hand-held stones not mounted on separate parts such as handles. The tools don't fall into distinct types with unique functions. There were no standardized bone tools, no bows and arrows. Some of the stone tools were undoubtedly used to make wooden tools, which rarely survive. One notable exception is a wooden thrusting spear eight feet long, found in the ribs of a long-extinct species of elephant at an archaeological site in Germany. Despite that (lucky?) success, Neanderthals were probably not very good at big-game hunting, because Neanderthal population densities (to judge from the number of their sites) were much lower than those of later Cro-Magnons, and because even anatomically more modern people living in Africa at the same time as the Neanderthals were undistinguished as hunters.

If you say "Neanderthal" to friends and ask for their first association, you'll probably get back the answer "caveman." While most excavated Neanderthal remains do come from caves, that's surely an artifact of preservation, since open-air sites would be eroded much more quickly. Among my hundreds of campsites in New Guinea, one was in a cave, and that's the only site where future archaeologists are likely to find my pile of discarded tin cans intact. The archaeologists will also be deceived into considering me a caveman. Neanderthals must have constructed some type of shelter against the cold climate in which they lived, but those shelters must have been crude. All that remains is a few piles of stones and some postholes, compared to the elaborate remains of houses built by the later Cro-Magnons.

The list of other quintessentially modern human things that Neanderthals lacked is a long one. They left no unequivocal art objects. They must have worn some clothing in their cold environment, but it had to be crude, as they lacked needles and other evidence of sewing. They evidently lacked boats, as no Neanderthal remains are known from Mediterranean islands or even from North Africa, just eight miles across the Straits of Gibraltar from Neanderthal-populated Spain. There was no long-distance overland trade: Neanderthal tools are made of stones available within a few miles of the site.

Today we take cultural differences among people inhabiting different areas for granted. Every human population alive today has its characteristic house style, implements, and art. If you were shown chopsticks, a Guinness beer bottle, and a blowgun and asked to associate one object each with China, Ireland, and Borneo, you'd have no trouble giving the right answers. No such cultural variation is apparent for Neanderthals, whose tools look much the same whether they come from France or Russia.

We also take cultural progress with time for granted. The wares from a Roman villa, medieval castle, and 1990 New York apartment differ obviously. In the year 2000 my sons will look with astonishment at the slide rule I used for calculations throughout the 1950s: "Daddy, are you *really* that old?" But Neanderthal tools from 100,000 and 40,000 years ago look essentially the same. In short, Neanderthal tools had no variation in either time or space to suggest that most human of characteristics, *innovation*. As one archaeologist put it, Neanderthals had "beautiful tools stupidly made." Despite Neanderthals' big brains, something was still missing.

Grandparenting, and what we consider old age, must also have been rare among Neanderthals. Their skeletons make clear that adults might live to their thirties or early forties, but not beyond forty-five. If we lacked writing *and* if none of us lived past forty-five, just think how the ability of our society to accumulate and transmit information would suffer.

I've had to mention all these subhuman qualities of Neanderthals, but there are three respects in which we can relate to their humanity. First, virtually all well-preserved Neanderthal caves have small areas of ash and charcoal indicating simple fireplaces. Hence, although Peking Man may have already used fire hundreds of thousands of years earlier, Neanderthals were the first people to leave undisputed evidence of fire's regular use. Neanderthals may also have been the first people who regularly buried their dead, but that's disputed, and whether it would imply religion is a matter of pure speculation. Finally, they regularly took care of their sick and aged. Most skeletons of older Neanderthals show signs of severe impairment, such as withered arms, healed but incapacitating broken bones, tooth loss, and severe osteoarthritis. Only care by young Neanderthals could have enabled such older Neanderthals to stay alive to the point of such incapacitation. After my long litany of what Neanderthals lacked, we've finally found something that lets us feel a spark of kindred spirit in these strange creatures of the last Ice Age—nearly human in form, and yet not really human in spirit.

Did Neanderthals belong to the same species as we do? That depends on whether we could and would have mated and reared a child with a Neanderthal man or woman, given the opportunity. Science-fiction novels love to imagine the scenario. You remember the blurb on many a back cover: "A team of explorers stumbles on a steep-walled valley in the center of deepest Africa, a valley that time forgot. In this valley they find a tribe of incredibly primitive people, living in ways that our Stone Age ancestors discarded thousands of years ago. Do they belong to the same species as we do? There's only one way to find out, but who among the intrepid explorers can bring himself [male explorers, of course] to make the test?" At this point one of the bone-chewing cavewomen suddenly is described as beautiful and sexy in a primitively erotic way, so that modern novel readers will find the brave explorer's dilemma believable: does he or doesn't he have sex with her?

Believe it or not, something like that experiment actually took place. It happened repeatedly around forty thousand years ago, at the time of the Great Leap Forward.

I MENTIONED that the Neanderthals of Europe and western Asia were just one of at least three human populations occupying different parts of the Old World around 100,000 years ago. A few fossils from eastern Asia suffice to show that people there differed from Neanderthals as well as from us moderns, but too few bones have been found to describe these Asians in more detail. The best-characterized contemporaries of the Neanderthals are those from Africa, some of whom were virtually modern in their skull anatomy. Does this mean that, 100,000 years ago in Africa, we have at last arrived at the watershed of human cultural development?

Surprisingly, the answer is still "no." The stone tools of these modern-looking Africans were very similar to those of the decidedly unmodern-looking Neanderthals, hence we refer to them as "Middle Stone Age Africans." They still lacked standardized bone tools, bows and arrows, nets, fishhooks, art, and cultural variation in tools from place to place. Despite their mostly modern bodies, these Africans were still missing something needed to endow them with full humanity. Once again, we face the paradox that mostly modern bones, and presumably mostly modern genes, aren't enough by themselves to produce modern behavior.

Some South African caves occupied around 100,000 years ago provide us with the first time point in human evolution when we have detailed information about what people actually were eating. Our confidence stems from the fact that the African caves are full of stone tools, animal bones with cut marks from stone tools, and human bones, but few or no bones of carnivores like hyenas. Thus, it's clear that people, not hyenas, brought the bones to the caves. Among the bones are many of seals and penguins, as well as shellfish such as limpets. That makes Middle Stone Age Africans the first people for whom there is even a hint that they exploited the seashore. However, the caves contain very few remains of fish or flying seabirds, undoubtedly because people still lacked the fishhooks and nets needed to catch fish and birds.

The mammal bones from the caves include those of quite a few medium-sized species, among which those of an antelope called eland predominate by far. Eland bones in the caves represent eland of all ages, as if people had somehow managed to capture a whole herd and kill every individual. At first, the relative abundance of eland among hunters' prey is surprising, since the caves' environment 100,000 years ago was much as it is today and since eland is now one of the least common large animals in the area. The secret to the hunters' success with eland probably lay in the fact that eland are rather tame, not dangerous, and easy to drive in herds. This suggests that hunters occasionally managed to drive a whole herd over a cliff, explaining why the distribution of eland ages among the cave kills is like that in a living herd. In contrast, more dangerous prey, such as Cape buffalo, pigs, elephants, and rhinos, yield a very different picture. Buffalo bones in the caves are mainly of very young or very old individuals, while pigs, elephants, and rhinos are virtually unrepresented.

Hence Middle Stone Age Africans can be considered big-game hunters, but only barely. They either avoided dangerous species entirely or confined themselves to weak old animals or babies. Those choices reflect sound prudence on the hunters' part, since their weapons were still spears for thrusting rather than bows and arrows. Along with drinking a strychnine cocktail, poking an adult rhino or Cape buffalo with a spear ranks as one of the most effective means of suicide that I know. Nor could the hunters have succeeded often at driving eland herds over cliffs, since elands weren't exterminated but continued to coexist with hunters. As with earlier peoples and modern Stone Age hunters, I suspect that plants and small game made up most of the diets of these not-so-great Middle Stone Age hunters. They were definitely more effective than chimpanzees, but not up to the skill of modern Bushmen and Pygmies.

The scene that the human world presented from around 100,000 to somewhat before fifty thousand years ago was this. Northern Europe, Siberia, Australia, oceanic islands, and the whole New World were still empty of people. In Europe and western Asia lived the Neanderthals; in Africa, people increasingly like us moderns in their anatomy; and in eastern Asia, people unlike either the Neanderthals or Africans but known from only a few bones. All three of these populations were, at least initially, still primitive in their tools, behavior, and limited innovativeness. The stage was set for the Great Leap Forward. Which among these three contemporary populations would take that leap?

THE EVIDENCE for an abrupt rise is clearest in France and Spain, in the Late Ice Age around forty thousand years ago. Where there had previously been Neanderthals, anatomically fully modern people (often known as Cro-Magnons, from the French site where their bones were first identified) now appear. Had one of those gentlemen or ladies strolled down the Champs Élysées in modern attire, he or she would not have stood out from the Parisian crowds in any way. As significant to archaeologists as the Cro-Magnons' skeletons are their tools, which are far more diverse in form and obvious in function than any in the earlier archaeological record. The tools suggest that modern anatomy had at last been joined by modern innovative behavior.

Many of the tools continued to be of stone, but they were now made from thin blades struck off larger stones, thereby yielding ten times more cutting edge from a given quantity of raw stone than obtainable previously. Standardized bone and antler tools appeared for the first time. So did unequivocal compound tools of several parts tied or glued together, such as spear points set in shafts or axe heads fitted onto wooden handles. Tools fall into many distinct categories whose function is often obvious, such as needles, awls, mortars and pestles, fishhooks, net sinkers, and rope. The rope (used in nets or snares) accounts for the frequent bones of foxes, weasels, and rabbits at Cro-Magnon sites, while the rope, fishhooks, and net sinkers explain the bones of fish and flying birds at contemporary South African sites.

Sophisticated weapons for safely killing dangerous large animals at a distance now appear—weapons such as barbed harpoons, darts,

spear throwers, and bows and arrows. South African caves occupied
by people now yield bones of such vicious prey as adult Cape buffalo
and pigs, while European caves are full of bones of bison, elk, reindeer, horse, and ibex. Even today, hunters armed with high-powered
telescopic rifles find it hard to bag some of these species, which must
have required highly skilled communal hunting methods based on
detailed knowledge of each species' behavior.

Several types of evidence testify to the effectiveness of Late Ice Age
people as big-game hunters. Their sites are much more numerous
than those of earlier Neanderthals or Middle Stone Age Africans,
implying more success at obtaining food. Numerous species of big
animals that had survived many previous ice ages became extinct
toward the end of the last Ice Age, suggesting that they were exterminated by human hunters' new skills. These likely victims, to be
discussed in later chapters, include the mammoths of North America,
Europe's woolly rhino and giant deer, southern Africa's giant buffalo
and giant Cape horse, and Australia's giant kangaroos. Evidently, the
most brilliant moment of our rise already contained the seeds of what
may yet prove a cause of our fall.

Improved technology now allowed humans to occupy new environments, as well as to multiply in previously occupied areas of
Eurasia and Africa. Australia was first reached by humans around
fifty thousand years ago, implying watercraft capable of crossing
water gaps as wide as sixty miles between eastern Indonesia and
Australia. The occupation of northern Russia and Siberia by at least
twenty thousand years ago depended on many advances: sewn clothing, whose existence is reflected in eyed needles, cave paintings of
parkas, and grave ornaments marking outlines of shirts and trousers;
warm furs, indicated by fox and wolf skeletons minus the paws
(removed in skinning and found in a separate pile); elaborate houses
(marked by postholes, pavements, and walls of mammoth bones),
with elaborate fireplaces; and stone lamps to hold animal fat and light
the long Arctic nights. The occupation of Siberia and Alaska in turn
led to the occupation of North America and South America around
eleven thousand years ago.

Whereas Neanderthals obtained their raw materials within a few
miles of home, Cro-Magnons and their contemporaries throughout
Europe practiced long-distance trade, not only for raw materials of
tools but also for "useless" ornaments. Tools of high-quality stone

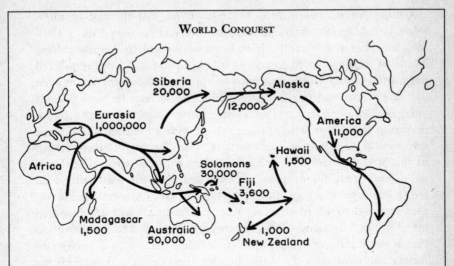

WORLD CONQUEST

Eurasia 1,000,000

Siberia 20,000

Alaska

12,000

America 11,000

Africa

Solomons 30,000

Hawaii 1,500

Fiji 3,600

Madagascar 1,500

Australia 50,000

1,000 New Zealand

Figure 3. This map illustrates stages in the spread of our ancestors from their African origins to populate the world. Numbers stand for estimated number of years before the present. Further discoveries of older archaeological sites may well show that some regions, such as Siberia or the Solomons, were colonized earlier than the estimated dates shown here.

such as obsidian, jasper, and flint are found hundreds of miles from where those stones were quarried. Baltic amber reached southeast Europe, while Mediterranean shells were carried to inland parts of France, Spain, and the Ukraine. I saw very similar patterns in modern Stone Age New Guinea, where cowry shells prized as decorations were traded up to the highlands from the coast, bird-of-paradise plumes were traded back down to the coast, and obsidian for stone axes was traded out from a few highly valued quarries.

The evident aesthetic sense reflected in Late Ice Age trade in ornaments relates to the achievements for which we most admire the Cro-Magnons: their art. Best known, of course, are the rock paintings from caves like Lascaux, with stunning polychrome depictions of now-extinct animals. But equally impressive are the bas reliefs, necklaces and pendants, fired-clay ceramic sculptures, Venus figurines of women with enormous breasts and buttocks, and musical instruments ranging from flutes to rattles.

Unlike Neanderthals, few of whom lived past the age of forty, some Cro-Magnon skeletons indicate survival to sixty. Many Cro-Magnons, but few Neanderthals, lived to enjoy their grandchildren. Those of us accustomed to getting our information from the printed page or television will find it hard to appreciate how important even just one or two old people are in preliterate society. In New Guinea villages it often happens that younger men lead me to the oldest person in the village when I stump them with a question about some uncommon bird or fruit. For example, when I visited Rennell Island in the Solomons in 1976, many islanders told me what wild fruits were good to eat, but only one old man could tell me what other wild fruits could be eaten in an emergency to avoid starvation. He remembered that information from a cyclone that had hit Rennell in his childhood (around 1905), destroying gardens and reducing his people to a state of desperation. One such person in a preliterate society can thus spell the difference between death and survival for the whole society. Hence the fact that some Cro-Magnons survived twenty years longer than any Neanderthal probably played a big role in Cro-Magnon success. Living to an older age required not just improved survival skills but also some biological changes, possibly including the evolution of human female menopause.

I've described the Great Leap Forward as if all those advances in tools and art appeared simultaneously forty thousand years ago. In fact, different innovations appeared at different times. Spear throwers appeared before harpoons or bows and arrows, while beads and pendants appeared before cave paintings. I've also described the changes as if they were the same everywhere, but they weren't. Among Late Ice Age Africans, Ukrainians, and French, only the Africans made beads out of ostrich eggs, only the Ukrainians built houses out of mammoth bones, and only the French painted woolly rhinos on cave walls.

These variations of culture in time and space are totally unlike the unchanging monolithic Neanderthal culture. They constitute the most important innovation that came with our rise to humanity: namely, the capacity for innovation itself. To us today, who can't picture a world in which Nigerians and Latvians in 1991 have virtually the same possessions as each other and as Romans in 50 B.C., innovation is utterly natural. To Neanderthals, it was evidently unthinkable.

Despite our instant sympathy with Cro-Magnon art, their stone tools and hunter-gatherer life-style make it hard for us to view them as other than primitive. Stone tools evoke cartoons of club-waving cavemen uttering grunts as they drag a woman off to their cave. But we can form a more accurate impression of Cro-Magnons if we imagine what future archaeologists will conclude after excavating a New Guinea village site from as recently as the 1950s. The archaeologists will find a few simple types of stone axes. Virtually all other material possessions were made of wood and will have perished. Nothing will remain of the multistory houses, beautifully woven baskets, drums and flutes, outrigger canoes, and world-quality painted sculpture. There will be no trace of the village's complex language, songs, social relationships, and knowledge of the natural world.

New Guinea material culture was until recently "primitive" (i.e., Stone Age) for historical reasons, but New Guineans are fully modern humans. New Guineans whose fathers lived in the Stone Age now pilot airplanes, operate computers, and govern a modern state. If we could carry ourselves back forty thousand years in a time machine, I suspect that we would find Cro-Magnons to be equally modern people, capable of learning to fly a jet plane. They made stone and bone tools only because no other tools had yet been invented; that's all they had the opportunity to learn.

IT USED TO BE ARGUED that Neanderthals evolved into Cro-Magnons within Europe. That possibility now seems increasingly unlikely. The last Neanderthal skeletons from somewhat after forty thousand years ago were still "full-blown" Neanderthals, while the first Cro-Magnons appearing in Europe at the same time were already anatomically fully modern. Since anatomically modern people were already present in Africa and the Near East tens of thousands of years earlier, it seems much more likely that anatomically modern people invaded Europe from that direction than that they evolved within Europe.

What happened when invading Cro-Magnons met the resident Neanderthals? We can be certain only of the end result: within a short time, no more Neanderthals. The conclusion seems to me inescapable that Cro-Magnon arrival somehow caused Neanderthal extinction. Yet many archaeologists recoil at this conclusion and invoke environmental changes instead. For example, the *Encyclopedia Britannica*'s fifteenth edition concludes its entry for Neanderthals with the sentence "The disappearance of the Neanderthals, although it cannot yet be fixed in time, was probably the result of being creatures of an interglacial period unable to avoid the ravages of another Ice Age." In fact, Neanderthals thrived during the last Ice Age, and suddenly disappeared over thirty thousand years after its start and an equal time before its end.

My guess is that events in Europe at the time of the Great Leap Forward were similar to events that have occurred repeatedly in the modern world, whenever a numerous people with more advanced technology invades the lands of a much less-numerous people with less-advanced technology. For instance, when European colonists invaded North America, most North American Indians proceeded to die of introduced epidemics; most of the survivors were killed outright or driven off their land; some of the survivors adopted European technology (horses and guns) and resisted for some time; and many of the remaining survivors were pushed onto lands that Europeans did not want, or else intermarried with Europeans. The displacement of Aboriginal Australians by European colonists, and of southern African San populations (Bushmen) by invading iron-age Bantu speakers, followed a similar course.

By analogy, I guess that Cro-Magnon diseases, murders, and displacements did in the Neanderthals. If so, then the Cro-Magnon-Neanderthal transition was a harbinger of what was to come, when the victors' descendants began squabbling among themselves. It may at first seem paradoxical that Cro-Magnons prevailed over the far more muscular Neanderthals, but weaponry rather than strength would have been decisive. Similarly, it's not gorillas that are now threatening to exterminate humans in central Africa, but vice versa. People with huge muscles require lots of food, and they thereby gain no advantage if slimmer, smarter people can use tools to do the same work.

Like the Great Plains Indians, some Neanderthals may have learned some Cro-Magnon ways and resisted for a while. This is the only sense I can make of a puzzling culture called the Châtelperronian, which coexisted in western Europe along with a typical Cro-

Magnon culture (the so-called Aurignacian culture) for a short time after Cro-Magnons arrived. Châtelperronian stone tools are a mixture of typical Neanderthal and Cro-Magnon tools, but the bone tools and art typical of Cro-Magnons are usually lacking. The identity of the people who produced Châtelperronian culture was debated by archaeologists until a skeleton unearthed with Châtelperronian artifacts at Saint-Césaire in France proved to be Neanderthal. Perhaps, then, some Neanderthals managed to master some Cro-Magnon tools and hold out longer than their fellows.

What remains unclear is the outcome of the interbreeding experiment posed in science-fiction novels. Did some invading Cro-Magnon men mate with some Neanderthal women? No skeletons that could reasonably be considered Neanderthal-Cro-Magnon hybrids are known. If Neanderthal behavior was as relatively rudimentary, and Neanderthal anatomy as distinctive as I suspect, few Cro-Magnons may have wanted to mate with Neanderthals. Similarly, although humans and chimps continue to coexist today, I'm not aware of any matings. While Cro-Magnons and Neanderthals weren't nearly as different, the differences may still have been a mutual turnoff. And if Neanderthal women were geared for a twelve-month pregnancy, a hybrid fetus might not have survived. My inclination is to take the negative evidence at face value, to accept that hybridization occurred rarely if ever, and to doubt that living people of European descent carry any Neanderthal genes.

So much for the Great Leap Forward in western Europe. The replacement of Neanderthals by modern people occurred somewhat earlier in eastern Europe, and still earlier in the Near East, where possession of the same area apparently shifted back and forth between Neanderthals and modern people from ninety thousand to sixty thousand years ago. The slowness of the transition in the Near East, compared to its speed in western Europe, suggests that the anatomically modern people living around the Near East before sixty thousand years ago had not yet developed the modern behavior that ultimately let them drive out the Neanderthals.

Thus, we have a tentative picture of anatomically modern people arising in Africa over a hundred thousand years ago, but initially making the same tools as Neanderthals and having no advantage over them. By perhaps sixty thousand years ago, some magic twist of behavior had been added to the modern anatomy. That twist (of which more in a moment) produced innovative, fully modern people who proceeded to spread westward from the Near East into Europe, quickly supplanting Europe's Neanderthals. Presumably, those modern people also spread east into Asia and Indonesia, supplanting the earlier people there, of whom we know little. Some anthropologists think that skull remains of those earlier Asians and Indonesians show traits recognizable in modern Asians and Aboriginal Australians. If so, the invading moderns may not have exterminated the original Asians without issue, as they did the Neanderthals, but instead interbred with them.

Two million years ago, several protohuman lineages had coexisted side by side until a shakedown left only one. It now appears that a similar shakedown occurred within the last sixty thousand years, and that all of us alive in the world today are descended from the winner of that shakedown. What was the last missing ingredient whose acquisition helped our ancestor to win?

THE IDENTITY of the ingredient that produced the Great Leap Forward poses an archaeological puzzle without an accepted answer. It doesn't show up in fossil skeletons. It may have been a change in only 0.1 percent of our DNA. What tiny change in genes could have had such enormous consequences?

Like some other scientists who have speculated about this question, I can think of only one plausible answer: the anatomical basis for spoken complex language. Chimpanzees, gorillas, and even monkeys are capable of symbolic communication not dependent on spoken words. Both chimpanzees and gorillas have been taught to communicate by means of sign language, and chimpanzees have learned to communicate via the keys of a large computer-controlled console. Individual apes have thus mastered "vocabularies" of hundreds of symbols. While scientists argue over the extent to which such communication resembles human language, there is little doubt that it constitutes a form of symbolic communication. That is, a particular sign or computer key symbolizes a particular something else.

Primates can use not just signs and computer keys, but also sounds, as symbols. For instance, wild vervet monkeys have a natural form of symbolic communication based on grunts, with slightly different grunts to mean "leopard," "eagle," and "snake." A month-old chimpanzee named Viki, adopted by a psychologist and his wife and reared virtually as their daughter, learned to "say" approximations of four words: "papa," "mama," "cup," and "up." (The chimp breathed rather than spoke those words.) Given this capability for symbolic communication using sounds, why have apes not gone on to develop much more complex natural languages of their own?

The answer seems to involve the structure of the larynx, tongue, and associated muscles that give us fine control over spoken sounds. Like a Swiss watch, all of whose many parts have to be well designed for the watch to keep time at all, our vocal tract depends on the precise functioning of many structures and muscles. Chimps are thought to be physically incapable of producing several of the commonest human vowels. If we too were limited to just a few vowels and consonants, our own vocabulary would be greatly reduced. For example, take this paragraph, convert all vowels other than "a" or "i" to either of those two, convert all consonants other than "d" or "m" or "s" to one of those three, and then see how much of the paragraph you can still understand.

That's why it's plausible that the missing ingredient may have been some modifications of the protohuman vocal tract to give us finer control and permit formation of a much greater variety of sounds. Such fine modifications of muscles need not be detectable in fossil skulls.

It's easy to appreciate how a tiny change in anatomy resulting in capacity for speech would produce a huge change in behavior. With language, it takes only a few seconds to communicate the message, "Turn sharp right at the fourth tree and drive the male antelope toward the reddish boulder, where I'll hide to spear it." Without language, that message could not be communicated at all. Without language, two protohumans could not brainstorm together about how to devise a better tool, or about what a cave painting might mean. Without language, even one protohuman would have had difficulty thinking out for himself or herself how to devise a better tool.

I don't suggest that the Great Leap Forward began as soon as the mutations for altered tongue and larynx anatomy arose. Given the right anatomy, it must have taken humans thousands of years to perfect the structure of language as we know it—to arrive at the concepts of word order and case endings and tenses, and to develop vocabulary. In Chapter 8 I'll consider some possible stages by which our language might have become perfected. But if the missing ingredient did consist of changes in our vocal tract that permitted fine control of sounds, then the capacity for innovation would follow eventually. It was the spoken word that made us free.

This interpretation seems to me to account for the lack of evidence for Neanderthal-Cro-Magnon hybrids. Speech is of overwhelming importance in the relations between men and women and their children. That's not to deny that mute or deaf people learn to function well in our culture, but they do so by learning to find alternatives for a spoken language that already exists. If Neanderthal language was much simpler than ours or nonexistent, it's not surprising that Cro-Magnons didn't choose to marry Neanderthals.

I'VE ARGUED that we were fully modern in anatomy and behavior and language by forty thousand years ago, and that a Cro-Magnon could have been taught to fly a jet airplane. If so, why did it take so long after the Great Leap Forward for us to invent writing and build the Parthenon? The answer may be similar to the explanation why the Romans, great engineers that they were, didn't build atomic bombs. To reach the point of building an A-bomb required two thousand years of technological advances beyond Roman levels, such as the invention of gunpowder and calculus, the development of atomic theory, and the isolation of uranium. Similarly, writing and the Parthenon depended on tens of thousands of years of cumulative developments after the arrival of Cro-Magnons—developments that included the bow and arrow, pottery, domestication of plants and animals, and many others.

Until the Great Leap Forward, human culture had developed at a snail's pace for millions of years. That pace was dictated by the slow pace of genetic change. After the leap, cultural development no longer depended on genetic change. Despite negligible changes in our anatomy, there has been far more cultural evolution in the past forty thousand years than in the millions of years before. Had a visitor from Outer Space come to the Earth in Neanderthal times, humans would not have stood out as unique among the world's species. At most, the visitor might have mentioned humans along with beavers, bowerbirds, and army ants as examples of species with curious behavior. Would the visitor have foreseen the change that would soon make us the first species, in the history of life on Earth, capable of destroying all life?

PART TWO

AN ANIMAL
WITH A STRANGE
LIFE CYCLE

We've just traced our evolutionary history through the appearance of humans with fully modern anatomy and behavioral capabilities. But that background doesn't prepare us to go straight on to consider the development of human cultural hallmarks, such as language and art. That's because we took up only the evidence of bones and tools. Yes, our evolution of large brains and upright posture was prerequisite to language and art, but wasn't enough by itself. Human bones alone don't guarantee humanity. Instead, our rise to humanity also required drastic changes in our life cycle, which will be the subject of Part Two.

For any species one can describe what biologists term its "life cycle." That means traits such as the number of offspring produced per litter or birth, the parental care (if any) that offspring receive from the mother or father, social relations between adult individuals, how a male and female select each other to mate with, frequency of sexual relations, menopause (if any), and life expectancy.

We take the forms of these traits as they exist in humans for granted, as the norm. But our life cycle is actually bizarre by animal standards. All the traits that I've just mentioned vary greatly between species, and we are extreme in most respects. To mention only some obvious examples, most animals produce litters much larger than one baby at a time, most animal fathers provide no parental care, and few other animal species live even a small fraction of three score years and ten.

Of these exceptional features of ours, some are shared by apes, suggesting that we merely retained traits already acquired by our apelike ancestors. For instance, apes too usually give birth to one baby at a time and live for several decades. Neither of these things is true of the other animals most familiar (but less closely related) to us, such as cats, dogs, songbirds, and goldfish.

In others of these respects, we're greatly different even from apes. Here are some obvious differences whose functions are well understood. Human babies continue to have all food brought to them by their parents even after weaning, whereas weaned apes gather their own food. Most human fathers as well as mothers, but only chimpanzee mothers, are closely involved in caring for their young. Like seagulls but unlike apes or most other mammals, we live in dense breeding colonies of nominally monogamous couples, some of whom also pursue extramarital sex. All these traits are as essential as large brain cases for the survival and education of human offspring. That's because our elaborate, tool-dependent methods of obtaining food make weaned human infants incompetent to feed themselves. Our infants first require a long period of food provisioning, training, and protection—an investment much more taxing than that facing the ape mother. Hence human fathers who want their offspring to survive to maturity have generally assisted their mates with more than just sperm, the sole parental input of an orangutan father.

Our life cycle also differs from that of wild apes in more subtle respects, whose functions are nevertheless still discernible. Many of us live longer than most wild apes: even hunter-gatherer tribes include some old individuals who are enormously important as repositories of experience. Men's testes are much larger than those of gorillas but smaller than those of chimps, for reasons that this book will explain. We regard human female menopause as inevitable, and I'll show why it makes good sense for humans, but it's almost unprecedented among other mammals. The closest mammalian parallel is among some tiny mouselike marsupials in Australia, and it's their males, not their females, that undergo menopause. Our longevity, testis size, and menopause were also prerequisites to our humanity.

Still other features of our life cycle differ far more drastically from those of apes than do our testes, yet the functions of those remaining novel features of ours remain hotly debated. We are unusual in having sex mainly in private and for fun, rather than mainly in public and only when the female is able to conceive. Ape females advertise the time when they are ovulating; human females conceal it even from themselves. While anatomists understand the value of men's moderate testis size, an explanation for men's relatively enormous penis still escapes us. Whatever their explanation, all these features, too, are part of what defines humanity. Certainly, it is hard to picture how fathers and mothers could cooperate harmoniously at rearing their children if human females resembled some primate females in having their genitalia turn bright red at the time of ovulation, becoming sexually receptive only at that time, flaunting their red badge of receptivity, and proceeding to have sex in public with any passing male.

Thus, human society and child rearing rest not only on the skeletal changes mentioned in Part One, but also on these remarkable new features of our life cycle. Unlike the case with our skeletal changes, however, we can't follow through our evolutionary history the timing of each of these life-cycle changes, because they leave no direct fossil imprint. As a result, they receive only brief attention in paleontology texts despite their importance. Archaeologists have recently discovered a Neanderthal hyoid bone, one of the key pieces of our speech-producing equipment, but as yet no trace of a Neanderthal penis. We don't know whether *Homo erectus* was already on the road to evolving a preference for having sex in private, in addition to having evolved his and her well-documented large brain. We can't even prove through fossils, as we can for our large brain size, that we rather than living apes are the ones whose life cycles diverged most from the ancestral condition. Instead, we have to be content with merely inferring that conclusion from the fact that our life cycles are exceptional compared not just to living apes but also to other primates, suggesting that we were the ones who did more changing.

Darwin established in the mid-nineteenth century that animals' anatomy has evolved through natural selection. Within this century, biochemists have similarly traced how animals' chemical makeup has evolved through natural selection. But so has animals' behavior, including reproductive biology and sexual habits in particular. Life-cycle traits have some genetic basis and vary quantitatively among individuals of the same species. For instance, some women are genetically predisposed to give birth to twins, while we're all aware that genes for long life span run in some families more than in others. Life-cycle traits affect our success in passing on our genes, through affecting our success in wooing mates, conceiving, rearing babies, and surviving as adults. Just as natural selection tends to adapt an animal's anatomy to its ecological niche and vice versa, so natural selection also tends to mold animals' life cycles. Those individuals leaving the most numerous surviving offspring promote their genes for life-cycle traits as well as for bones and chemical makeup.

A difficulty with this reasoning is that it seems as if some of our traits, such as menopause and aging, would reduce (rather than enhance) our output of offspring and shouldn't have resulted from natural selection. It often proves profitable to try to understand these paradoxes through the concept of "trade-offs." In the animal world there's nothing that's free or pure good. Everything involves costs as well as benefits, by using space, time, or energy that could have been devoted to something else. You might otherwise have thought that women who never underwent menopause would leave more descendants than women designed as they actually are. But we'll see that consideration of the hidden costs of forgoing menopause suggests why evolution didn't design these strategies into us. The same considerations illuminate such painful questions as why we grow old and die, and whether we're better off (even in a narrow evolutionary sense) in being faithful to our spouses or in pursuing extramarital affairs.

I've been assuming in this discussion that our distinctively human life-cycle traits have some genetic basis. The comments that I made in Chapter 1 about the function of genes in general apply here as well. Just as our height and most of our other observable traits aren't influenced by just a single gene, there surely isn't a single gene specifying menopause or monogamy. In fact, we know little about the genetic bases of human life-cycle traits, though selective breeding experiments in mice and sheep have illuminated the genetic control of their testis size. Enormous cultural influences obviously operate on our motivation for providing child care or seeking extramarital sex, and there is no reason to believe that genes contribute significantly to differences among individual people in these traits. However, genetic differences between humans and the other two chimpanzee species probably do contribute to the consistent differences in many life-cycle traits between all human populations and all chimpanzee populations. There is no human society, regardless of its cultural practices, whose men have chimpanzee-sized testes and whose women forgo menopause. Among those 1.6 percent of our genes that differ between us and chimps and that have any function, a significant fraction is likely to be involved in specifying traits of our life cycle.

In our discussion of our uniquely human life cycle, we'll begin by taking up the distinctive features of human social organization and of sexual anatomy, physiology, and behavior. As already mentioned, features that make us strange among animals include our societies of nominally monogamous couples, our genital anatomy, and our constant and generally private pursuit of sex. Our sex lives are reflected not only in our genitalia but also in the relative sizes of men's and women's bodies (much more equal than are the bodies of male and female gorillas or orangutans). We'll see how some of these familiar distinctive features have known functions, while others continue to defy understanding.

No honest discussion of the human life cycle could get away with noting that we're nominally monogamous and just leaving it at that. Pursuit of extramarital sex is obviously greatly influenced by each individual's particular upbringing and by the norms of the society in which the individual lives. Despite all that cultural influence, we're left with having to explain the facts that *both* the institution of marriage *and* the occurrence of extramarital sex have been reported from all human societies; but that extramarital sex is unknown in gibbons, although they do practice "marriage" (i.e., lasting male-female pairing to rear offspring); and that the question of extramarital sex is meaningless for chimpanzees because they don't practice "marriage." Thus, an adequate discussion of our uniquely human life cycle must account for our combination of marriage with extramarital sex. As I'll show, animal precedents exist to help us make evolutionary sense of our combination: men and women tend to differ in their attitudes toward extramarital sex much as geese and ganders do.

We'll then turn to another distinctive human life-cycle trait: how we select our sex partners, marital or otherwise. That problem scarcely arises for baboon troops, in which there is little selection: any male tries to mate with each female as she comes into heat. While common chimpanzees do some picking of their sex partners, they're still much less selective and much more promiscuously baboonlike than are humans. Mate selection is a decision of major consequence in the human life cycle, because married couples share parental responsibilities as well as sexual involvement. Precisely because care of human children demands such heavy and prolonged parental investment, we have to select our coinvestor much more carefully than does a baboon. Nevertheless, it turns out that we can find animal precedents for our procedure in choosing sex partners, by going beyond primates to rats and birds.

Those mate-selection criteria of ours prove relevant to the vexed question of human racial variation. Humans native to different parts of the globe vary conspicuously in external appearance, as do gorillas, orangutans, and most other animal species occupying a sufficiently extensive geographic range. Some of the geographic variation in our appearance surely reflects natural selection molding us to local climate, just as weasels in areas with winter snow develop white fur in winter for better camouflage and survival. But I'll argue that our visible geographic variability arose mainly through sexual selection, as a result of our mate-choice procedures.

To bring the discussion of our life cycle to an end, I'll ask why our lives themselves have to come to an end. Aging is another feature of our life cycle so familiar that we take it for granted: of course we'll all grow old and eventually die. So will all individuals of all animal species, but different species age at very different rates. Among animals we are relatively long-lived and became even more so around the time that Cro-Magnons replaced Neanderthals. Our longevity has been important for our humanity, by permitting effective transmission of learned skills between generations. But even humans grow old. Why is aging inevitable, given our extensive capacity for biological self-repair?

Here, more than anywhere else in this book, the importance of thinking in terms of evolutionary trade-offs becomes clear. As measured by leaving increased numbers of offspring, it paradoxically just wouldn't pay us to make the increased investment in self-repair mechanisms required to live longer. We'll see that the trade-off concept also illuminates the puzzle of menopause: a shutdown of childbearing, paradoxically programmed by natural selection so that women can leave more surviving children.

CHAPTER 3

The Evolution
of Human Sexuality

No WEEK PASSES WITHOUT PUBLICATION OF YET ANOTHER BOOK
about sex. Our desire to read about sex is surpassed only by our desire
to practice it. Hence, you might suppose, the basic facts of human
sexuality must be familiar to lay people and understood by scientists.
Just test your own grasp of sex by trying to answer these five easy
questions:

Among the various ape species and man, which has by far the
biggest penis, and what for?

Why should men be bigger than women?

How can men get away with having much smaller testes than
do chimpanzees?

Why do humans copulate in private, while all other social animals do it in public?

Why don't women resemble almost all other female mammals
in having easily recognized days of fertility, with sexual receptivity confined to those days?

If you jumped at "the gorilla" as your answer to the first question, put on a dunce cap; the correct answer is man. If you gave any intelligent answers to the next four questions, publish them; scientists are still debating rival theories.

These five questions illustrate how hard it is to explain the most obvious facts of our sexual anatomy and physiology. Part of the problem is our hang-ups about sex: scientists did not begin to study it seriously until recently, and they still have trouble being objective. Another difficulty is that scientists can't do controlled experiments on sexual practices of us humans, as they can on our cholesterol intake or toothbrushing habits. Finally, sex organs don't exist in isolation: they are adapted to their owners' social habits and life cycle, which are in turn adapted to food-gathering habits. In our own case that means, among other things, that evolution of human sex organs has been intertwined with that of human tool use, large brains, and child-rearing practices. Our progress from being just another species of big mammal to being uniquely human therefore depended on the remodeling not just of our pelvises and skulls, but also of our sexuality.

GIVEN KNOWLEDGE of how an animal feeds, a biologist can often predict the animal's mating system and genital anatomy. In particular, if we want to understand how human sexuality got to be the way it is, we have to begin by understanding the evolution of our own diet and our society. From the vegetarian diet of our ape ancestors, we diverged within the last several million years to become social carnivores as well as vegetarians. Yet our teeth and claws remained those of apes, not of tigers. Our hunting prowess depended instead on large brains: by using tools and operating in coordinated groups, our ancestors were able to hunt successfully despite their deficient anatomical equipment, and they regularly shared food with each other. Our ability to gather roots and berries as well came to depend on tools and thus also to require large brains.

As a result, human children took years to acquire the information and the practice needed to be efficient hunter-gatherers, just as they still take years to learn how to be farmers or computer programmers today. During those many years after weaning, our children are still too ignorant and helpless to acquire their own food; they depend entirely on us parents to bring food to them. These habits are so natural to us that we forget that baby apes gather food as soon as they are weaned.

The reasons that human infants are totally incompetent at food gathering are actually twofold: mechanical and mental. First, making and wielding the tools used to obtain food requires fine finger coordination that children take years to develop. Just as my four-year-old sons still can't tie their own shoelaces, four-year-old hunter-gatherer children can't sharpen a stone axe or build a dugout canoe. Second, we depend on much more brainpower than do other animals in acquiring food, because we have a much more varied diet and more varied and complicated food-gathering techniques. For instance, New Guineans with whom I work typically have separate names for about a thousand different species of plants and animals living in the vicinity. For each of those species they know something about its distribution and life history, how to recognize it, whether it is edible or otherwise useful, and how best to capture or harvest it. All this information takes years to acquire.

Weaned human infants can't support themselves because they lack these mechanical and mental skills. They need adults to teach them, and they also need adults to feed them for the decade or two that they are being taught. As is true of so many other human hallmarks, these problems of ours too have animal precedents. In lions and many other species, the young must be trained to hunt by their parents. Chimpanzees too have a varied diet, employ varied foraging techniques, and assist their young in obtaining food, while common (but not pygmy) chimps make some use of tools. Our distinction isn't absolute but one of degree: for us the necessary skills and hence the parental burden are far greater than for lions or chimpanzees.

The resulting parental burden makes parental care by the father as well as the mother important for a child to survive. Orangutan fathers provide their offspring with nothing beyond their initial donation of semen; gorilla, chimpanzee, and gibbon fathers go beyond that to offer protection; but hunter-gatherer human fathers provide some food and much teaching as well. Human food-gathering habits require a social system in which a male retains his relationship with a female after fertilizing her, in order to assist in rearing the resulting child. Otherwise, the child would be less likely to survive, and the father less likely to pass on his genes. The orangutan system, in which the father departs after copulation, wouldn't work for us.

But the chimpanzee system, in which several adult males are likely to copulate with the same estrous female, also wouldn't work for us. The result of that system is that a chimpanzee father has no idea which infants in the troop he has sired. For the chimp father that's no loss, as his exertions on behalf of troop infants are modest. For the human father, however, who will contribute significantly to the care of what he thinks is his child, he had better have some confidence in his paternity—e.g., through having been the exclusive sexual partner of the child's mother. Otherwise, his child-care contribution may help pass on some other man's genes.

Confidence in paternity would be no problem if humans, like gibbons, were scattered over the landscape as separate couples, so that each female would only rarely encounter a male other than her consort. But there are compelling reasons why almost all human populations have consisted of groups of adults, despite the paranoia about paternity that this causes. Among the reasons: much human hunting and gathering involves cooperative group efforts among men, women, or both; much of our wild food occurs in scattered but concentrated patches, able to sustain many people; and groups offer better protection against predators and aggressors, especially against other humans.

In short, the social system we evolved to accommodate our un-apelike food habits seems utterly normal to us, but is bizarre by ape standards and is virtually unique among mammals. Adult orangutans are solitary; adult gibbons live as separate monogamous, male-female pairs; gorillas live in polygamous harems, each consisting of several adult females and usually one dominant adult male; common chimpanzees live in fairly promiscuous communities consisting of scattered females plus a group of males; and pygmy chimpanzees form even more promiscuous communities of both sexes. But our societies, like our food habits, resemble those of lions and wolves: we live in bands containing many adult males *and* many adult females. Furthermore, we diverge from even lions and wolves in how those societies are organized: our males and females are paired off with each other. In contrast, any male lion within a lion pride can and regularly does mate with any of the pride's lionesses, making paternity unidentifiable. Our peculiar societies instead have their closest parallels in colonies of seabirds, like gulls and penguins, which also consist of male-female pairs.

At least officially, human pairing is more or less monogamous in most modern political states, but is "mildly polygynous" among most surviving hunter-gatherer bands, which are better models for how mankind lived over the last million years. (This description omits consideration of extramarital sex, through which we become effectively more polygamous and whose scientifically fascinating aspects I'll discuss in the next chapter.) By "mildly polygynous," I mean that most hunter-gatherer men can support only a single family, but a few powerful men have several wives. Polygyny on the scale of elephant seals, among which powerful males have dozens of wives, is impossible for hunter-gatherer men, because they differ from elephant seals in having to provide child care. The big harems for which some human potentates are famous didn't become possible until the rise of agriculture and centralized government let a few princes tax everyone else in order to feed the royal harem's babies.

Now let's see how this social organization shapes the bodies of men and women. Take first the fact that adult men are slightly bigger than similar-aged women (about 8 percent taller and 20 percent heavier, on the average). A zoologist from Outer Space would take one look at my five-foot-eight wife next to me (five-feet-ten), and would instantly guess that we belonged to a mildly polygynous species. How, you may ask, can one possibly guess mating practices from relative body size?

It turns out that, among polygynous mammals, average harem size increases with the ratio of the male's body size to the female's body size. That is, the biggest harems are typical of species in which males are much larger than females. For example, males and females are the same size among gibbons, which are monogamous; male gorillas, with a typical harem of three to six females, weigh nearly double each female; but the average harem is forty-eight wives for the southern elephant seal, whose three-ton male dwarfs his seven hundred-pound wives. The explanation is that, in a monogamous species, every male can win a female, but in a very polygynous species most males languish without any mates, because a few dominant males have succeeded in rounding up all the females into their harems. Hence, the bigger the harem, the fiercer is the competition among males and the more important it is for a male to be big, since the bigger male usually wins the fights. We humans, with our slightly bigger males and slight polygyny, fit this pattern. (However, at some point in human evolution, male intelligence and personality came to count for more than size: male basketball players and sumo wrestlers don't tend to have more wives than male jockeys or coxswains.)

Because competition for mates is fiercer in polygynous than monogamous species, the polygynous species also tend to have more marked differences between males and females in other respects besides body size. These differences are the secondary sexual characteristics that play a role in attracting mates. For instance, males and females of the monogamous gibbons look identical at a distance, while male gorillas (befitting their polygyny) are easily recognized by their crested heads and silver-haired backs. Here too, our anatomy reflects our mild polygyny. The external differences between men and women aren't nearly as marked as sex-related differences in gorillas or orangutans, but the zoologist from Outer Space could probably still distinguish men and women by the body and facial hair of men, men's unusually large penises, and the large breasts of women even before first pregnancy (in this we are unique among primates).

PROCEEDING NOW to the genitalia themselves, the combined weight of the testes in the average man is about 1½ ounces. This may boost the macho man's ego when he reflects on the slightly lower testis weight in a 450-pound male gorilla. But wait: our testes are dwarfed by the 4-ounce testes of a 100-pound male chimpanzee. Why is the gorilla so economical, and the chimp so well-endowed, compared to us?

The Theory of Testis Size is one of the triumphs of modern physical anthropology. By weighing the testes of thirty-three primate species, British scientists identified two trends: species that copulate more often need bigger testes; and promiscuous species in which several males routinely copulate in quick sequence with one female need especially big testes (because the male that injects the most semen has the best chance of being the one to fertilize the egg). When fertilization is a competitive lottery, large testes enable a male to enter more sperm in the lottery.

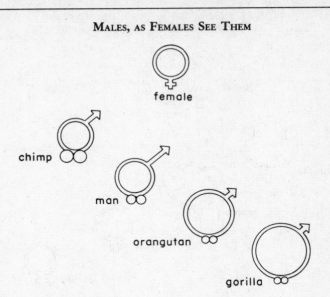

MALES, AS FEMALES SEE THEM

female

chimp

man

orangutan

gorilla

Figure 4. Humans and great apes differ with respect to the relative body size of males and females, penis length, and testis size. The main circles represent the body size of the male of each species relative to that of the female of the same species. Female body size is arbitrarily shown as the same for all species at upper right. Thus, chimps of both sexes weigh about the same; men are slightly larger than women; but male orangutans and gorillas are much bigger than females. The arrows on the male symbols are proportional to the length of the erect penis, while the twin circles represent testis weight relative to that of the body. Men have the longest penises, chimps the largest testes, and orangutans and gorillas the shortest penises and smallest testes.

Here's how these considerations account for the differences in testis size among the great apes and humans. A female gorilla does not resume sexual activity until three or four years after giving birth, and she is receptive for only a couple of days a month until she becomes pregnant again. Even the successful male gorilla with a harem of several females experiences sex as a rare treat: if he is lucky, a few times a year. His relatively tiny testes are quite adequate for

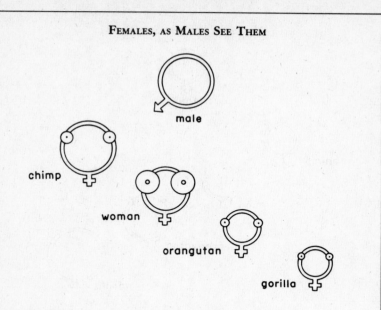

FEMALES, AS MALES SEE THEM

Figure 5. Human females are unique in their breasts, which are considerably larger than those of apes even before the first pregnancy. The main circles represent female body size relative to male body size of the same species.

those modest demands. The sex life of a male orangutan may be somewhat more demanding, but not much. However, each male chimp in a promiscuous troop of many females lives in sexual nirvana, with nearly daily opportunities to copulate for a common chimp and several daily copulations for the average pygmy chimp. That, plus his need to outdo other male chimps in semen output if he is to fertilize the promiscuous females, explains his need for gigantic testes. We humans make do with medium-sized testes, because the average man copulates more often than gorillas or orangutans but less often than chimps. In addition, the typical woman in a typical menstrual cycle does not force several men into sperm competition to fertilize her.

Thus, primate testis design well illustrates the principles of trade-offs and evolutionary cost/benefit analyses explained on pages 62–65.

Each species has testes big enough to do their job, but not unnecessarily larger ones. Bigger testes would just entail more costs without proportional benefits, by diverting space and energy from other tissues and increasing the risk of testicular cancer.

From this triumph of scientific explanation we descend to a glaring failure: the inability of twentieth-century science to formulate an adequate Theory of Penis Length. The length of the erect penis averages 1¼ inches in a gorilla, 1½ inches in an orangutan, 3 inches in a chimp, and 5 inches in a man. Visual conspicuousness varies in the same sequence: a gorilla's penis is inconspicuous even when erect because of its black color, while the chimp's pink erect penis stands out against the bare white skin behind it. The flaccid penis is not even visible in apes. Why does the human male need his relatively enormous, attention-getting penis, which is larger than that of any other primate? Since the male ape successfully propagates his kind with much less, does not the human penis represent largely wasted protoplasm that would be more valuable if devoted, say, to the cerebral cortex or improved fingers?

Biologist friends to whom I pose this conundrum usually think of distinctive features of human coitus in which they suppose a long penis might somehow be useful: our frequent use of the face-to-face position, our acrobatic variety of coital positions, and the leisurely duration of our coital bouts. None of these explanations survives close scrutiny. The face-to-face position is also a preferred one for orangutans and pygmy chimps, and is used occasionally by gorillas. Orangs vary face-to-face copulation with dorsoventral and sideways positions, and do it while hanging from branches of trees: surely that demands more penile acrobatics than our comfortable boudoir exercises. Our mean duration of coitus (about four minutes for Americans) is much longer than for gorillas (one minute), pygmy chimps (fifteen seconds), or common chimps (seven seconds), but shorter than for orangutans (fifteen minutes) and lightning-fast compared to the twelve-hour-long copulations of marsupial mice.

Since these facts make it unlikely that special features of human coitus demand a large penis, a popular alternative theory is that the human penis has also become an organ of display, like a peacock's tail or a lion's mane. This theory is reasonable but begs the question: what type of display, and to whom?

Proud male anthropologists unhesitatingly answer: an attractive display, to women. But the anthropologists' answer represents mere wishful thinking. Many women say that they are turned on by a man's voice, legs, and shoulders more than by the sight of his penis. A telling point is that the women's magazine *Viva* initially published photos of nude men but dropped them after surveys showed lack of female interest. When *Viva*'s nude men disappeared, the number of female readers increased, and the number of male readers decreased. Evidently the male readers were the ones buying *Viva* for its nude photos. While we can agree that the human penis is an organ of display, the display is intended not for women but for fellow men.

Other facts confirm the role of a large penis as a threat or status display toward other men. Recall all the phallic art created by men for men, and the widespread obsession of men with their penis size. Evolution of the human penis was effectively limited by the length of the female vagina: a man's penis would damage a woman if it were significantly larger. However, I can guess what the penis would look like if this practical constraint were removed and if men could design themselves. It would resemble the penis sheaths (phallocarps) used as male attire in some areas of New Guinea where I do fieldwork. Phallocarps vary in length (up to two feet), diameter (up to 4 inches), shape (curved or straight), angle made with the wearer's body, color (yellow or red), and decoration (e.g., a tuft of fur at the end). Each man has a wardrobe of several sizes and shapes from which to choose each day, depending on his mood that morning. Embarrassed male anthropologists interpret the phallocarp as something used for modesty or concealment, to which my wife had a succinct answer on seeing a phallocarp: "The most immodest display of modesty I've ever seen!"

Thus, astonishing as it seems, important functions of the human penis remain obscure. Here is a rich field for research.

PASSING NOW from anatomy to physiology, we are immediately confronted by our sexual-activity pattern, which must be considered freakish by the standards of other mammal species. Most mammals are sexually inactive most of the time. They copulate only when the female is in estrus—i.e., when she is ovulating and capable of being fertilized. Female mammals apparently "know" when they are ovulating, for they solicit copulation then by presenting their genitals toward males. Lest a male miss the point, many female primates go further: the area around the vagina, plus in some species the buttocks and breast, swells up and turns red, pink, or blue. This visual advertisement of female availability affects male monkeys in the same way that the sight of a seductively dressed woman affects male humans. In the presence of females with brightly swollen genitals, male monkeys stare much more often at the female's genitals, develop higher testosterone levels, attempt to copulate more often, and penetrate more quickly and after fewer pelvic thrusts than in the presence of females not displaying their wares.

Human sexual cycles are quite different. The human female maintains her sexual receptivity more or less constantly, instead of having it sharply confined to a short estrus phase. Indeed, despite numerous studies aimed at settling whether a woman's receptivity varies at all through her cycle, there is still no agreement about the answer—nor about the cycle phase when receptivity is maximal if it does vary.

So well concealed is human ovulation that we did not have accurate scientific information on its timing until around 1930. Before that, many physicians thought that women could conceive at any point in their cycle, or even that conception was most likely at the time of menstruation. In contrast to the male monkey, who has only to scan his surroundings for brightly swollen lady monkeys, the unfortunate human male has not the faintest idea which ladies around him are ovulating and capable of being fertilized. A woman herself may learn to recognize sensations associated with ovulation, but it is often tricky, even with the help of thermometers and ratings of vaginal mucus quality. Furthermore, today's would-be mother, who tries in such ways to sense ovulation in order to achieve (or avoid) fertilization, is responding by cold-blooded calculation to hard-won, modern book knowledge. She has no other choice; she lacks the innate, hot-blooded sense of sexual receptivity that drives other female mammals.

Our concealed ovulation, constant receptivity, and brief fertile period in each menstrual cycle ensure that most copulations by humans are at the wrong time for conception. To make things worse, menstrual-cycle length varies more between women, or from cycle to cycle in a given woman, than for other female mammals. As a result, even young newlyweds who omit contraception and make love at maximum frequency have only a 28 percent probability of conception per menstrual cycle. Animal breeders would be in despair if a prize cow had such low fertility, but in fact they can schedule a *single* artificial insemination so that the cow has a 75 percent chance of being fertilized!

Whatever the main biological function of human copulation, it isn't conception, which is just an occasional by-product. In these days of growing human overpopulation, one of the most ironic tragedies is the Catholic Church's claim that human copulation has conception as its natural purpose, and that the rhythm method is the only proper means of birth control. The rhythm method would be terrific for gorillas and most other mammal species, but not for us. In no species besides humans has the purpose of copulation become so unrelated to conception, or the rhythm method so unsuited for contraception.

For animals, copulation is a dangerous luxury. While occupied *in acto flagrante,* an animal is burning up valuable calories, neglecting opportunities to gather food, vulnerable to predators eager to eat it, and vulnerable to rivals eager to usurp its territory. Hence copulation is something to be accomplished in the minimum time required to do the job of fertilization. In contrast, human sex, as a device to achieve fertilization, would have to be rated a huge waste of time and energy, an evolutionary failure. Had we retained a proper estrus cycle like other mammals, the wasted time could have been diverted by our hunter-gatherer ancestors to butchering more mastodons. By this results-oriented view of sex, any hunter-gatherer band whose females advertised their estrus period could thereby have fed more babies and outcompeted neighboring bands.

Thus, the most hotly debated problem in the evolution of human reproduction is to explain why we ended up with concealed ovulation, and what good all our mistimed copulations do for us. For scientists, it's no answer just to say that sex is fun. Sure, it's fun, but evolution made it that way. If we weren't getting big benefits from our mistimed copulations, mutant humans who had evolved not to enjoy sex would have taken over the world.

Related to this paradox of concealed ovulation is the paradox of concealed copulation. All other group-living animals have sex in public, whether they are promiscuous or monogamous. Paired seagulls mate in the middle of the colony; an ovulating female chimpanzee may mate consecutively with five males in each other's presence. Why are we unique in our strong preference for copulating in private?

Biologists are currently arguing over at least six different theories to explain the origin of concealed ovulation and concealed copulation in humans. Interestingly, the debate proves to be a Rorschach test for the gender and outlook of the scientists involved. Here are the theories and their proponents:

1. *Theory preferred by many traditional male anthropologists.* According to this view, concealed ovulation and copulation evolved in order to enhance cooperation and reduce aggression among male hunters. How could cavemen bring off the precise teamwork needed to spear a mammoth if they had been fighting that morning for the public favors of a cavewoman in estrus? The implicit message of this theory: women's physiology is important chiefly for its effect on bonds between men, the real movers of society. However, one can broaden this theory to make it less blatantly sexist: visible estrus and sex would disrupt human society by affecting female/female and male/female as well as male/male bonds.

To illustrate this broadened version of the prevalent theory, consider the following scene from an imaginary soap opera, showing what life would be like for us modern hunter-gatherers if we did not have concealed ovulation and private copulation. Our soap opera stars Bob and Carol and Ted and Alice and Ralph and Jane. Bob, Alice, Ralph, and Jane work together in an office where the men hunt contracts and the women gather accounts payable. Ralph is married to Jane. Bob's wife is Carol, and Alice's husband is Ted. Carol and Ted work elsewhere.

One morning, Alice and Jane both discover on awakening that they have turned bright red in order to advertise impending ovulation and sexual receptivity. Alice and Ted make love at home before they go off in their separate directions to work. Jane and Ralph go together to work, where they copulate occasionally on the office sofa in the presence of their coworkers.

Bob cannot help lusting for Alice and Jane when he sees them bright red and sees Jane and Ralph copulating. He is unable to concentrate on his work. He repeatedly propositions Jane and Alice.

Ralph drives Bob away from Jane.

Alice is faithful to Ted and rejects Bob, but the hassle also interferes with her work.

All day, Carol in her office elsewhere is seething with jealousy at the thought of Alice and Jane, because Carol knows that Alice and Jane are bright red and attractive to Bob, while she (Carol) is not.

As a result, the office succeeds in bagging few contracts and accounts. In the meantime, other offices, where ovulation is concealed and where copulation is private, prosper. Eventually, Bob, Alice, Ralph, and Jane's office goes extinct. The only offices that survive are those with concealed ovulation and copulation.

This parable suggests that the traditional theory, by which concealed ovulation and copulation evolved to promote cooperation within human societies, is plausible. Unfortunately, there are other, equally plausible theories that I'll now explain more briefly.

2. *Theory preferred by many other traditional male anthropologists.* Concealed ovulation and copulation cement the bonds between a particular man and woman, thereby laying the foundations of the human family. A woman remains sexually attractive and receptive so that she can satisfy a man sexually all the time, bind him to her, and reward him for his help in rearing her baby. The sexist message: women evolved to make men happy. Left unexplained by this theory is the question why pairs of gibbons, whose unflinching devotion to monogamy should make them role models for the Moral Majority, remain constantly together despite having sex only every few years.

3. *Theory of a more modern male anthropologist.* (Donald Symons). Symons noted that a male chimpanzee who kills a small animal is more likely to share the meat with an estrous female than with a nonestrous female. This suggested to Symons that human females might have evolved a constant state of estrus in order to ensure a frequent meat supply from male hunters by rewarding them with sex. As an alternative theory, Symons noted that women in most hunter-gatherer societies have little say in selection of a husband. The societies are male-dominated, and male clans just suit themselves by exchanging daughters in marriage. However, by being constantly attractive, even a woman wed to an inferior male could privately seduce a superior male and secure his genes for her children. Symons's theories, while still male-oriented, at least represent

a step forward in that he views women as cleverly pursuing their own goals.

4. *Theory produced jointly by a male biologist and a female biologist* (Richard Alexander and Katherine Noonan). If a man could recognize signs of ovulation, he could use that knowledge to fertilize his wife by copulating with her only while she's ovulating. He could then safely neglect her the rest of the time and go off and philander, secure in the knowledge that the wife he left behind was unreceptive, if not already fertilized. Hence women evolved concealed ovulation to force men into a permanent marriage bond, by exploiting male paranoia about fatherhood. Not knowing the time of ovulation, a man must copulate often with his wife to have a chance of fertilizing her, and that leaves him less time to develop dalliances with other women. The wife benefits, but so does the husband. He gains confidence in his paternity of his children, and he need not worry that his wife will suddenly attract many competing men by turning bright red on a particular day. At last, we have a theory seemingly grounded in sexual equality.

5. *Theory of a female sociobiologist* (Sarah Hrdy). Hrdy was impressed by the frequency with which many primates—including not only monkeys but also baboons, gorillas, and common chimps—kill infants not their own. The bereaved mother is thereby induced to come into estrus again and often mates with the murderer, thus increasing his output of progeny. (Such violence has been common in human history: male conquerors kill the vanquished men and children but spare the women.) As a countermeasure, Hrdy reasoned, women evolved concealed ovulation in order to manipulate men by confusing the issue of paternity. A woman who distributed her favors widely would thereby enlist many men to help feed (or at least not to kill) her infant, since many men could suppose themselves to be the infant's father. Whether this theory is right or wrong, we must applaud Hrdy's overturning of conventional masculine sexism and her transferring of sexual power to women.

6. *Theory of another female sociobiologist* (Nancy Burley). The average seven-pound human newborn weighs double a newborn gorilla, but the two-hundred-pound gorilla mother dwarfs the average

human mother. Because the human newborn is so much larger in relation to its mother than are newborn apes, birth is exceptionally painful and dangerous in humans. Until the advent of modern medicine, women often died in childbirth, whereas I have never heard of such a fate's befalling a female gorilla or chimpanzee. Once humans had evolved enough intelligence to associate conception with copulation, estrous women could have chosen to avoid copulating at the time of ovulation, and could have thereby spared themselves the pain and peril of childbirth. But such women would have left fewer descendants than did women who could not detect their ovulation. Thus, where male anthropologists saw concealed ovulation as something evolved by women for men (Theories 1 and 2), Nancy Burley sees it as a trick that women evolved to deceive themselves.

WHICH OF THESE SIX THEORIES for the evolution of concealed ovulation is correct? Not only are biologists uncertain; it is only in recent years that the question has begun to receive serious attention. This dilemma exemplifies a pervasive problem in establishing causation in evolutionary biology, as well as in history, psychology, and many other fields where one can't manipulate variables to perform controlled experiments. Such experiments would afford the most convincing way to demonstrate cause or function. If we could remodel one tribe of people so that all women advertised their day of ovulation, we could then see whether cooperation within or between couples broke down, or whether the women used their new knowledge to avoid becoming pregnant. In the absence of such experiments, we can never be certain what human society would really be like today without concealed ovulation.

If it's hard to determine the function of things happening today under our eyes, how much harder must it be to determine functions in the vanished past! We know that human bones and tools were different hundreds of thousands of years ago, when concealed ovulation may have been evolving. Probably human sexuality, including the function of concealed ovulation, may also have been different then, in ways now hard for us to picture. Interpretation of our past runs the constant risk of degenerating into mere "paleopoetry": stories that we spin today, stimulated by a few bits of fossil bone, and expressing like Rorschach tests our own personal prejudices, but devoid of any claim to validity about the past.

Nevertheless, having mentioned six plausible theories, I can't just walk away from the problem without attempting some synthesis. Here again, we come up against another pervasive problem in establishing causation. It's rare for complex phenomena like concealed ovulation to be influenced by only a single factor. It would be as silly to seek a single cause of concealed ovulation as to claim that there was a single root cause of World War I. Instead, there were many somewhat independent factors in the period 1900–14 pushing toward war, others pushing toward peace. War finally broke out when the net weight of factors tipped toward war. But that doesn't excuse going to the opposite extreme, and "explaining" complex phenomena by an unweighted laundry list encompassing every conceivable factor.

As a first step toward pruning down our laundry list of six theories, let's realize that, whatever factors caused our distinctive sexual habits to evolve in the distant past, they wouldn't be persisting today if there weren't some factors still sustaining them. But the factors responsible for their initial appearance needn't have been the same as the ones now operative. In particular, although the factors behind Theories 3, 5, and 6 may have been major ones long ago, they don't seem to be so now. Only a minority of modern women use sex to obtain food or other resources from many men, or to confound paternity and induce many men simultaneously to support the woman's child. Postulates of their former role are paleopoetry, albeit plausible paleopoetry. Let's just content ourselves with trying to understand why concealed ovulation and frequent private copulation might make sense now. At least our guesses can be guided by introspection about ourselves plus observations of others.

The factors behind Theories 1, 2, and 4 seem to me still operative today, and to be facets of the same paradoxical feature of human social organization. That paradox is that a man and woman desirous for their child (and genes) to survive must cooperate with each other for a long time to rear their child, *but* must also cooperate economically with other couples living close by. It's obvious that regular sexual relations between a man and woman intensifies their connection, compared to their connections with other women and men whom they see daily but with whom they are not involved sexually. Concealed ovulation and constant receptivity advance this "new" function of sex (new by the standards of most mammals) as a social cement, not just a device for fertilization. This function is not, as in the traditional male-chauvinist version of Theories 1 and 2, a sop thrown by a cold, calculating woman to a sex-starved man, but instead an inducement for both sexes. Not only have all signs of female ovulation vanished, but the act of sex itself takes place privately, to emphasize the distinction between sexual and nonsexual partners within the same close group. As for the objection that gibbons remain monogamously involved without the reward of constant sex, that's easy to explain: each gibbon couple has minimal social and no economic involvement with other gibbon couples.

Human testis size also seems to me an outcome of that same basic paradox of human social organization. While our testes are larger than a gorilla's, because we have frequent sex for fun, they're still smaller than a chimpanzee's, because we're more monogamous. The oversized human penis may have evolved as an arbitrary sexual display symbol, as arbitrary as a lion's mane or a woman's enlarged breasts. Why weren't lionesses the ones to develop enlarged breasts, lions an oversized penis, and men a mane? If they had, those permuted signals could have functioned equally well. That it didn't come out that way may be just an accident of evolution, a result of each species' and sex's relative ease of evolving those various structures.

But there's still something basic missing from our discussion so far. I've talked about an idealized form of human sexuality: monogamous couples (plus a few polygynous households), husbands confident in their paternity of their wives' children, and husbands helping their wives with child rearing rather than neglecting the kids in order to philander. As justification for discussing this fictitious ideal, I maintain that actual human practice is much closer to this ideal than to baboon or chimpanzee practice. But the ideal is still fictitious. Any social system with rules of conduct is open to the risk of individuals' cheating when they find the advantages of cheating to outweigh the burden of sanctions. The question is thus a quantitative one: does cheating become so regular that the whole system collapses, or does cheating occur but not so often as to destroy the system, or is cheating vanishingly rare? As translated for human sexuality, that question becomes: are 90 percent, 30 percent, or 1 percent of human babies fathered extramaritally? Let's now confront that question and its consequences.

CHAPTER 4

The Science of Adultery

PEOPLE HAVE MANY REASONS TO LIE WHEN ASKED WHETHER THEY HAVE committed adultery. That's why it's notoriously difficult to get accurate scientific information about this important subject. One of the few existing sets of hard facts emerged as a totally unexpected by-product of a medical study, performed nearly half a century ago for a different reason. That study's findings have never been revealed until now.

I recently learned those facts from the distinguished medical scientist who ran the study. (Since he does not wish to be identified in this connection, I shall refer to him as Dr. X.) In the 1940s Dr. X was studying the genetics of human blood groups, which are molecules that we acquire only by inheritance. Each of us has dozens of blood-group substances on our red blood cells, and we inherit each substance either from our mother or from our father. The study's research plan was straightforward: go to the obstetrics ward of a highly respectable U.S. hospital; collect blood samples from one thousand newborn babies and their mothers and fathers;

identify the blood groups in all the samples; and then use standard genetic reasoning to deduce the inheritance patterns.

To Dr. X's shock, the blood groups revealed nearly 10 percent of those babies to be the fruits of adultery! Proof of the babies' illegitimate origin was that they had one or more blood groups lacking in both alleged parents. There could be no question of mistaken maternity: the blood samples were drawn from an infant and its mother soon after the infant emerged from the mother. A blood group present in a baby but absent in its undoubted mother could only have come from its father. Absence of the blood group from the mother's husband as well showed conclusively that the baby had been sired by some other man, extramaritally. The true incidence of extramarital sex must have been considerably higher than 10 percent, since many other blood-group substances now used in paternity tests were not yet known in the 1940s, and since most bouts of intercourse do not result in conception.

At the time that Dr. X made his discovery, research on American sexual habits was virtually taboo. He decided to maintain a prudent silence, never published his findings, and it was only with difficulty that I got his permission to mention his results without betraying his name. However, his results were later confirmed by several similar genetic studies whose results did get published. Those studies variously showed between about 5 and 30 percent of American and British babies to have been adulterously conceived. Again, the proportion of the tested couples of whom at least the wife had practiced adultery must have been higher, for the same two reasons as in Dr. X's study.

We can now answer the question whether extramarital sex is for humans a rare aberration, a frequent exception to a "normal" pattern of marital sex, or so frequent as to make a sham of marriage. The middle alternative proves to be the correct one. Most fathers really are raising their own children, and human marriage is not a sham. We're not just promiscuous chimpanzees pretending to be otherwise. Yet it's also clear that extramarital sex is an integral, albeit unofficial, part of the human mating system. Adultery has also been observed in many other animal species whose societies resemble ours in being based on male and female coparents with a lasting bond. Since such lasting bonds don't characterize common chimpanzee or pygmy chimpanzee society, it's meaningless to talk of adultery in chimps. We must have reinvented it after our chimp-like ancestors had rendered it obsolete. Thus, we can't discuss human sexuality, and its role in our rise to humanity, without carefully considering the science of adultery.

Most of our information about adultery's incidence has come from researchers asking people about their sex lives, rather than from blood-grouping their babies. Since the 1940s, the myth that marital infidelity is rare in the U.S. has been publicly exploded by a long succession of surveys, beginning with the Kinsey report. Nevertheless, even though this is the supposedly liberated 1990s, we are still profoundly ambivalent about adultery. It is thought of as exciting: no television soap opera could attract many viewers without it. It has few rivals as a basis of jokes. Yet, as Freud pointed out, we often use humor to deal with things intensely painful. Throughout history, adultery has had few rivals as a cause of murder and human misery. In writing about this subject, it's impossible to remain completely serious, but it's also impossible not to be revolted at the sadistic institutions by which societies have attempted to deal with extramarital sex.

WHAT MAKES A MARRIED PERSON decide to seek or avoid adultery? Scientists have theories to explain many other things, so it should not be surprising that there is also a theory of extramarital sex (abbreviated EMS, and not to be confused with premarital sex = PMS, in turn not to be confused with premenstrual syndrome = PMS). With many species of animals the problem of EMS never arises, because they don't opt for marriage in the first place. For instance, a female Barbary macaque in heat copulates promiscuously with every adult male in her troop and averages one copulation per seventeen minutes. However, some mammals and most bird species do opt for "marriage." That is, a male and a female form a lasting pair bond to devote care or protection to their joint offspring. Once there is marriage, there is also the possibility of what sociobiologists euphemistically term "the pursuit of a mixed reproductive strategy" (abbreviated MRS). In plain English, that means being married while simultaneously seeking extramarital sex.

Married animals vary enormously in the degree to which they mix their reproductive strategies. There appears to be no recorded instance of EMS in the little apes called gibbons, while snow geese indulge regularly. Human societies similarly vary, but I suspect that none comes close to the faithful gibbons. To explain all this variation, sociobiologists have found it useful to apply the reasoning of game theory. That is, life is considered an evolutionary contest whose winners are those individuals leaving the largest number of surviving offspring.

Contest rules are set by the ecology and reproductive biology of the particular species. The problem is then to figure out which strategy is most likely to win the contest: rigid fidelity, pure promiscuity, or a mixed strategy. But I must make one thing clear right at the outset. While this sociobiological approach certainly proves useful for understanding adultery in animals, its relevance for human adultery is an explosive issue to which I shall return.

The first thing one realizes when one starts to think about the contest is that the best game strategy differs between males and females of the same species. This is because of two profound differences between the reproductive biology of males and females: in the minimum necessary reproductive effort, and in the risk of being cuckolded. Let's consider these differences, which are painfully familiar to humans.

For men, the minimum effort needed to sire an offspring is the act of copulation, a brief expenditure of time and energy. The man who sires a baby by one woman one day is biologically capable of siring a baby by another woman the next day. For women, however, the minimum effort consists of copulation plus pregnancy plus (throughout most of human history) several years spent nursing—a huge commitment of time and energy. Thus, a man potentially can sire far more offspring than can a woman. A nineteenth-century visitor who spent a week at the court of the Nizam of Hyderabad, a polygamous Indian potentate, reported that four of the Nizam's wives gave birth within eight days, and that nine more births were anticipated for the following week. The record lifetime number of offspring for a man is 888, sired by Emperor Moulay Ismail the Bloodthirsty of Morocco, while the corresponding record for a woman is only sixty-nine (a nineteenth-century Moscow woman specializing in triplets). Few women have topped twenty children, whereas some men easily do so in polygynous societies.

As a result of this biological difference, a man stands to gain much more from EMS or polygamy than does a woman—if one's sole criterion is number of offspring born. (To female readers about to stop reading in outrage, or to male readers about to cheer, I warn you now: keep reading, there is much more to the question of EMS.) For human EMS the statistical evidence is naturally hard to come by, but for human polygamy it is readily available. In the sole polyandrous society for which I could find data, the Tre-ba of Tibet, women with two husbands average *fewer* children, not more children, than women with one husband. In contrast, nineteenth-century American Mormon men realized big benefits from polygyny: men with one wife averaged only 7 children, but men with two wives averaged 16 children, and those with three wives averaged 20. Polygynous Mormon men as a group averaged 2.4 wives and 15 children, while polygynous Mormon church leaders in particular averaged 5 wives and 25 children. Similarly, among the polygynous Temne people of Sierra Leone, a man's average number of children increases from 1.7 to 7 as his number of wives increases from 1 to 5.

The other sexual asymmetry relevant to mating strategy involves confidence that one really is the biological parent of one's putative offspring. A cuckolded animal, deceived into rearing offspring not its own, has thereby lost the evolutionary game while advancing the victory of another player, the real parent. Barring a switch of babies in the hospital nursery, women cannot be "cuckolded": they see their baby emerge from them. Nor can there be cuckoldry of males in animal species practicing external fertilization (i.e., fertilization of eggs outside the female's body). For instance, some male fish watch a female shed eggs, then immediately deposit sperm on the eggs and scoop them up to care for them, secure in their paternity. However, men and other male animals practicing internal fertilization— fertilization of eggs inside the female's body—can readily be cuckolded. All that the putative father knows for sure is that his sperm went into the mother, and eventually an offspring came out. Only observation of the female throughout her whole fertile period can absolutely exclude the possibility that some other male's sperm also entered and did the actual fertilizing.

An extreme solution to this simple asymmetry is the one formerly adopted by southern India's Nayar society. Among the Nayar, women freely took many lovers simultaneously or in sequence, and husbands accordingly had no confidence in paternity. To make the best of a bad situation, a Nayar man did not live with his wife or care for his supposed children, but he instead lived with his sisters and cared for his sisters' children. At least, those nieces and nephews were sure to share one-quarter of his genes.

Bearing in mind these two basic facts of sexual asymmetry, we can now examine what is the best game strategy, and when EMS pays. Let's examine three game plans of increasing complexity.

Game Plan 1. A man should always seek EMS, because he has so little to lose and so much to gain. Consider the hunter-gatherer conditions prevailing throughout most of human evolution, under which a woman could at best rear about four children in the course of her life. Through one dalliance, her otherwise faithful husband could increase his lifetime reproductive output from four to five: an enormous increase of 25 percent, for only a few minutes' work. What's wrong with this dazzlingly naïve reasoning?

Game Plan 2. A moment's reflection should expose a basic flaw of Game Plan 1: it considers only the potential benefits of EMS to a man and ignores his potential costs. Obvious costs would include: the risk of detection and injury or murder by the husband of the woman sought as EMS partner; the risk that one's own wife will desert; the risk of being cuckolded by one's wife while one is off seeking EMS; and the risk that one's legitimate children will suffer through one's neglect of them. Thus, according to Game Plan 2, the would-be Casanova, like a sophisticated investor, should seek to maximize his gains while minimizing his losses. What reasoning could be more impeccably judicious?

Game Plan 3. The man silly enough to be satisfied with Game Plan 2 has obviously never approached a woman with an offer of EMS or PMS. Worse yet, the silly man has never even thought about the statistics of human heterosexual intercourse, which dictate that, for every bout of EMS by a man, there must be one bout of EMS (or at least PMS) by a woman. Game Plans 1 and 2 share the flaw that they ignore considerations of the woman's strategy, without which any male strategy is doomed to failure. So an improved Game Plan 3 must combine a male strategy and a female strategy. But, since one husband suffices to realize a woman's maximum reproductive potential, what could possibly attract a woman to EMS or PMS? This question puzzles the current generation of theoretical sociobiologists with a purely intellectual interest in EMS, just as it has taxed the ingenuity of would-be male adulterers throughout human history.

To PROCEED FURTHER with our theoretical exploration of Game Plan 3, we need rigorous empirical data on EMS. Because surveys of people's sexual habits are notoriously unreliable, let's first turn to some recently published studies of birds that nest as mated pairs in large colonies. These, rather than our closest relatives the apes, are the animals whose mating system most closely resembles our own. Compared to us, birds have the disadvantage that one can't ask them about their motives for EMS, but this is no great loss, as our answers are often lies anyway. The great virtue of colonial birds for EMS research is that one can band the birds in a colony, then sit nearby for hundreds of hours and determine exactly who does what with whom. I am unaware of equivalent information for a large human population.

Important recent observations of adultery among birds were made on five species of herons, gulls, and geese. All five nest in dense colonies composed of nominally monogamous male-female pairs. One parent alone is incapable of rearing a chick, as an unguarded nest is likely to be destroyed while the parent is off food-gathering. Nor is a male capable of feeding or guarding two families simultaneously. Thus, among the ground rules of mating strategy for these colonial birds are the following: polygamy is forbidden; copulation with or by an unmated female is pointless, unless she soon acquires a mate to care for the resulting offspring: but surreptitious fertilization by one male of another male's mate is a viable strategy.

The first study involved great blue herons and great egrets at Hog Island, Texas. In these species the male builds a nest and stays there to court visiting females. Eventually a male and female accept each other and copulate about twenty times. The female then lays eggs and goes off to spend most daylight hours feeding, while the male remains to guard the nest and eggs. During the first day or two after pairing, the male often resumes courting any passing female as soon as his mate leaves to feed, but EMS does not result. Instead, the male's halfway-unfaithful behavior seems to constitute "divorce insurance" that lines up a backup mate for him in case his own mate deserts (she does desert him in up to 20 percent of the pairings reported). The passing "backup" female pursues the courtship out of ignorance: she is seeking a mate and has no way of knowing that the male is already mated, until his spouse returns (which she does at frequent intervals) and drives them off. Eventually, the male gains complete confidence that he will not be deserted, and he ceases to court any passing females.

In the second study, of little blue herons in Mississippi, behavior that might have originated as divorce insurance took a more serious turn. Sixty-two cases of EMS were documented, mostly between a female on her nest and a male from the neighboring nest while the female's mate was busy finding food. Most females initially resisted but then ceased resisting, and some females engaged in more EMS than marital sex. To reduce his own risk of being cuckolded, the adulterous male did his feeding as quickly as possible, returned often to his own nest to guard his mate, and traveled no further than neighboring nests to seek EMS. EMS was usually timed to occur when the female chosen had not yet completed egg laying and could still be fertilized. However, adulterous copulations were quicker than marital copulations (eight versus twelve seconds), hence possibly less effective at fertilizing, and nearly half of all nests involved in EMS were subsequently abandoned.

Among herring gulls in Lake Michigan, 35 percent of mated males were observed to engage in EMS. This percentage is nearly the same as the value of 32 percent reported for young American husbands in a study published by Playboy Press in 1974. But there is a big difference between gulls and humans in female behavior. Whereas Playboy Press reported EMS for 24 percent of young American wives, every mated female gull virtuously rejected adulterous male advances and never solicited the neighboring male in her own mate's absence. Instead, all cases of male EMS involved unmated female gulls practicing PMS. To decrease his own risk of being cuckolded, the male spent more time chasing intruders away from his nest when his mate was fertile than when she was not fertile. As for how the male induced his mate to remain faithful during the time that he was off seeking EMS, his secret—like that of some married men similarly pursuing a mixed reproductive strategy—consisted of feeding her diligently and copulating often whenever she was receptive.

Our final set of rigorous data involves snow geese breeding in Manitoba. Just as I already explained for little blue herons, EMS in snow geese mainly involves a male approaching an initially resisting female on a neighboring nest in the absence of her mate. The mate's absence is usually because he himself is off seeking EMS. It may seem as if the male thereby loses as much as he gains, but a male goose is not so dumb. As long as the female is still laying eggs, her mate remains to guard her. (A nesting female is propositioned fifty times less often in her mate's presence than in his absence.) Only after the female has finished laying does her mate go off on EMS quests, with his paternity assured at home.

Such bird studies illustrate the value of a scientific approach to adultery. They have revealed a series of sophisticated strategies by which adulterous male birds try to have it both ways, so as to obtain confidence of paternity at home while sowing their seed abroad. The strategies include wooing unmated females for "divorce insurance," as long as one feels unsure of one's wife's fidelity; guarding one's fertile spouse; feeding her copiously and copulating with her often, to induce her to remain faithful in one's absence; and coveting one's neighbor's spouse at a time when she is fertile and one's own spouse is no longer fertile. However, not even these applications of the scientific method in all its power sufficed to clarify what, if anything, female birds gain from EMS. One possible answer is that female herons weighing desertion of their mates may use EMS to shop around for a new mate. Another is that some unmated female gulls in colonies with a deficit of males may get fertilized by PMS, and then try to rear the chicks with the help of another, similar female.

The chief limitation of these colonial bird studies is that the females often seem to be unwilling participants in EMS. For understanding of a more active female role, we have no choice but to turn to human studies, riddled as they are with problems of cultural variation, observer bias, and dubiously reliable survey responses.

SURVEYS COMPARING MEN WITH WOMEN in various cultures scattered around the world typically purport to find the following differences: men are more interested in EMS than are women; men are more interested than women in seeking a variety of sexual partners for the sake of variety itself; women's motives for EMS are more likely to be marital dissatisfaction and/or a desire for a lasting new relationship; and men are less selective in taking on casual female sexual partners than women are in taking on casual male partners. For example, among the New Guinea highlanders with whom I work, a man will say he seeks EMS because sex with his own wife (or even wives, in the case of polygynous men) inevitably becomes boring, while a woman who seeks EMS does so mainly because her husband cannot satisfy her sexually (e.g., because of old age). In the questionnaires that several hundred young Americans filled out for a computer dating service, women expressed stronger partner preferences than men did in almost every respect: intelligence, status, dancing ability, religion, race, etc. The only category in which men were more selective than women was physical attractiveness. After dates the men and women filled out "debriefing" questionnaires, with the result that two and a half times as many men as women expressed strong romantic attraction to their computer-selected partner. Thus, the women were choosier, the men more undiscriminating, in their reactions to partners.

Obviously, we are on shaky ground if we expect an honest answer when we ask people their attitudes about EMS. However, people also express their attitudes in laws and behaviors. In particular, some widespread hypocritical and sadistic features of human societies stem from two fundamental difficulties that men face in seeking EMS. First, a man who pursues an MRS is trying to have it both ways: he wishes to obtain sex with other men's wives, while denying sex with his own wife (or wives) to other men. Thus, some men inevitably gain at the expense of other men. Second, as we have discussed, there is a realistic biological basis for men's widespread paranoia about being cuckolded.

Adultery laws provide a clear example of how men have dealt with these dilemmas. Until recently, essentially all such laws—Hebraic, Egyptian, Roman, Aztec, Moslem, African, Chinese, Japanese, and others—were asymmetrical. They existed to secure a married man's confidence in his paternity of his children, and for no other purpose. Hence these laws define adultery by the marital status of the participating woman; that of the participating man is irrelevant. EMS by a married woman is considered an offense against her husband, who is commonly entitled to damages, often including violent revenge or else divorce with refund of the bride price. EMS by a married man is not considered an offense against his wife. Instead, if his partner in adultery is married, the offense is against her husband; if she is unmarried, the offense is against her father or brothers (because her value as prospective bride is reduced).

No criminal law against male infidelity even existed until a French law of 1810, and that law only forbade a married man to keep a concubine in his conjugal house against his wife's wishes. Viewed from the perspective of human history, the absence or near-symmetry of modern western adultery laws is a novelty that only appeared in the last 150 years. Even today, prosecutors, judges, and juries in the United States and England often reduce a homicide charge to manslaughter of the lowest degree, or else acquit altogether, when a husband kills an adulterous wife or her lover caught in the act.

Perhaps the most elaborate system to uphold confidence of paternity was that maintained by Chinese emperors of the T'ang Dynasty. For each of the emperor's hundreds of wives and concubines, a team of court ladies kept records on dates of menstruation, so that the emperor could copulate with that wife on a date likely to result in fertilization. Dates of copulation were also recorded, and as an auxiliary form of record keeping, were commemorated by an indelible tattoo on the woman's arm and by a silver ring on her left leg. It goes without saying that equal thoroughness was applied to excluding men other than the emperor from the harem.

Men of other cultures have resorted to less complicated but even more repulsive means of ensuring paternity. These measures limit sexual access to wives, or else to daughters or sisters who would command a high bride price if delivered as proven virgin goods. Relatively mild measures include close chaperoning or virtual imprisonment of women. Similar purposes are served by the code of "honor and shame" widespread in Mediterranean countries. (Translation: EMS for me but not for you; only the latter is a shame to *my* honor.) Stronger measures include the barbaric mutilations euphemistically and misleadingly termed "female circumcision." These consist of removal of the clitoris or most of the external female genitalia to reduce female interest in sex, marital or extramarital. Men bent on total certainty invented infibulation: suturing a woman's labia majora nearly shut, so as to make intercourse impossible. An infibulated wife can be deinfibulated for childbirth or for reinsemination after each child is weaned, and can be reinfibulated when the husband takes a long trip. Female circumcision and infibulation are still practiced in twenty-three countries today, from Africa through Saudi Arabia to Indonesia.

When adultery laws, imperial records, and coercive restraint still fail to ensure paternity, murder is available as a last resort. The role of sexual jealousy as one of the commonest causes of homicide emerges from studies in many American cities and in many other countries. Usually, the murderer is a husband while the victim is his adulterous wife or her lover; or else the lover kills the husband. The table on the facing page gives some actual numbers for murders committed in Detroit in 1972. Until the formation of centralized political states provided soldiers with loftier motives, sexual jealousy also loomed large in human history as a cause of war. It was Paris's seduction (abduction, rape) of Menelaus's wife Helen that provoked the Trojan War. In the modern New Guinea highlands, only disputes over ownership of pigs rival disputes over sex at triggering war.

Asymmetric adultery laws, tattooing of wives after insemination, virtual imprisonment of women, genital mutilation of women: these behaviors are unique to the human species, defining humanity as much as does invention of the alphabet. More exactly, they are new means to the old evolutionary goal of males' promoting their genes. Some of our other means to this goal are ancient ones shared with many animals, including jealous murder, infanticide, rape, intergroup warfare, and adultery itself. Human male infibulators stitch the vagina closed; some male animals achieve the same result by cementing a female's vagina after copulating with her.

Sociobiologists have had considerable success at understanding the marked differences among animal species in details of these practices. As a result of recent research, it is no longer controversial to conclude that natural selection caused animals to evolve behaviors, as well as anatomical structures, that tend to maximize the number of their descendants. Few scientists doubt that natural selection molded human anatomy. However, no theory has caused such bitter divisions among my fellow biologists today as the claim that natural selection likewise molded our social behavior. Most of the human behaviors discussed in this chapter are considered barbaric by modern western society. Some biologists are outraged not only by the behaviors themselves but also by sociobiological explanations for the evolution of the

*Breakdown of murders caused by sexual jealousy
in the U.S. city of Detroit in 1972*

Total: 58 murders

47 murders precipitated by jealous man:
    16 cases: jealous man killed the unfaithful woman
    17 cases: jealous man killed the rival man
    9 cases: jealous man was killed by the accused woman
    2 cases: jealous man was killed by the accused woman's
        relatives
    2 cases: jealous man killed unfaithful homosexual male
        lover
    1 case: jealous man killed innocent bystander
        accidentally

11 murders precipitated by jealous woman:
    6 cases: jealous woman killed the unfaithful man
    3 cases: jealous woman killed the rival woman
    2 cases: jealous woman was killed by the accused man

behaviors. To "explain" a behavior seems uncomfortably close to defending it.

Like nuclear physics and all other knowledge, sociobiology is available for abuse. People have never lacked pretexts to justify the abuse or killing of other people, but ever since Darwin formulated his theory of evolution, evolutionary reasoning has also been abused as such a pretext. Sociobiological discussions of human sexuality can be seen as seeking to justify men's abuse of women, analogous to the biological justifications advanced for whites' treatment of blacks or Nazis' treatment of Jews. In the critiques that some biologists have directed at sociobiology, two fears recur: that a demonstrated evolutionary basis for a barbaric behavior would seem to justify it; and that a demonstrated genetic basis for the behavior would imply the futility of attempts at change.

In my view, neither fear is warranted. As for the first fear, one can seek to understand how something arose, regardless of whether one considers that something admirable or abominable. Most books analyzing the motives of murderers are not written in an effort to justify murder, but instead to understand its causes as a way of preventing it. As for the second fear, we are not mere slaves to our evolved characteristics, not even to our genetically acquired ones. Modern civilization is fairly successful at thwarting ancient behaviors like infanticide. One of the main objectives of modern medicine is to thwart the effects of our harmful genes and microbes, despite our having come to understand why it's natural for those genes and microbes to tend to kill us. Thus, the case against infibulation does not collapse even if the practice can be shown to be genetically advantageous to male infibulators. Instead, we condemn it because we hold the mutilation of one person by another to be ethically loathsome.

While sociobiology is thus useful for understanding the evolutionary context of human social behavior, this approach still shouldn't be pushed too far. The goal of all human activity can't be reduced to the leaving of descendants. Once human culture was firmly in place, it acquired new goals. Many people debate today whether to have children, and many decide that they prefer to devote their time and energy to other activities. I claim only that evolutionary reasoning is valuable for understanding the origin of human social practices; I don't claim that it's the only way to understand their current forms.

In short, we evolved, like other animals, to win at the contest of leaving as many descendants as possible. Much of the legacy of that game strategy is still with us. But we have also chosen to pursue ethical goals, which can conflict with the goals and methods of our reproductive contest. Having that choice among goals represents one of our most radical departures from other animals.

CHAPTER 5

How We Pick Our Mates
and Sex Partners

ARE THERE ANY UNIVERSAL STANDARDS OF HUMAN BEAUTY AND SEX appeal accepted by peoples as different in appearance as Chinese, Swedes, and Fijians? If not, do we inherit our particular taste in marriage partner through our genes, or do we learn it by looking at other members of our society? How, really, do we pick our sex partners and spouses?

It may be surprising to realize that this problem is one that arose anew during the evolution of the human species—or at least became much more important for us than for the other two chimpanzees. As we've already discussed, our familiar human mating system, based ideally on couples' maintaining ongoing involvement, is a human innovation. Pygmy chimps are the opposite of sexually selective: females mate in sequence with many males, and there is much sexual activity between females and between males as well. Common chimps are not so completely promiscuous—a male and female may sometimes go off and "consort" with each other for a few days—but they still rank as promiscuous by human standards. However, humans are much more selective sexually, since rearing a human child is difficult (at least for hunter-gatherers) without a father's help, and since sex becomes part of the cement that differentiates coparents from other men and women frequently encountered. Choosing a mate or sex partner is not so much a human invention as a reinvention of something practiced by many other (nominally) monogamous animals with lasting pair bonds, and lost by our chimpanzeelike ancestors. Those choosy animals include many bird species, plus our distant ape relatives the gibbons.

We just saw in the last chapter that this ideal depiction of a human society based on monogamous couples coexists with a good deal of extramarital sex. Sex appeal plays an even greater role in our choices of extramarital partners than of spouses, with adulterous women tending to be more finicky than adulterous men. Thus, selection of sex partners, married or otherwise, is another important piece of what defines humanity. It's as basic to our rise from chimpanzee status as is our remodeled pelvis. We'll see that much of what we think of as human racial variation may have arisen as a by-product of the beauty standards by which we pick our bedmates.

IN ADDITION to this theoretical interest, the question how we make these choices is of much personal interest. It preoccupies most of us for much of our lives. Those of us who are still unattached spend daily hours dreaming about whom we will consort with or marry. The question becomes more intriguing when we compare what turns on different people within the same culture. Think of the men or women that you find sexually attractive. If you're a man, for instance, do you prefer women who are blond or brunette, flat-chested or buxom, and with big or small eyes? If you're a woman, do you like men who are bearded or smooth-shaven, tall or short, and smiling or scowling? Probably you don't just go for anyone, but only certain types attract you. Everyone can name friends who got divorced, then chose a second spouse who was the exact image of the first one. A colleague of mine went through a long series of plain, slim, brown-haired, round-faced girlfriends, until he finally found one he got along with and married her. Whatever your own preference, you'll have noticed that some of your friends have completely different tastes.

The particular ideal that each of us pursues is an example of what are called "search images." (A search image is a mental picture against which we compare objects and people around us in order to be able to recognize something quickly, like a Perrier bottle amidst all the other bottled waters on the supermarket shelf, or one's child at a playground with other kids.) How do we develop our private search image for a mate? Do we seek someone familiar and similar to us, or are we more turned on by someone exotic? Would most European men really marry a Polynesian woman if given the chance? Do we seek someone complementary to us so as to fulfill our needs? For instance, there undoubtedly are some dependent men who marry a mothering woman, but how typical are such pairings?

Psychologists have tackled this question by examining many married couples, measuring everything conceivable about their physical appearance and other characteristics, and then trying to make sense out of who married whom. A simple numerical way of describing the result is by means of a statistical index called the correlation coefficient. If you line up one hundred husbands in order of their ranking for some characteristic (say, their height), and if you also line up their one hundred wives with respect to the same characteristic, the correlation coefficient describes whether a man tends to be at the same position in the husbands' line-up as his wife is in the wives' line-up. A correlation coefficient of plus 1 would mean perfect correspondence: the tallest man marries the tallest woman, the thirty-seventh-tallest man marries the thirty-seventh-tallest woman, and so on. A correlation coefficient of minus 1 would mean perfect matching by opposites: the tallest man marries the shortest woman, the thirty-seventh-tallest man marries the thirty-seventh-shortest woman, and so on. Finally, a correlation coefficient of zero would mean that husbands and wives assort completely randomly by height: a tall man is as likely to marry a short woman as a tall woman. These examples are for height, but correlation coefficients can also be calculated for anything else, such as income and IQ.

If you measure enough things about enough couples, here is what you'll find. Not surprisingly, the highest correlation coefficients—typically around +0.9—are for religion, ethnic background, race, socioeconomic status, age, and political views. In other words, most husbands and wives prove to be of the same religion, ethnic background, and so on. Perhaps you also won't be surprised that the next-highest correlation coefficients, usually around +0.4, are for measures of personality and intelligence, such as extroversion, neatness, and IQ. Slobs tend to marry slobs, though the chances of a slob marrying a compulsively neat person aren't as low as the chances of a political reactionary marrying a left-winger.

What about matching of husbands and wives for physical characteristics? The answer isn't one that would leap out at you immediately if you just looked at a few married couples. That's because we don't select our own mates for their bodies as carefully as we select the mates of our show dogs, race horses, and beef cattle. But we select nevertheless. If you measure enough couples, the answer that finally emerges is unexpectedly simple: *on the average,* spouses resemble each other slightly but significantly in almost every physical feature examined.

That's true of all obvious traits you'd first think of when asked to design your ideal beloved—his or her height, weight, hair color, eye color, and skin color. But it's also true of an astonishing variety of other traits that you probably wouldn't have mentioned in your description of the perfect sex partner. Those other traits include ones as diverse as breadth of nose, length of earlobe or middle finger, circumference of wrist, distance between eyes, and lung volume! Experimenters have made this finding for people as diverse as Poles in Poland, Americans in Michigan, and Africans in Chad. If you don't believe it, try noting eye colors (or measuring ear lobes) the next time you are at a dinner party with many couples, and then get your pocket calculator to give you the correlation coefficient.

Coefficients for physical traits are on the average $+0.2$—not so high as for personality traits ($+0.4$) or religion ($+0.9$), but still significantly higher than zero. For a few physical traits the correlation is even higher than 0.2—e.g., an astonishing 0.61 for length of middle finger. At least unconsciously, people care more about their spouse's middle-finger length than about his or her hair color and intelligence!

In short, like tends to marry like. Among the obvious explanations that contribute to these results, one is propinquity: we tend to live in neighborhoods defined by socioeconomic status, religion, and ethnic background. For instance, in large American cities one can point to the rich neighborhoods and the poor neighborhoods, and also to the

Jewish section, Chinese section, Italian section, black section, and so on. We meet people of the same religion when we go to church, and we tend to meet people of similar socioeconomic status or political views in many of our daily activities. Since those contacts give us far more opportunities to meet people like us than unlike us in these respects, of course we're more likely to marry someone of our religion, socioeconomic status, and so on. But we don't live in neighborhoods grouped by length of earlobe, so there must be some other reason why spouses tend to be matched in that respect as well.

Another obvious reason why likes tend to marry likes is that marriage isn't just a choice; it's a negotiation. We don't go out searching until we find a person with the right eye color and length of middle finger, then announce to that person, "You are marrying me." For most of us, marriage results from a proposal rather than a unilateral announcement, and the proposal is the culmination of some sort of negotiation. The more similar a man and woman are in political views, religion, and personality, the smoother will be the negotiation. The match in personality traits is on the average closer for married couples than for dating couples, closer for happily than unhappily married couples, and closer for couples who stay married than for those who get divorced. But this still doesn't explain spousal resemblance in earlobe length, which is only rarely cited as a factor in divorce.

The remaining factor deciding whom you'll marry, besides propinquity and smoothness of negotiation, is surely sexual attraction based on physical appearance. That in itself is no surprise. Most of us are aware of our preferences in obvious visible features like height, build, and hair color. What's initially surprising is the importance of so many other physical traits that we usually don't consciously notice, such as earlobes, middle fingers, and interocular distances. Nevertheless, all those other traits contribute unconsciously to the snap decisions we make when we're introduced to someone and a voice inside tells us, "She's my type!"

Here's an example. When my wife and I were introduced to each other, I instantly found Marie attractive and vice versa. In retrospect, I can understand why: we're both brown-eyed, similar in height and build and hair color, and so on. But, on the other hand, I also had a sense that there was something about Marie that didn't quite match my ideal, even though I couldn't figure out what exactly it was. Not until Marie and I first went to a ballet together did I solve the puzzle. I lent Marie my opera glasses, and when she passed them back to me, I found that she had pushed the eyepieces so close together that I couldn't see through them until I had spread them apart again. I then realized that Marie has more close-set eyes than I do, and that most women whom I had pursued before had wide-set eyes like my own. Thanks to Marie's earlobes and other merits, I've been able to make peace with my and her mismatched interocular distances. Nevertheless, the episode with the opera glasses made me appreciate for the first time that I have always found wide-set eyes a turn-on, even though I hadn't been explicitly aware of it.

So we tend to marry someone who looks like us. But—wait a minute. The men who look most similar to a woman are the men who share half of her genes: her father or brother! Similarly, the best-matched mate for a man would be his mother or sister! Yet most of us obey the incest taboo and certainly do not marry our parent or sibling of the opposite sex.

Instead, I'm saying that people tend to marry a person who *looks like* the parent or sibling of the opposite sex. That's because we begin already as children to develop our search image of a future sex partner, and that image is heavily influenced by the people of opposite sex whom we see most often. For most of us that's our mother (or father) and sister (or brother), plus close childhood friends. Our behavior is summed up by a popular song of the 1920s,

> *I want a girl*
> *Just like the girl*
> *That married dear old Dad . . .*

AT THIS POINT, you are probably turning to your spouse or significant other, pulling out your tape measure, and discovering gross mismatch between your and his (or her) earlobes. Or perhaps you've pulled out a photo of your mother or sister, and you detect not the faintest resemblance when you hold it beside your spouse. If your wife isn't a dead ringer for your mother, don't stop reading, and conversely don't get worried that you should see a psychiatrist about your pathological search image. After all, remember:

1. Studies consistently show that factors like religion and personality influence our choice of spouse much more strongly than does physical appearance. I'm merely saying that physical traits have *some* influence. In fact, I'd predict much higher correlation coefficients for physical traits between casual sex partners than between spouses. That's because we can select casual sex partners solely on the basis of physical attraction, without regard to religion or political views. This prediction awaits testing.

2. Remember also that your search image could have been influenced by any of the people of opposite sex that you regularly saw around you as you were growing up. That includes playmates and siblings as well as parents. Perhaps your spouse resembles the little girl next door, rather than Mother.

3. Finally, remember that lots of independent physical traits enter into our search images, so most of us end up with a mild average resemblance to our spouses in many traits, rather than with a very close resemblance in a few traits. This idea is known as the "buxom redhead theory." If a man's mother and sister were both buxom redheads, he might grow up to consider buxom redheads very exciting. But redheads are relatively rare, and buxom redheads still rarer. Furthermore, the man's preference even in a casual sex partner is likely to depend on some other physical traits as well, and his preference in a wife will certainly depend on her views about children, politics, and money. As a result, among a group of sons of buxom redheads, a few lucky ones will find a girl like Mother in those two respects, some will have to settle for buxom nonredheads, others for nonbuxom redheads, and most for run-of-the-mill nonbuxom brunettes.

You may also be objecting at this point that my argument applies only to societies where spouses pick each other. As friends from India and China are quick to remind me, that's a peculiar custom of the twentieth-century U.S. and Europe. It wasn't true of the U.S. and Europe in the past, and it's still not true of most of the world today, where marriages are instead arranged by the families involved. The bride and groom often aren't even introduced until the wedding day. How could my argument possibly apply to such marriages?

Of course it couldn't, if one is talking just about legal marriages. But my argument would still apply to choice of extramarital sex partners, who may father a nontrivial fraction of children, just as blood-group studies proved for American and British children. In fact, I'd expect that if extramarital fathering is frequent even in societies where a woman already exercises her sexual preferences in choosing a husband, it may be even more frequent in societies with arranged marriages, where a woman's choice can only be expressed extramaritally.

It's NOT JUST THE CASE, then, that Fijian men prefer Fijian women over Swedish women, and vice versa: our search images are much more specific. However, these insights still leave questions unanswered. Did I inherit or learn my search image for someone like Mother? If I were offered the choice of sex with my sister or a strange woman, I would certainly reject the offer of my sister and probably my first cousin, but would I prefer my second cousin over a strange woman (because the cousin probably resembles me more)? There are some crucial experiments that would settle these questions—e.g., keeping a man in a large cage with his female first, second, third, fourth, and fifth cousins, counting how many times he had sex with each, and repeating the experiment with many men (or women) and their cousins. Alas, such experiments are hard to do in humans, but they've been done for several animal species, with instructive results. I'll give just three examples: the cousin-loving quail, and the perfumed mice and rats. (We can't use our closest relatives the chimpanzees for these examples, since they're so unselective.)

Consider first the case of Japanese quail, which normally grow up with their biological parents and siblings. However, it's also possible to "cross-foster" quail by switching eggs between quail mothers and their nests before the eggs hatch. In that way, a baby quail may be reared by foster parents and grow up with "pseudo-siblings"—i.e., litter mates among whom the baby hatched but to whom the baby isn't genetically related.

The preferences of male quail have been tested by putting a male in a cage with two females and observing with which female the male spent more time or copulated. When a male was given a choice between females that he had never seen before (although some were his relatives that had been separated from him before hatching), he preferred his first cousin to his third cousin or an unrelated female, but he also preferred his first cousin to his sister. Evidently, male quail as they grow up learn the appearance of their sisters (or mother) with whom they are reared, then seek a mate that is very similar but not *too* similar. In fancy technical language, biologists term this the Principle of Optimal Intermediate Similarity. Like other things in life, inbreeding seems to be good in moderation—a little inbreeding, but not too much. For instance, among unrelated females a male prefers an unfamiliar one over a familiar one with whom he grew up (a "pseudo-sister," who pushes the male's not-too-much-incest button).

Mice and rats similarly learn in childhood what to look for in mates, but they choose by smell more than by appearance. When infant female mice were reared by parents sprayed repeatedly with Parma Violet perfume, the females on reaching adulthood sought out Parma Violet–scented males in preference to unscented males. In another experiment, infant male rats were reared by mother rats whose nipples and vagina were sprayed with lemon odor, then the male on reaching adulthood was put in a cage with a lemon-smelling or unscented female rat. Each such encounter was videotaped and played back to note the times of key events. It turned out that males with scented mothers mounted and ejaculated more quickly when placed with a scented female than an unscented one, while the reverse was true for males with unscented mothers. For example, sons of scented mother rats were so excited by a scented sex partner that they ejaculated in only 11½ minutes, while they took over 17 minutes to ejaculate with an unscented female. But sons of unscented mother rats took over 17 minutes with the *scented* partner and only 12 minutes with the *unscented* partner. Obviously, the males had learned to be sexually excited by their mother's smell (or lack of smell); they did not inherit the knowledge.

WHAT DO THESE EXPERIMENTS on quail, mice, and rats show? The message is clear: animals of those species learn to recognize their parents and siblings as they grow up, then are programmed to seek out an individual fairly similar to the parent or sibling of opposite sex—but not Mother or Sister herself. They may *inherit* some search image of what constitutes a rat, but they evidently *learn* their search image of who in particular is a beautiful, eligible rat.

We can immediately appreciate what experiments are needed to get unequivocal proof of this theory for humans. We should take an average happy family, spray Father every day with Parma Violet, spray Mother's nipples daily with lemon oil while she is nursing, and then wait twenty years to see whom the sons and daughters marry. Alas, we would be frustrated by the many obstacles to establishing Scientific Truth for humans. But some observations and accidental experiments still let us tiptoe toward the truth.

Take the incest taboo. Scientists debate whether the taboo itself in humans is instinctive or learned. Given that we somehow acquire an incest taboo, do we learn to whom to apply it, or do we inherit that information in our genes? Normally we grow up with our closest relatives (parents and siblings), so our subsequent avoidance of them as sex partners could equally well be genetic or learned. But adoptive brothers and sisters also tend to avoid incest, suggesting learned avoidance.

This conclusion is strengthened by an interesting set of observations made in Israeli kibbutzim—the collective settlements whose members house, school, and care for all their children together as a large group. Kibbutz children live from birth until young adulthood in intimate association with each other, like a gigantic family of brothers and sisters. If propinquity were the main factor influencing whom we marry, most kibbutz children should marry within the kibbutz. In fact, a study of 2,769 marriages contracted by kibbutz-reared children turned up only thirteen between children from the same kibbutz; all the other children married outside the kibbutz on reaching maturity.

Even those thirteen cases turned out to be the exceptions that proved the rule: all involved couples one of whom had moved into that kibbutz only after the age of six! Among children reared in the same peer group since birth, there were not only no marriages, but also no cases of adolescent or adult heterosexual activity at all. This is astonishing restraint on the part of nearly three thousand young men and women who enjoyed daily opportunities for sexual involvement with each other, and who had far fewer opportunities for involvement with outsiders. It illustrates dramatically that the period between birth and age six is a critical time for formation of our sexual preferences. We *learn,* however unconsciously, that our intimate associates from that period are ineligible as sex partners when we become mature.

We also appear to learn the part of our search image that tells us whom to seek, not just the part that tells us whom to avoid. For instance, a friend of mine who is 100 percent Chinese herself happened to grow up in a community in which every other family was white. Eventually she moved as an adult to an area with many Chinese men, and for some time she dated both Chinese and white men, but came to realize that it was the whites who attracted her. She has been married twice, both times to white men. Her own experiences led her to ask her Chinese women friends about their backgrounds. It turned out that most of her friends reared in white enclaves also ended up marrying white men, while those reared in Chinese neighborhoods married Chinese men—although all had plenty of men of both types from whom to choose during their young adult years. Hence those who surround us as we grow up, though ineligible themselves as eventual mates, nevertheless shape our beauty standards and search images.

Think to yourself: what sort of men or women do you find physically attractive, and where did you develop that taste? I'd guess that most people, like me, can trace their preferences to the appearance of parents or siblings or childhood friends. So don't be discouraged by all those old generalizations about sex appeal—"Gentlemen prefer blondes," "Men seldom make passes at girls who wear glasses," etc. Each such "rule" applies only to some of us, and there are plenty of men out there whose mothers were myopic brunettes. Fortunately for my wife and me—both of us brunettes requiring glasses, born of brunette glasses-wearing parents—beauty is in the eye of the beholder.

CHAPTER 6

Sexual Selection,
and the
Origin of Human Races

WHITE MAN! LOOKIM THIS-FELLER LINE THREE-FELLER MAN. This-feller number-one he belong Buka Island, na nother-feller number-two he belong Makira Island, na this-feller number-three he belong Sikaiana Island. Yu no savvy? Yu no enough lookim straight? I think, eye-belong-yu he bugger-up finish?"

No, damn it, my eyes-belong-me were not ruined beyond repair. It was my first visit to the Solomon Islands in the Southwest Pacific, and I told my scornful guide through the medium of pidgin English that I saw perfectly well the differences between those three men in a row over there. The first one had jet-black skin and frizzy hair, the second had much lighter skin and frizzy hair, and the third had straighter hair and more slanted eyes. The only thing the matter with me was that I had no experience of what people from each particular Solomon island looked like. By the end of my first trip through the Solomons, I too could identify people to islands by their skin and hair and eyes.

In those variable features, the Solomons are a microcosm of humanity. Simply by looking at a person, even laymen can often tell what part of the world that person comes from, and trained anthropologists may be able to "place" him or her to the right part of the right country. For example, given one person each from Sweden, Nigeria, and Japan, none of us would have any trouble deciding at a glance which person was from which country. The most visibly variable features in clothed people are of course skin color, the color and form of the eyes and hair, body shape, and (in men) the amount of facial hair. If the people to be identified were undressed, we might also notice differences in amount of body hair, the size and shape and color of a woman's breasts and nipples, the form of her labia and buttocks, and the size and angle of a man's penis. All those variable features contribute to what we know as human racial variation.

Those geographic differences among humans have long fascinated travelers, anthropologists, bigots, and politicians, as well as the rest of us. Since scientists have solved so many arcane questions about obscure unimportant species, surely you might expect them to have answered one of the most obvious questions about ourselves: "Why do people from different areas look different?" Our understanding of how humans came to differ from other animals would remain incomplete if we didn't also consider how, in the process, human populations acquired their most visible differences from each other. Nevertheless, the subject of human races is so explosive that Darwin excised all discussion of it from his famous 1859 book *On the Origin of Species*. Even today, few scientists dare to study racial origins, lest they be branded racists just for being interested in the subject.

But there's another reason why we don't understand the significance of human racial variation: it proves to be an unexpectedly difficult problem. Twelve years after Darwin wrote his book attributing the origin of species to natural selection, he wrote another book 898 pages long, attributing the origin of human races to our sexual preferences, which I described in the last chapter, and entirely rejecting a role of natural selection. Despite that verbal overkill, many readers were unconvinced. To this day, Darwin's theory of sexual selection (as he called it) remains controversial. Instead, modern biologists generally invoke natural selection to explain the visible differences among human races—especially the differences in skin color, whose relation to sun exposure seems obvious. However, biologists can't even agree on why natural selection led to dark skin in the tropics. I shall explain why I believe natural selection to have played only a secondary role in our racial origins, and why Darwin's preference for sexual selection seems to me correct. I thus consider visible human racial variation to be largely a by-product of the remodeled human life cycle.

FIRST, TO PLACE MATTERS IN PERSPECTIVE, let's realize that racial variation is not at all confined to humans. Most animal and plant species with sufficiently wide distributions, including all higher ape species except the geographically localized pygmy chimp, also vary geographically. So marked is variation in some bird species, such as North America's white-crowned sparrow and Eurasia's yellow wagtail, that experienced bird-watchers can identify an individual bird's approximate birthplace by its plumage pattern.

Variation in apes encompasses many of the same characteristics that vary geographically in humans. For example, among the three recognized races of gorillas, western lowland gorillas have the smallest bodies and rather gray or brown hair, while mountain gorillas have the longest hair, and eastern lowland gorillas share black hair with mountain gorillas. Races of white-handed gibbons similarly vary in hair color (variously black, brown, reddish, or gray), hair length, tooth size, protrusion of the jaws, and protrusion of the bony ridges over the eyes. All these traits that I have just mentioned as varying among gorilla or gibbon populations also differ among human populations.

How does one decide whether recognizably distinct animal populations from different localities constitute different species, or belong instead to the same species and just constitute different races (also known as subspecies)? As I have explained, the distinction is based on interbreeding under normal circumstances: members of the same species may interbreed normally if given the opportunity, while members of different species don't. (But closely related species that wouldn't normally interbreed in the wild, like lions and tigers, may do so if a male of one is caged with a female of the other and given no other choice.) By this criterion, all living human populations belong to the same species, since some interbreeding has occurred whenever humans from different regions have come into contact— even people as dissimilar in appearance as African Bantus and Pygmies. With humans as with other species, populations may intergrade into each other, and it becomes arbitrary which populations to group as races. By the same criterion of interbreeding, the large gibbons known as siamangs are a species distinct from the smaller gibbons, since both occur together in the wild without hybridizing. This is also the criterion for considering Neanderthals possibly as a distinct species from *Homo sapiens,* since hybrid skeletons have not been identified despite apparent Cro-Magnon-Neanderthal contact.

Racial variation has characterized humans for at least the past several thousand years, and possibly much longer. Already around 450 B.C., the Greek historian Herodotus described the Pygmies of West Africa, the black-skinned Ethiopians, and a blue-eyed, red-haired tribe in Russia. Ancient paintings, mummies from Egypt and Peru, and bodies of people preserved in European peat bogs confirm that people several thousand years ago differed in their hair and faces much as they do today. Origins of modern races can be pushed back still further, to at least ten thousand years ago, since fossil skulls of that age from various parts of the world differ in many of the same respects that modern skulls from the same regions differ. More controversial are the studies of some anthropologists, contested by others, reporting continuity of racial skull characteristics for hundreds of thousands of years. If those studies are correct, then some of the human racial variation that we see today may predate the Great Leap Forward, and may have gone back to *Homo erectus* times.

Now LET'S TURN to the question whether natural selection or sexual selection has made the larger contribution to those visible geographic differences of ours. Take first the arguments about natural selection, the selection of traits that enhance survival. No scientist denies today that natural selection does account for many of the differences between species, like the fact that lions have paws with claws while we have grasping fingers. No one denies either that natural selection explains some geographic variation ("racial variation") within some animal species. For instance, Arctic weasels that live in areas covered by winter snow change color from brown in summer to white in winter, while more southerly weasels stay brown all year. That racial difference enhances survival, because white weasels against a brown background would be glaringly conspicuous to their prey but are camouflaged against snow.

By the same token, natural selection surely explains *some* geographic variation in humans. Many African blacks but no Swedes have the sickle-cell hemoglobin gene, because the gene protects against malaria, a tropical disease that would otherwise kill many Africans. Other localized human traits that surely evolved through natural selection include the big chests of Andean Indians (good for extracting oxygen from thin air at high altitudes), the compact shapes of Eskimos (good for conserving heat), the slender shapes of southern Sudanese (good for losing heat), and the slitlike eyes of north Asians (good for protecting eyes against cold and against sun glare off the snow). All these examples are easy to understand.

Can natural selection similarly explain the racial differences that we think of first, those in skin color and eye color and hair? If so, one might expect that the same trait (e.g., blue eyes) would reappear in different parts of the world with similar climates, and that scientists would agree on what the trait is good for.

Seemingly the simplest trait to understand is skin color. Our skins run the spectrum from various shades of black, brown, copper, and yellowish to pink with or without freckles. The usual story to explain this variation by natural selection goes as follows. People from sunny Africa have blackish skins. So too (supposedly) do people from other sunny places, like southern India and New Guinea. Skins are said to get paler as one moves north or south from the equator, until one reaches northern Europe, with the palest skins of all. Obviously, dark skins evolved in those people who were exposed to much sunlight. That's just like the skins of whites tanning under the summer sun (or in tanning parlors!), except that tanning is a reversible response to sun rather than a permanent genetic one. It's equally obvious what good a dark skin does in sunny areas: it protects against sunburn and skin cancer. Whites who spend lots of time outdoors in the sun tend to get skin cancer, and they get it on exposed parts of their body like their head and hands. Doesn't that all make sense?

Unfortunately, it's really not so simple at all. To begin with, skin cancer and sunburn cause little debilitation and few deaths. As agents of natural selection, they have an utterly trivial impact compared to infectious diseases of childhood. Many other theories have instead been proposed to explain the supposed pole-to-equator gradient in skin color.

One favorite competing theory notes that the sun's ultraviolet rays promote vitamin D formation in a layer of our skin beneath the main pigmented layer. Thus, people in sunny tropical areas might have evolved dark skin to protect them against the risk of kidney disease caused by too much vitamin D, while people in Scandinavia with its long, dark winters evolved pale skins to protect them against the risk of rickets caused by too little vitamin D. Two other popular theories: dark skins are to protect our internal organs against overheating by the tropical sun's infrared rays; or—just the opposite—dark skins help keep tropical people warm when the temperature drops. And if those four theories aren't enough for you, consider four more: that dark skins provide camouflage in the jungle, or that pale skins are less sensitive to frostbite, or that dark skins protect against beryllium poisoning in the tropics, or that pale skins cause deficiency of another vitamin (folic acid) in the tropics.

With at least eight theories in the running, we can hardly claim to understand why people from sunny climates have dark skins. That in itself doesn't refute the idea that, somehow, natural selection caused the evolution of dark skins in sunny climates. After all, dark skins could have multiple advantages, which scientists may sort out some-day. Instead, the heaviest objection to any theory based on natural selection is that the association between dark skins and sunny cli-mates is a very imperfect one. Native peoples had very dark skins in some areas receiving relatively little sunlight, like Tasmania, while skin color is only medium in sunny areas of tropical Southeast Asia. No American Indians have black skins, not even in the sunniest parts of the New World. When one takes cloud cover into account, the world's most dimly lit areas, receiving a daily average of under 3½ hours of sunlight, include parts of equatorial West Africa, South China, and Scandinavia, inhabited respectively by some of the world's blackest, yellowest, and palest peoples! Among the Solomon Islands, all of which share a similar climate, jet-black people and lighter people replace each other over short distances. Evidently, sunlight has not been the sole selective factor that influenced skin color.

The first response of anthropologists to these objections is to raise a counterobjection: the time factor. This argument tries to explain away the cases of pale-skinned people in the tropics by claiming that those particular peoples migrated to the tropics too recently to have evolved black skins. For example, the ancestors of American Indians may have reached the New World only eleven thousand years ago:

perhaps that hasn't been long enough to evolve black skins in the tropical Americas. But if you're going to evoke the time factor to explain away objections to the climate theory of skin color, then you also have to consider the time factor for peoples who supposedly support that theory.

One of the prime supports of the climate theory is the pale skins of Scandinavians, living in the cold, dark, foggy North. Unfortunately, Scandinavians have been in Scandinavia for an even shorter time than American Indians have been in the Amazon. Until about nine thousand years ago, Scandinavia was covered by an ice sheet and could hardly have supported any people, pale-skinned or dark-skinned. Modern Scandinavians reached Scandinavia only around four or five thousand years ago, as a result of the expansion of farmers from the Near East and of Indo-European speakers from southern Russia. Either Scandinavians acquired their pale skins long ago in some other area with a different climate, or else they acquired them in Scandinavia within half the time that Indians have spent in the Amazon without becoming dark-skinned.

The sole people in the world about whom we can be certain that they spent the last ten thousand years in the same location were the natives of Tasmania. Lying south of Australia, at the temperate latitude of Chicago or Vladivostok, Tasmania used to be connected to Australia until it was cut off by rising sea level ten thousand years ago and became an island. Since modern Tasmanian natives didn't have boats capable of going more than a few miles, we know that they were derived from colonists who walked out to Tasmania at the time of its connection to Australia, and who remained there continuously until they were exterminated by British colonists in the nineteenth century. If any people had enough time for natural selection to match their skin color to their local temperate-zone climate, it was the Tasmanians. Yet they had blackish skins, supposedly adapted to the equator.

If the case for natural selection of skin color seems weak, that for hair color and eye color is virtually nonexistent. There are no consistent correlations with climate, and not even any halfway-plausible theories for the supposed advantage lent by each color type. Blond hair is common in cold, wet, dimly lit Scandinavia and also among Aborigines of the hot, dry, sunny desert of central Australia. What do those two areas have in common, and how does being blond help both Swedes and Aborigines to survive? Blue eyes are common in Scandinavia and supposedly help their owners see farther in dim, misty light. But that speculation is unproven, and all my friends in the even dimmer, mistier mountains of New Guinea see just fine with their dark eyes.

The racial traits for which it seems most absurd to seek an explanation based on natural selection are our variable genitalia and secondary sex characteristics. Are hemispherical breasts an adaptation to summer rainfall and conical breasts an adaptation to winter fog, or vice versa? Do the protruding labia minora of Bushman women protect them against pursuing lions, or reduce their water losses in the Kalahari Desert? You surely don't think that men with hairy chests can thereby keep warm while going shirtless in the Arctic, do you? If you do think so, then please explain why women don't share hairy chests with men, since women also have to keep warm.

FACTS SUCH AS THESE were what made Darwin despair of imputing human racial variation to his own concept of natural selection. He finally dismissed the attempt with a succinct statement: "Not one of the external differences between the races of man are of any direct or special service to him." When Darwin came up with a theory that he preferred, he termed it "sexual selection" to contrast with natural selection, and he devoted an entire book to explaining it.

The basic notion behind this theory is easily grasped. Darwin noted many animal features that had no obvious survival value but that did play an obvious role in securing mates, either by attracting an individual of the opposite sex or by intimidating a rival of the same sex. Familiar examples are the tails of male peacocks, the manes of male lions, and the bright red buttocks of female baboons in estrus. If an individual male is especially successful at attracting females or intimidating rival males, that male will leave more descendants and will tend to pass on his genes and traits—as a result of sexual selection, not natural selection. The same argument applies to female traits as well.

For sexual selection to work, evolution must produce two changes simultaneously: one sex must evolve some trait, and the other sex must evolve in lockstep a liking for that trait. Female baboons could hardly afford to flash red buttocks if the sight revolted male baboons to the point of their becoming impotent. As long as the female has it and the male likes it, sexual selection could lead to any arbitrary trait, just as long as it doesn't impair survival too much. In fact, many traits produced by sexual selection do seem quite arbitrary. A visitor from Outer Space who had yet to see humans could have no way of predicting that men rather than women would have beards, that the beards would be on the face rather than above the navel, and that women would not have red and blue buttocks.

That sexual selection really can work was proved by an elegant experiment carried out by the Swedish biologist Malte Andersson on the long-tailed widowbird of Africa. In this species the male's tail in the breeding season grows to twenty inches long, while the female's tail is only three inches. Some males are polygamous and acquire up to six mates, at the expense of other males who get none. Biologists had guessed that a long tail served as an arbitrary signal by which males attracted females to join their harem. Hence Andersson's test was to cut off parts of the tails from nine males so that their tails were only six inches long. He then glued those cut segments to the tails of nine other males to give them thirty-inch tails, and he waited to see where the females built their nests. It turned out that the males with the artificially lengthened tails attracted on the average over four times as many mates as the males with artificially shortened tails.

Perhaps our first reaction to Andersson's experiment is: Those dumb birds! Imagine a female selecting a particular male to father her offspring merely because his tail is longer than other males' tails! But before we get too smug, let's consider again what we learned in the last chapter about how we humans select our own mates. Are our criteria such good indicators of genetic worth? Don't some men and women set disproportionate value on the size or form of certain body parts, which are really nothing more than arbitrary signals for sexual selection? Why did we evolve to pay any attention at all to a beautiful face, which is useless to its owner in the struggle for survival?

In animals some of the traits that vary racially are ones produced by sexual selection. For instance, lions' manes vary in length and in color. Similarly, snow geese occur in two color phases, a blue phase commoner in the western Arctic and a white phase commoner in the eastern Arctic. Birds of each phase prefer a mate of the same phase. Could human breast shape and skin color similarly be the outcome of sexual preferences that vary arbitrarily from area to area?

After 898 pages of his book Darwin convinced himself that the answer to this question was a resounding "yes." He noted that we pay inordinate attention to breasts, hair, eyes, and skin color in selecting our mates and sex partners. He noted also that people in different parts of the world define beautiful breasts, hair, eyes, and skin by what is familiar to them. Thus, Fijians, Hottentots, and Swedes each grow up with their own learned, arbitrary beauty standards, which tend to maintain each population in conformity with those standards, since individuals deviating too far from the standards would find it harder to obtain a mate.

Darwin died before his theory could be tested against rigorous studies of how people actually do select their mates. Such studies have proliferated in recent decades, and I summarized the results in the preceding chapter. There I showed that people tend to marry individuals who resemble themselves in every conceivable character, including hair and eye and skin color. To explain that seeming narcissism of ours, I reasoned that we develop our beauty standards by imprinting on the people we see around us in childhood—especially on our parents and siblings, the people we see the most. But our parents and siblings are also the people to whom we bear the strongest physical resemblance, since we share their genes. Thus, if you're a fair-skinned, blue-eyed blond who grew up in a family of fair-skinned blue-eyed blonds, that's the sort of person whom you'll consider most beautiful and will seek as a mate.

To test that imprinting theory of human mate choice rigorously, one would have to do experiments like shipping some Swedish babies to adoptive parents in New Guinea, or painting some Swedish parents permanently black. Then, after waiting twenty years for the babies to grow up, one could study whether they preferred Swedes or New Guineans as sex partners. Alas, once again, the Search for Truth about humans founders on practical problems. But such tests can be performed with full experimental rigor on animals.

Take snow geese, for example, with their blue and white color phases. Do white geese learn or inherit their preference in the wild for white geese over blue ones? Canadian biologists hatched gosling eggs in an incubator, then put the goslings into a nest of goose "foster parents." When those goslings grew up, they chose mates with the color of the foster parents. When goslings were reared in a large mixed flock of both blue and white birds, they showed no preference between blue and white prospective mates on reaching adulthood. Finally, when the biologists dyed some white parents pink, their offspring came to prefer pink-dyed geese. Thus, geese do not inherit but learn a color preference, by imprinting on their parents (and on their siblings and playmates).

How, THEN, do I think that people in different parts of the world evolved their differences? Our insides remained invisible to us and were molded only by natural selection, with such results as that tropical Africans but not Swedes got the antimalarial defense of the sickle-cell hemoglobin gene. Many visible features of our outsides also got molded by natural selection. But, just as in animals, sexual selection had a big effect in molding the external traits by which we pick our mates.

For us humans those traits are especially the skin, eyes, hair, breasts, and genitals. In each part of the world those traits evolved in lockstep with our imprinted aesthetic preferences to reach different, somewhat arbitrary end points. Which particular human population ended up with any given eye or hair color may have been partly an accident of what biologists term the "founder effect." That is to say, if a few individuals colonize an empty land and their descendants then multiply to fill the land, the genes of those few founding individuals may still dominate the resulting population many generations later. Just as some birds of paradise ended up with yellow plumes and others with black plumes, so some human populations ended up with yellow hair and others with black hair, some with blue eyes and others with green eyes, some with orange nipples and others with brown nipples.

I don't mean thereby to claim that climate has nothing whatsoever to do with skin color. I acknowledge that tropical peoples tend on the average to have darker skins than temperate-zone peoples, though there are many exceptions, and that this is probably due to natural selection, though we're unsure of the exact mechanism. Instead, I'm saying that sexual selection has been strong enough to render the correlation between skin color and sun exposure quite imperfect.

If you're still skeptical about how traits and aesthetic preferences can evolve together to different arbitrary end points, just think about our changing fashion preferences. When I was a schoolboy in the early 1950s, women rated men with crew cuts and clean-shaven faces as handsome. Since then, we've seen a parade of men's fashions, including beards, long hair, earrings, purple-dyed hair, and the Mohawk hair style. A man daring to flaunt any of those fashions in the 1950s would have revolted the girls and enjoyed zero mating success. That's not because crew cuts were better adapted to the atmospheric conditions of Stalin's last years, while a purple Mohawk has higher survival value in our post-Chernobyl era. Instead, men's appearances and women's tastes changed together, and the changes occurred far more rapidly than evolutionary changes in skin color, since no gene mutations were required. Either women came to like crew cuts because good men had them, or men adopted crew cuts because good women liked them, or something of both happened. The same goes for women's appearances and men's tastes.

To a zoologist, the visible geographic variability that sexual selection produced in humans is impressive. I've argued that much of our variability is a by-product of a distinctive feature of the human life cycle: our choosiness with respect to our spouses and sex partners. I don't know of any other wild animal species in which eye color of different populations can be green, blue, gray, brown, or black, while skin color varies geographically from pale to black and hair is either red, yellow, brown, black, gray, or white. There may be no limits, except those imposed by evolutionary time, on the colors with which sexual selection can adorn us. If humanity survives another twenty thousand years, I predict that there will be women with naturally green hair and red eyes—and men who think such women are the sexiest.

CHAPTER 7

Why
Do We Grow Old
and Die?

DEATH AND AGING CONSTITUTE A MYSTERY THAT WE OFTEN ASK about as children, deny in youth, and reluctantly come to accept as adults. I scarcely reflected on aging when I was a college student. Now that I am fifty-four years old, I find it decidedly more interesting. Life expectancy among U.S. white adults is presently about seventy-eight years for men, eighty-three for women. But few of us will survive to one hundred. Why is it so easy to live to eighty, so hard to live to one hundred, and almost impossible to live to 120? Why do humans with access to the best medical care, and animals kept in a cage with plenty of food and no predators, inevitably grow infirm and die? It's the most obvious feature of our life cycle, but there's nothing obvious about what causes it.

In the bare fact of our aging and dying we resemble all other animals. In the details, however, we've improved considerably over the course of our evolutionary history. Not a single individual of any ape species has been recorded as achieving the current life expectancy of U.S. whites, and only exceptional apes reach their fifties. Evidently, we age more slowly than do our closest relatives. Some of that slowdown may have developed recently, around the time of the Great Leap Forward, since quite a few Cro-Magnons lived into their sixties while few Neanderthals passed forty.

Slow aging is as crucial to the human life-style as are marriage, concealed ovulation, and the other life-cycle features that we've been discussing in the preceding chapters. That's because our life-style depends on transmitted information. As language evolved, far more information became available to us to pass on than previously. Until the invention of writing, old people acted as the repositories of that transmitted information and experience, just as they continue to do in tribal societies today. Under hunter-gatherer conditions, the knowledge possessed by even one person over the age of seventy could spell the difference between survival and starvation for a whole clan. Our long life span, therefore, was important for our rise from animal to human status.

Obviously, our ability to survive to a ripe old age depended ultimately upon advances in culture and technology. It's easier to defend yourself against a lion if you're carrying a spear than just a hand-held stone, and easier yet with a high-powered rifle. However, advances in culture and technology alone would not have been enough, unless our bodies had also become redesigned to last longer. No caged ape in a zoo, enjoying all the benefits of modern human technology and veterinary care, reaches eighty. We'll see in this chapter that our biology became remolded to the increased life expectancy that our cultural advances made possible. In particular, I'd guess that Cro-Magnon tools weren't the sole reason why Cro-Magnons lived on the average longer than Neanderthals. Instead, around the time of the Great Leap Forward our biology must also have changed so that we aged more slowly. That may even have been the time when menopause, the concomitant of aging that paradoxically functions to let women live longer, evolved.

THE WAY IN WHICH SCIENTISTS THINK about aging depends on whether they are interested in so-called proximate explanations or ultimate explanations. To appreciate this difference, consider the question, "Why do skunks smell bad?" A chemist or molecular biologist would answer, "It's because skunks secrete chemical compounds with certain particular molecular structures. Owing to the principles of quantum mechanics, those structures result in bad smells. Those particular chemicals would smell bad no matter what the biological function of their bad smell was."

But an evolutionary biologist would instead reason, "It's because skunks would be easy victims for predators if they didn't defend themselves with bad smells. Natural selection made skunks evolve to secrete bad-smelling chemicals; those skunks with the worst smells survived to produce the most baby skunks. The molecular structure of those chemicals is a mere incidental detail; any other bad-smelling chemicals would suit skunks equally well."

The chemist has offered a proximate explanation: i.e., the mechanism immediately responsible for the observation that was to be explained. The evolutionary biologist has instead offered an ultimate explanation: the function or chain of events that caused that mechanism to be present. The chemist and the evolutionary biologist would each dismiss the other's answer as not being "the real explanation."

Similarly, studies of aging are pursued independently by two groups of scientists who scarcely communicate with each other. One group seeks a proximate explanation, the other an ultimate explanation. Evolutionary biologists try to understand how natural selection could ever permit aging to occur, and they think that they've found an answer to this question. Physiologists inquire instead into the cellular mechanisms underlying aging, and admit that they don't yet have an answer. But I'll argue that aging can't be understood unless we seek both explanations simultaneously. In particular, I expect that the evolutionary (ultimate) explanation will help us find the physiological (proximate) explanation of aging that has so far eluded scientists.

BEFORE I CAN PURSUE THIS REASONING, I must anticipate objections of my physiologist friends. They tend to believe that something about our physiology somehow makes aging inevitable, and that evolutionary considerations are irrelevant. For instance, one such theory attributes aging to the progressive difficulties that our immune system is said to face in distinguishing our own cells from foreign cells. Physiologists subscribing to this view make an implicit assumption:

that natural selection couldn't lead to an immune system without that fatal defect. Is this belief warranted?

To evaluate this objection, let's consider biological repair mechanisms, because aging may be thought of simply as unrepaired damage or deterioration. Our first association to the word "repair" is likely to be to those repairs that cause us the most frustration—car repairs. Our cars tend to grow old and die, but we spend money to postpone their inevitable fate. Similarly, we are unconsciously but constantly repairing ourselves too, at every level from molecules to tissues or whole organs. Our own self-repair mechanisms, like those we lavish on our cars, are of two sorts: damage control and regular replacement.

An automotive example of damage control is that we replace a car's fender only if it's bashed in; we don't routinely replace the fenders at every regular oil change. The most visible example of damage control applied to our bodies is wound healing, by which we repair damage to our skin. Many animals can achieve more spectacular results: lizards regenerate severed tails, starfish and crabs their limbs, sea cucumbers their intestines, and ribbon worms their poison stylets. At the invisible molecular level our genetic material, DNA, is repaired exclusively by damage control: we have enzymes that recognize and fix damaged sites in the DNA helix while ignoring intact DNA.

The other type of repair, regular replacement, is also familiar to every car owner: we periodically change the oil, air filter, and ball bearings to eliminate slight wear, without waiting for the car to break down first. In the biological world, teeth are similarly replaced on a prescheduled basis: humans go through two sets, elephants six sets, and sharks an indefinite number, during their lifetimes. Though we humans go through life with the same skeleton with which we were born, lobsters and other arthropods regularly replace their exoskeletons by molting them and growing new ones. Still another highly visible example of scheduled repair is the continual growth of our hair: no matter how short we cut it, its growth will replace the cut portion.

Regular replacement also goes on at a microscopic or submicroscopic level. We constantly replace many of our cells: about once every few days for the cells lining our intestine, once every two months for the cells lining the urinary bladder, and once every four months for our red blood cells. At the molecular level, our protein molecules are subject to continuous turnover at a rate characteristic of each particular protein; we thereby avoid the accumulation of damaged molecules. If you compare your beloved's appearance today with a photo taken a month ago, he (or she) may look the same, but many of the individual molecules forming that beloved body are different. While all the king's horses and men couldn't put Humpty-Dumpty together again, Nature is taking us apart and putting us back together every day.

In such ways, much of an animal's body can be repaired as needed, or is regularly replaced anyhow, but the details of how much is replaceable vary greatly with the part and with the species. There is nothing physiologically inevitable about the limited repair capabilities of us humans. Since starfish can regrow amputated limbs, why can't we? What prevents us from having six sequential sets of teeth like an elephant, rather than just baby teeth and adult teeth? With four more natural sets, we wouldn't need fillings, crowns, and dentures as we got older. Why don't we protect ourselves against arthritis?—all we'd need is to replace our joints periodically, as crabs do. Why don't we guard against heart disease by periodically replacing our hearts, as ribbon worms replace their poison stylets? One might suppose that natural selection would favor the man or woman who didn't die of heart disease around age eighty but continued to live and produce babies at least until age two hundred. Why, for that matter, can't we repair or replace everything in our bodies?

The answer surely has something to do with the expense of repair. Here again, the analogy of car repair is helpful. If the boasts of the Mercedes-Benz company are to be believed, their cars are so well built that, even should you do no maintenance whatsoever—not even lubrication or oil changes—your Mercedes will still run for years. At the end of that time, of course, it will fall apart from accumulated irreversible damage. Hence Mercedes owners generally do choose to service their cars regularly. My Mercedes-owning friends tell me that Mercedes service is very expensive: hundreds of dollars every time they drive into the shop. Nevertheless, they consider the expense worth it: a serviced Mercedes lasts much longer than an unserviced Mercedes, and it's much cheaper to service your old Mercedes regularly than to discard it and buy a new one every few years.

That's how Mercedes owners reason in Germany and the United States. But suppose you were living in Port Moresby, the capital of Papua New Guinea, automobile-accident capital of the world, where any car is likely to be totaled within a year no matter how you maintain it. Many car owners in New Guinea don't go to the expense of maintaining their car: they use the saved money to help buy the inevitable next car.

By analogy, how much an animal "should" invest in biological repairs depends on the expense of the repairs, and on a comparison of the animal's expected life span with and without the repairs. But such "should" questions belong to the realm of evolutionary biology, not physiology. Natural selection tends to maximize one's rate of producing offspring that survive to leave offspring of their own. Evolution can thus be regarded as a strategy game, in which the individual whose strategy leaves the most descendants wins. The type of reasoning used in game theory is therefore helpful in understanding how we came to be the way we are.

THIS PROBLEM OF LIFE SPAN, and of investment in biological repair, is in turn one of an even broader class of evolutionary problems addressed by game theory: the mystery of what sets the maximum limit on any advantageous trait. There are lots of other biological traits, besides life span, that beg the question why natural selection hasn't made them longer or bigger or faster or made more of them. For instance, people who are big or smart or can run fast have obvious advantages over small, stupid, slow people—especially throughout most of human evolution, when we were still fending off lions and hyenas. Why didn't we evolve to become on the average even bigger, smarter, and faster than we now are?

The complication that makes these evolutionary design problems less simple than they might at first seem is this: natural selection acts on whole individuals, not on single parts of an individual. It's you, not your big brain or fast legs, that does or doesn't survive and leave offspring. Increasing one part of an animal's body may be beneficial in some obvious respect but harmful in other respects. For instance, that one larger part might not fit in well with other parts of the same animal, or it might drain off energy from other parts.

To evolutionary biologists, the magic word that expresses this complication is "optimize." Natural selection tends to mold each trait to the size, speed, or number that maximizes the survival and reproductive success of the whole animal, given the animal's basic design. Each trait in itself doesn't tend toward a maximal value. Instead, each trait converges on some optimal intermediate value, neither too big nor too small. The whole animal is thereby more successful than it would be if that trait were bigger or smaller.

Should this reasoning about animals seem abstract, think instead of our everyday machines. Essentially the same principles apply to engineering design, of machines by humans, as to evolutionary design, of animals by natural selection. For example, consider my pride and joy among my machines, my 1962 Volkswagen Beetle, the only car I've ever owned. (Car buffs will remember 1962 as the year that Volkswagen introduced the big rear window in the Beetle.) On a smooth, level highway with an assisting tailwind, my VW can go 65 mph. To BMW owners, that may sound distinctly submaximal. Why don't I junk my puny 4-cylinder 40-horsepower engine, install instead the 12-cylinder 296-horsepower engine from my neighbor's BMW 750 IL, and roar off at 180 mph down the San Diego Freeway?

Well, even I, dodo about cars that I am, know that that wouldn't work. To begin with, that huge BMW engine wouldn't fit into my VW's engine compartment, which would need enlarging. Then the BMW engine is meant to go in front, but the VW engine compartment is in back, so I'd have to change the gear box and transmission and other things. I'd also have to change the shock absorbers and brakes, designed to smooth the ride and stop a car at 65 mph but not at 180 mph. By the time I had finished modifying my VW to take the BMW engine, there wouldn't be much remaining from my original Beetle. And the modifications would have cost me a big pile of money. I'd guess that my puny 40-HP engine is optimal, in the sense that I couldn't increase my cruising speed without sacrificing other performance features of my car—as well as sacrificing other money-requiring features of my life-style.

While the marketplace *eventually* eliminates engineering monstrosities like a VW with a BMW engine, all of us can think of monstrosities that took quite a while to eliminate. To those of you who share my fascination with naval warfare, British battle cruisers are a good example. Before and during World War I, the British navy launched thirteen warships called battle cruisers, designed to be as large and with as many big guns as battleships but much faster. By maximizing speed and firepower, the battle cruisers immediately caught the public imagination and became a propaganda sensation. However, if you take a 28,000-ton battleship, keep the weight of the big guns nearly constant, and greatly increase the weight of the engines while still maintaining total weight around 28,000 tons, you have to skimp on the weight of some other parts. The battle cruisers skimped especially on weight of armor, but also on weight of small guns, internal compartments, and antiaircraft defense.

The results of this suboptimal overall design were inevitable. In 1916 H.M.S. *Indefatigable, Queen Mary,* and *Invincible* all blew up almost as soon as they were hit by German shells at the Battle of Jutland. H.M.S. *Hood* blew up in 1941, a mere eight minutes after entering battle with the German battleship *Bismarck.* H.M.S. *Repulse* was sunk by Japanese bombers a few days after the Japanese attack on Pearl Harbor, thereby acquiring the dubious distinction of being the first large warship to be destroyed from the air while in combat at sea. Faced with this stark evidence that some spectacularly maximal parts don't make an optimal whole, the British navy let its program of building battle cruisers go extinct.

In short, engineers can't tinker with single parts in isolation from the rest of a machine, because each part costs money, space, and weight that might have gone into something else. Engineers instead have to ask what *combination* of parts will optimize a machine's effectiveness. By the same reasoning, evolution can't tinker with single traits in isolation from the rest of an animal, because every structure, enzyme, or piece of DNA consumes energy and space that might have gone into something else. Instead, natural selection favors that combination of traits that maximizes the animal's reproductive output. Both engineers and evolutionary biologists have to evaluate the trade-offs involved in increasing anything: its costs as well as the benefits that it would bring.

AN OBVIOUS DIFFICULTY in applying this reasoning to our life cycles is that they have many features seeming to reduce, not to maximize, our ability to produce offspring. Growing old and dying are just one example; other examples are human female menopause, bearing one baby at a time, producing babies only once every year or so at most, and not even starting to produce babies until age twelve to sixteen.

Wouldn't natural selection favor the woman who entered puberty at age five, completed gestation in three weeks, regularly bore quintuplets, never underwent menopause, put lots of biological energy into repair of her body, lived to two hundred, and thereby left hundreds of offspring?

But posing the question in that form pretends that evolution can change our bodies one piece at a time, and ignores the hidden costs. For instance, a woman certainly couldn't reduce the length of pregnancy to three weeks without changing anything else about herself or her baby. Remember that we only have a finite amount of energy available to us. Even people doing hard exercise and eating rich food—lumberjacks or marathon runners in training—can't metabolize much more than about six thousand calories per day. How should we allocate those calories between repairing ourselves and rearing babies, if our goal is to raise as many babies as possible?

At the one extreme, if we put all our energy into babies and devoted no energy to biological repair, our bodies would age and disintegrate before we could rear our first baby. At the other extreme, if we lavished all our available energy on keeping our bodies in shape, we might live a long time but would have no energy left for the exhausting process of making and rearing babies. What natural selection must do is to adjust an animal's relative expenditures of energy on repair and on reproduction, so as to maximize its reproductive output, averaged over its lifetime. The answer to that problem varies among animal species, depending on factors such as their risk of accidental death, their reproductive biology, and the costs of various types of repairs.

This perspective can be employed to make testable predictions about how animals should differ in their repair mechanisms and rates of aging. In 1957 the evolutionary biologist George Williams cited some striking facts about aging that become comprehensible only from an evolutionary perspective. Let's consider several of Williams's examples and reexpress them in the physiological language of biological repair, by taking slow aging as an indication of good repair mechanisms.

The first example concerns the age at which an animal first breeds and produces offspring. That age varies enormously among species: few humans are so precocious as to produce babies before the age of twelve years, while any self-respecting mouse a mere two months old can already make baby mice. Animals belonging to a species whose age of first breeding is late, like us, need to devote much energy to repair, in order to ensure that they survive to that reproductive age. So we expect investment in repair to increase with age at first reproduction.

For instance, correlated with our having a much later age of first reproduction than do mice, we humans age far more slowly than mice and are presumed to repair our bodies much more effectively. Even with plenty of food and the best medical care, a mouse is lucky to reach its second birthday, while we would be unlucky not to reach our seventy-second birthday. The evolutionary reason: a human who invested no more of his/her energy in repair than does a mouse would be dead long before reaching puberty. Hence it is more worthwhile to repair a human than a mouse.

What might that postulated extra energy expenditure of ours actually consist of? At first, our human repair capabilities seem unimpressive. We can't regrow an amputated arm, and we don't regularly replace our skeleton, in the way that some short-lived invertebrates do. However, such spectacular but infrequent replacements of a whole structure probably aren't the biggest items in an animal's repair budget. Instead, the biggest expense is all that invisible replacement of so many of your cells and molecules, day after day. Even if you spend all day every day just lying in bed, you need to eat about 1,640 calories per day if you're a man (1,430 for a woman) just to maintain your body. Much of that maintenance metabolism goes to our invisible scheduled replacement. And so I'd guess that where we cost more than a mouse is in putting a bigger fraction of our energy into self-repair, and a smaller fraction into other purposes like keeping warm or caring for babies.

The second example I'll discuss involves the risk of irreparable injury. Some biological damage is potentially reparable, but there is also damage that is guaranteed to be fatal (e.g., being eaten by a lion). If you're likely to be eaten by a lion tomorrow, there's no point in paying a dentist to start expensive orthodontic work on your teeth today. You'd do better to let your teeth rot and start having babies immediately. But if an animal's risk of death from irreparable accidents is low, then there is a potential payoff, in the form of increased life span, from putting energy into expensive repair mechanisms that retard aging. This is the reasoning by which

Mercedes owners decide to pay for lubrication of their cars in Germany and the United States but not in New Guinea.

Biological analogies are that the risk of death from predators is lower for birds than for mammals (because birds can escape by flying), and lower for turtles than for most other reptiles (because turtles are protected by a shell). Thus, birds and turtles stand to gain a lot from expensive repair mechanisms, compared to flightless mammals and shell-less reptiles that will soon be eaten by predators anyway. Indeed, if one compares longevities of well-fed pets protected from predators, birds do live longer (i.e., age more slowly) than similar-sized mammals, and turtles live longer than similar-sized shell-less reptiles. The bird species best protected from predators are seabirds like petrels and albatrosses that nest on remote oceanic islands free of predators. Their leisurely life cycles rival our own. Some albatrosses don't even breed until they're ten years old, and we still don't know how long they live: the birds themselves last longer than the metal rings that biologists began putting on their legs a few decades ago in order to keep track of their ages. In the ten years that it takes an albatross to start breeding, a mouse population could have gone through sixty generations, most of which would already have succumbed to predators or old age.

As our third example, let's compare males and females of the same species. We expect more potential payoff from repair mechanisms, and lower rates of aging, in that sex with the lower accidental mortality rate. For many or most species, males suffer greater accidental mortality than females, partly because males put themselves at greater risk by fighting and bold displays. This is certainly true of human males today and has probably been so throughout our history as a species: men are the sex most likely to die in wars against men of other groups, and in individual fights within a group. Also, in many species the males are bigger than the females, but studies of red deer and of New World blackbirds show that males are thereby more likely than females to die when food becomes scarce.

Correlated with this greater accidental death rate of men, men also age faster and have a higher nonaccidental death rate than women. At present, women's life expectancy is about six years greater than that of men; some of this difference is because more men than women are smokers, but there is a sex-linked difference in life expectancy even among nonsmokers. These differences suggest that evolution has programmed us so that women put more energy into self-repair, while men put more energy into fighting. Expressed another way, it just isn't worth as much to repair a man as it is to repair a woman. But I don't mean to denigrate male fighting, which serves a useful evolutionary purpose for a man: to gain wives and to secure resources for his children and his tribe, at the expense of other men and their children and tribe.

My REMAINING EXAMPLE of how some striking facts of aging become comprehensible only from an evolutionary perspective concerns the distinctively human phenomenon of survival past reproductive age, especially past female menopause. Since transmitting one's genes to the next generation is what drives evolution, other animal species rarely survive past reproductive age. Instead, Nature programs death to coincide with the end of fertility, because there is then no longer any evolutionary benefit to gain from keeping one's body in good repair. It's an exception in need of explanation to realize that women are programmed to live for decades after menopause, and that men are programmed to live to an age when most men are no longer busy siring babies.

But the explanation becomes apparent on reflection. The intense phase of parental care is unusually protracted in the human species and lasts nearly two decades. Even those older people whose own children have reached adulthood are tremendously important to the survival of not just their children but of their whole tribe. Especially in the days before writing, they acted as the carriers of essential knowledge. Nature has programmed us with the capacity to keep the rest of our bodies in reasonable repair even at an age when the female reproductive system itself has fallen into disrepair.

Conversely, though, we have to wonder why natural selection programmed female menopause into us in the first place. It too, like aging, can't be explained away as something physiologically inevitable. Most mammals, including human males plus chimps and gorillas of both sexes, merely experience a gradual decline and eventual cessation of fertility with age, rather than the abrupt shutdown of women's fertility. Why did that peculiar, seemingly counterproductive feature of ours evolve? Wouldn't natural selection favor the woman who remained fertile until the bitter end?

Human female menopause probably resulted from two other distinctively human characteristics: the exceptional danger that childbirth poses to the mother, and the danger that a mother's death poses to her offspring. Recall the enormous size of the human infant at birth relative to its mother: our big seven-pound babies emerging from hundred-pound mothers, compared to little four-pound gorilla babies emerging from two-hundred-pound gorilla mothers. As a result, childbirth is dangerous to women. Especially before the advent of modern obstetrics, women often died in childbirth, whereas mother gorillas and chimps virtually never do. A study of the outcome of labor in 401 pregnant rhesus monkeys recorded only a single maternal death.

Now recall the extreme dependence of human infants on their parents, especially on their mother. Because human infants develop so slowly and can't even feed themselves after weaning (unlike young apes), the death of a hunter-gatherer mother would have been likely to be fatal to her offspring up to a later age in childhood than for any other primate. A hunter-gatherer mother with several children was gambling the lives of those children at every subsequent childbirth. Since her investment in those prior children increased with their age, and since her own risk of death in childbirth also increased with her age, the odds of her gamble's paying off got worse and worse as she got older. When you already have three children alive but still dependent on you, why risk those three for a fourth?

Those worsening odds probably led through natural selection to menopausal shutdown of human female fertility, in order to protect a mother's prior investment in kids. But since childbirth carries no risk of death for fathers, men did not evolve menopause. Like aging, menopause illustrates how an evolutionary approach illuminates features of our life cycle that would otherwise make no sense. It's even possible that menopause evolved only within the past forty thousand years, when Cro-Magnons and other anatomically modern humans began frequently to survive to age sixty or more. Neanderthals and earlier humans usually died before age forty anyway, so that menopause would have brought their women no benefits if it were to occur at the same age as in modern *Femina sapiens*.

Thus, the longer life span of modern humans as compared to that of apes does not rest only on cultural adaptations, such as tools to acquire food and deter predators. It also rests on the biological adaptations of menopause and increased investment in self-repair. Whether those biological adaptations developed especially at the time of the Great Leap Forward or earlier, they rank among the life-history changes that permitted the rise of the third chimpanzee to humanity.

THE LAST CONCLUSION that I wish to draw from an evolutionary approach to aging is that it undermines the approach that has long dominated the physiological study of aging. The gerontological literature is obsessed with a search for The Cause of Aging—preferably a single cause, certainly not more than a few major causes. Within my own lifetime as a biologist, hormonal changes, deterioration in the immune system, and neural degeneration have vied in popularity for the title of The Cause, without compelling support having been adduced to date for any of the candidates. But evolutionary reasoning suggests that this search will remain futile. There *should* not be just one, or even a few, dominant physiological mechanisms of aging. Instead, natural selection should act to match rates of aging in all physiological systems, with the result that aging involves innumerable simultaneous changes.

The basis of this prediction is as follows. There's no point in doing expensive maintenance on one piece of the body if other pieces are deteriorating more rapidly. Conversely, there's no point in permitting a few systems to deteriorate long before all the others, because the cost of extra repairs on just those few systems would have bought a big increase in life expectancy. Natural selection doesn't make such pointless mistakes. By analogy, Mercedes owners shouldn't install cheap ball bearings when they are lavishing expense on all other parts of the car. Had they been so foolish, they could have doubled the lifetime of their costly car just by spending a few more dollars for better ball bearings. But it wouldn't pay either to go to the expense of installing diamond ball bearings, when all the rest of the car would have rusted away before those ball bearings wore out. Thus, the optimal strategy for Mercedes owners, and for us, is to repair all parts of our cars or bodies at rates such that everything finally collapses all at once.

It seems to me that this depressing prediction is borne out, and that this evolutionary ideal of simultaneous total collapse describes the fates of our bodies better than does the physiologists' long-sought single Cause of Aging. Signs of aging can be found wherever one looks for them. Already I am conscious in myself of tooth wear, considerable decreases in muscle performance, and significant losses in hearing, vision, smell, and taste. For all these senses, the acuity of women is greater than that of equal-aged men, whatever the age group compared. Ahead for me lies the familiar litany: weakening of the heart, hardening of the arteries, increasing brittleness of bones, decreases in kidney filtration rates, lower resistance of the immune system, and loss of memory. The list could be extended almost indefinitely. Evolution seems indeed to have arranged things so that all our systems deteriorate, and that we invest in repair only as much as we are worth.

From a practical standpoint, this conclusion is disappointing. If there had been one dominant cause of aging, curing that cause would have provided us with a fountain of youth. This thought, operating at a time when aging was thought to be largely a hormonal phenomenon, inspired some attempts at miraculous rejuvenation of old people by hormonal injections or implantation of young gonads. Such an attempt was the subject of Sir Arthur Conan Doyle's story "The Adventure of the Creeping Man," in which the aged Professor Presbury becomes infatuated with a young woman, desperately wants to rejuvenate himself, and instead is found climbing up a vine like a monkey after midnight. The great Sherlock Holmes discovers the reason: the professor has been seeking youth by injecting himself with the serum of langur monkeys.

I could have warned Professor Presbury that his myopic obsession with proximate causation would lead him astray. Had he thought of ultimate evolutionary causation, he would have realized that natural selection would never permit us to deteriorate through a single mechanism with one simple cure. Perhaps it's just as well. Sherlock Holmes worried greatly what would happen if such an elixir of life was found. "There is danger there—a very real danger to humanity. Consider, Watson, that the material, the sensual, the worldly would all prolong their worthless lives. . . . It would be the survival of the least fit. What sort of cesspool may not our poor world become?"

Holmes would be relieved to know that his worries now appear unlikely to materialize.

PART THREE

UNIQUELY
HUMAN

Parts One and Two discussed the biological foundations of our unique cultural traits. We saw that those foundations include our familiar skeletal hallmarks, such as our large braincase and our adaptations for upright gait. They also include features of our soft tissues, behavior, and endocrinology concerned with reproduction and social organization.

However, if those genetically specified features were our sole distinctions, we wouldn't stand out among animals, and we wouldn't now be threatening the survival of ourselves and other species. Other animals, such as ostriches, walk erect on two legs. Others have relatively large brains, though not as large as ours. Others live monogamously in colonies (many seabirds), or are very long-lived (albatrosses and tortoises).

Instead, our uniqueness lies in the cultural traits that rest on those genetic foundations and that in turn give us our power. Our cultural hallmarks include spoken language, art, tool-based technology, and agriculture. But if we stopped there, we'd have a one-sided and self-congratulatory view of our uniqueness. The hallmarks I just mentioned are ones that we're proud of. Yet the archaeological record shows the introduction of agriculture to have been a mixed blessing, seriously harming many people while benefiting others. Chemical abuse is a wholly ugly human hallmark. At least it doesn't threaten our survival, as do two of our other cultural practices: genocide, and mass exterminations of other species. We're uncomfortable about whether to regard these as occasional pathological aberrations, or as features no less basic to humanity than the traits we're proudest of.

All of these cultural features that define humanity are seemingly absent in animals, even in our closest relatives. They must have arisen sometime after our ancestors parted company from the other chimpanzees around seven million years ago. Furthermore, while we have no way of knowing whether Neanderthals spoke or indulged in drug abuse and genocide, they certainly didn't have agriculture, art, or the capacity to build radios. Hence these latter traits must be very recent human innovations of the last few tens of thousands of years. But they couldn't have arisen from nothing. There had to have been animal precursors, if we could only recognize them.

For each of our defining cultural traits, we need to ask: What were those precursors? When in our ancestry did the trait approach its modern form? What were the early stages of its evolution like, and can those stages be traced archaeologically? We're unique on Earth, but how unique are we in the universe?

In this part we'll consider some of the above questions for our noble, two-edged, or only mildly destructive characteristics. We'll first take up the origin of spoken language, which I suggested earlier might have triggered the Great Leap Forward, and which anyone would list among our most important distinctions from animals. On first reflection, the task of tracing the development of human language appears plainly impossible. Language before the dawn of writing left no archaeological remains, unlike our first experiments in art, agriculture, and tools. There seems to be no surviving simple human language, no animal language, that could exemplify the early stages.

In fact, there are innumerable animal precursors: the vocal communication systems evolved by many species. We're just beginning to appreciate the sophistication of some of these systems. If they exemplify a first stage, results of recent experiments at teaching language to apes represent a second stage, by revealing apes' innate capacities. The progress by which children learn to speak may trace out further stages. We'll also see that there really are some simple languages that modern humans have unconsciously invented and that prove unexpectedly instructive.

Among our unique cultural traits, art is perhaps the noblest human invention. There seems to be a gulf separating human art, supposedly created just for pleasure and doing nothing to perpetuate our genes, from any animal behavior. Yet paintings and drawings created by captive apes and elephants, whatever the motives of those animal artists, look so similar to work of human artists that they have fooled experts and have been bought by art collectors. If one nevertheless dismisses those animal artworks as unnatural productions, what is one to say about the carefully arranged colored bowers of normal male bowerbirds? Those bowers play an unquestioned crucial role in passing on genes. I'll argue that human art also had that role originally, and often still does today. Since art, unlike language, does show up in archaeological deposits, we know that human art didn't proliferate until the time of the Great Leap Forward.

Agriculture, another human hallmark, has an animal precedent but not precursor in the gardens of leafhopper ants, which lie far off of our direct lineage. The archaeological record lets us date our "reinvention" of agriculture to a time long after the Great Leap Forward, within the last ten thousand years. That transition from hunting and gathering to agriculture is generally considered a decisive step in our progress, when we at last acquired the stable food supply and leisure time prerequisite to the great accomplishments of modern civilization. In fact, careful examination of that transition suggests another conclusion: for most people the transition brought infectious diseases, malnutrition, and a shorter life span. For human society in general it worsened the relative lot of women and introduced class-based inequality. More than any other milestone along the path from chimpanzeehood to humanity, agriculture inextricably combines causes of our rise and our fall.

Abuse of toxic chemicals is a widespread human hallmark documented only within the last five thousand years, though it may well go back much earlier into preagricultural times. Unlike agriculture, it doesn't even rank as a mixed blessing but as a pure evil threatening the survival of individuals, though not of our species. Like art, drug abuse seems at first to lack animal precedents or biological functions. I'll argue, however, that it fits into a broad class of animal structures or behaviors that are dangerous to their owners or practitioners, and whose function depends paradoxically on that danger.

While animal precursors can be identified for all of our hallmarks, they still rank as human hallmarks because we're unique on Earth in the extreme degree to which we've developed them. How unique are we in the universe? Once conditions suitable for life exist on a planet, how likely are intelligent, technologically advanced life forms to evolve? Was their emergence on Earth practically inevitable, and do they now exist on innumerable planets circling other stars?

There is no direct way to prove whether creatures capable of language, art, agriculture, or drug abuse exist elsewhere in the universe, because from Earth we can't detect the existence of those traits on planets of other stars. However, we might be able to detect high technology elsewhere in the universe if it included our own capacity to send out space probes and interstellar electromagnetic signals. I'll conclude this part by examining the ongoing search for extraterrestrial intelligent life. I'll argue that evidence from a quite different field—studies of woodpecker evolution on Earth—instructs us about the inevitability of evolving intelligent life, and therefore about our uniqueness, not only on Earth but also in the accessible universe.

CHAPTER 8

Bridges to
Human Language

HUMAN LANGUAGE ORIGINS CONSTITUTE THE MOST IMPORTANT MYS-tery in understanding how we became uniquely human. After all, language lets us communicate with each other far more precisely than can any animals. It lets us lay joint plans, teach one another, and learn from what others experienced elsewhere or in the past. With it, we can store precise representations of the world within our minds, and encode and process information far more efficiently than can any animals. Without language we could never have conceived and built Chartres Cathedral—or V-2 rockets. These are the reasons that I speculated that the Great Leap Forward (the stage in human history when innovation and art at last emerged) was made possible by the emergence of spoken language as we know it.

Between human language and the vocalizations of any animal lies a seemingly unbridgeable gulf. As has been clear since the time of Darwin, the mystery of human language origins is an *evolutionary* problem: how was this unbridgeable gulf nevertheless bridged? If we accept that we evolved from animals lacking human speech, then our language must have evolved and become perfected with time, along with the human pelvis, skull, tools, and art. There must once have been intermediate languagelike stages linking Shakespeare's sonnets to monkey's grunts. Darwin diligently kept notebooks on his children's linguistic development, and he reflected on the languages of "primitive" peoples, in the hope of solving this evolutionary mystery.

Unfortunately, the origins of language prove harder to trace than the origins of the human pelvis, skull, tools, and art. All those latter things may survive, and can be recovered and dated, but the spoken word vanishes in an instant. In frustration, I often dream of a time machine that would let me place tape recorders in ancient hominoid camps. Perhaps I'd discover that australopithecines uttered grunts little different from those of chimpanzees; that early *Homo erectus* used recognizable single words, progressing after a million years to two-word sentences; that *Homo sapiens* before the Great Leap Forward had gotten as far as longer strings of words, but still without much grammar; and that syntax and the full range of modern speech sounds arrived only with the Great Leap.

Alas, we have no such retrospective tape recorder, and no prospects for ever getting one. How can we hope to trace speech origins without such a magic time machine? Until recently, I would have said that it was hopeless to do more than speculate. In this chapter, however, I shall try to draw on two exploding bodies of knowledge that may let us start to build bridges across the seemingly unbridgeable gulf between animal and human sounds, starting from each of its opposite shores.

Sophisticated new studies of wild-animal vocalizations, especially those of our primate relatives, constitute the bridgehead on the animal shore of the gulf. It has always been obvious that animal sounds must have been ancestral to human speech, but only now are we beginning to sense how far animals have come toward inventing their own "languages." In contrast, it has been quite unclear where to locate the bridgehead on the human shore, since all existing human languages seem infinitely advanced over animal sounds. Recently, though, it has been argued that a numerous set of human languages neglected by most linguists really does exemplify two primitive stages on the human side of the causeway.

*   *   *

Many wild animals communicate with each other by sounds, of which bird songs and the barking of dogs are especially familiar to us. Most of us are within earshot of some calling animal most days of our lives. Scientists have been studying animal sounds for centuries. Despite this long history of intimate association, our understanding of these ubiquitous familiar sounds has suddenly exploded because of the application of new techniques: use of modern tape recorders to record animal calls, electronic analysis of the calls to detect subtle variations imperceptible to the unaided human ear, broadcasting recorded calls back to animals to observe how they react, and observing their reactions to electronically reshuffled calls. These methods are revealing animal vocal communication to be much more language-like than anyone would have guessed thirty years ago.

The most sophisticated "animal language" studied to date is that of a common cat-sized African monkey known as the vervet. Equally at home in trees and on the ground of savanna and rain forest, vervets are among the monkey species that visitors to East African game parks are most likely to see. They must have been familiar to Africans for the hundreds of thousands of years that we have existed as the species *Homo sapiens*. They may have reached Europe as pets over three thousand years ago, and they certainly have been familiar to European biologists exploring Africa since the nineteenth century. Many lay people who have never visited Africa still know vervets from zoo captives.

Like other animals, wild vervets regularly face situations in which efficient communication and representation would help them to survive. About three-quarters of wild vervet deaths are caused by predators. If you're a vervet, it's essential to know the differences between a martial eagle, one of the leading killers of vervets, and a white-backed vulture, an equally large soaring bird that eats carrion and is no danger to live monkeys. It's essential to act appropriately when the eagle appears, and to tell your relatives. If you fail to recognize the eagle, you die; if you fail to tell your relatives, they die, carrying your genes with them; and if you think it's an eagle when it's really just a vulture, you're wasting time on defensive measures while other monkeys are safely out there gathering food.

Besides these problems posed by predators, vervets have complex social relationships with each other. They live in groups and compete for territory with other groups. Hence it's also essential to know the difference between a monkey intruding from another group, an un-related member of your own group likely to steal food from you, and a close relative in your own group on whose support you can count. Vervets that get into trouble need ways of telling their relatives that they, and not some other monkey, are in trouble. It's also essential to know and communicate about sources of food: for instance, which of the thousand plant and animal species in the environment are good to eat, which are poisonous, and where and when the edible ones are likely to be found. For all these reasons, vervets would profit from efficient ways of communicating about and representing their world.

Despite these reasons, and despite our long and close association with vervets, we had no appreciation of their complex world knowledge and vocalizations until the mid-1960s. Since then, ob-servations of vervet behavior have revealed that they make finely graded discriminations among types of predators, and among each other. They adopt quite different defensive measures when threat-ened by leopards, eagles, and snakes. They respond differently to dominant and subordinate members of their own troop, differently again to dominant and subordinate members of rival troops, dif-ferently to members of different rival troops, and differently to their mothers, maternal grandmothers, siblings, and unrelated members of their own troop. They know who is related to whom: if an infant monkey calls, its mother turns toward it, but other vervet mothers turn instead toward that infant's mother to see what she will do. It's as if vervets had names for several predator species and several dozen individual monkeys.

The first clue to how vervets communicate this information came from observations that the biologist Thomas Struhsaker made on vervets in Kenya's Amboseli National Park. He noted that three types of predators triggered different defensive measures by vervets, and also triggered alarm calls sufficiently distinct for Struhsaker to hear the differences even without making any sophisticated electronic analysis. When vervets encounter a leopard or other species of large wild cat, male monkeys give a loud series of barks, females give a high-pitched chirp, and all monkeys within earshot may run up a tree. The sight of a martial or crowned eagle soaring overhead causes vervets to give a short cough of two syllables, whereupon listening monkeys look up into the air or run into a bush. A monkey who spots a python or other dangerous snake gives a "chuttering" call, and that stimulates other vervets in the vicinity to stand erect on their hind legs and look down (to see where the snake is).

Beginning in 1977, a husband-and-wife team named Robert Seyfarth and Dorothy Cheney proved by experiments that these calls really had the different functions suggested by Struhsaker's observations. Their procedure was as follows. First, they made a tape recording of a monkey giving a call whose apparent function Struhsaker had observed (say, the "leopard call"). Then, on a later day, after locating the same troop of monkeys, Cheney or Seyfarth hid the tape and loudspeaker equipment nearby in a bush, while the other investigator (Seyfarth or Cheney) started filming the monkeys with a movie or video camera. After fifteen seconds, Scientist Number 1 broadcast the tape, and Scientist Number 2 kept filming the monkeys for one minute to see whether the monkeys behaved appropriately for the call's suspected function (e.g., whether the monkeys ran up a tree on hearing a broadcast of the supposed leopard call). It turned out that playback of the "leopard call" really did stimulate the monkeys to run up a tree, while the "eagle call" and "snake call" similarly stimulated monkeys to the behaviors that seemed to be associated with these calls under natural conditions. Thus, the apparent association between the observed behaviors and the calls had not resulted from some coincidence, and the calls did have the functions suggested by observation.

The three calls that I've mentioned by no means exhaust a vervet's vocabulary. Besides those loud and frequently given alarm calls, there appear to be at least three fainter alarms that are given less frequently. One, triggered by baboons, causes listening vervets to become more alert. A second, given in response to mammals like jackals and hyenas that prey on vervets only infrequently, makes the monkeys watch the animal and perhaps move slowly toward a tree. The remaining faint alarm call responds to unfamiliar humans and results in the vervets' quietly moving toward a bush or the top of a tree. However, the postulated functions of these three fainter alarm calls remain unproved because they have not yet been tested by playback experiments.

Vervets also utter gruntlike calls when interacting with each other. Even to scientists who have spent years listening to vervets, all these social grunts sound the same. When the grunts are recorded and displayed as a frequency spectrum on the screen of a sound-analyzing instrument, they look the same. Only when the spectra were measured in elaborate detail could Cheney and Seyfarth detect (sometimes but not always!) average differences between the grunts given in four social contexts: when a monkey approaches a dominant monkey, when it approaches a subordinate monkey, when it watches another monkey, or when it sees a rival troop.

Broadcasts of grunts recorded in these four different contexts caused monkeys to behave in subtly different ways. For example, they looked toward the loudspeaker if the grunt had originally been recorded in the "approach dominant monkey" context, while they looked off in the direction toward which the call was being broadcast if it had originally been recorded in the "see rival troop" context. Further observations of the monkeys under natural conditions showed that the natural calls themselves had been eliciting these subtly different behaviors.

Evidently, vervets are much more finely attuned than we are to their calls. Merely listening to and watching vervets, without recording and playing back their calls, gave no hint that they had at least four distinct grunts—and possibly many more. As Seyfarth writes, "Watching vervets grunt to each other is really very much like watching humans engaged in conversation without being able to hear what they're saying. There aren't any obvious reactions or replies to grunts, so the whole system seems very mysterious—mysterious, that is, until you start doing playbacks." These discoveries illustrate how easy it is to underestimate the size of an animal's vocal repertoire.

THE VERVETS OF AMBOSELI thus have *at least* ten putative "words": their words for "leopard," "eagle," "snake," "baboon," "other predatory mammal," "unfamiliar human," "dominant monkey," "subordinate monkey," "watch other monkey," and "see rival troop." However, virtually every claim of any animal behavior suggestive of elements of human language is greeted with skepticism by many scientists, convinced of the linguistic gulf separating us from animals. Such skeptics consider it simpler to assume that humans are unique, and that the burden of proof should be borne by anyone who thinks otherwise. Any claim of languagelike elements for animals is considered a more complicated hypothesis, to be dismissed as unnecessary in the absence of positive proof. Yet the alternative hypotheses by

which the skeptics instead attempt to explain animal behaviors sometimes strike me as more complicated than the simple and often plausible explanation that humans are not unique.

It seems a modest claim to propose that the different calls that
vervets give in response to leopards, eagles, and snakes actually refer
to these animals or are intended as communications to other monkeys. However, skeptics were disposed to believe that only humans
could emit voluntary signals referring to external objects or events.
The skeptics suggested that the vervet alarm calls were just involuntary expressions of the monkey's emotional state ("I'm scared out
of my wits!") or of its intent ("I'm going to run up a tree"). After all,
those explanations apply to some of our own "calls." If I saw a
leopard coming at me, I too might scream even though there wasn't
anyone around to whom to communicate. We grunt as we throw
ourselves into some physical activities, like lifting a heavy object.

Suppose that zoologists from an advanced civilization in Outer
Space observed me to give a trisyllabic scream "argh, leopard" and to
climb a tree when I saw a leopard. The zoologists might well doubt
that my lowly species could express anything beyond grunts of emotion or intent—certainly not symbolic communications. To test their
hypothesis, the zoologists would resort to experiments and detailed
observations. If I gave the scream regardless of whether any other
human was in earshot, that would support the theory of a mere
expression of emotion or intent. If I gave the scream only in the
presence of another person, and only when approached by a leopard
but not by a lion, that would suggest a communication with a specific
external referent. And if I gave the scream to my son but remained
silent when I saw the leopard stalk a man with whom I had frequently been seen to fight, the visiting zoologist would feel certain
that a purposeful communication was involved.

Similar observations convinced earthling zoologists of the communicative role of vervet alarm calls. A solitary vervet chased by a
leopard for nearly an hour remained silent throughout the whole
ordeal. Mother vervets give more alarm calls when accompanied by
their own offspring than by unrelated monkeys. Vervets occasionally
give the "leopard call" when no leopard is present but when their
troop is fighting with another troop and losing the fight. The fake
alarm sends all combatants scrambling for the nearest tree and
thereby serves as a deceptive "time-out." Hence the call is clearly a

voluntary communication, not an automatic expression of fear at the sight of a leopard. Nor is the call a mere reflex grunt given in the act of climbing a tree, since a calling monkey may either climb a tree, jump out of a tree, or do nothing, depending on the circumstances.

As for the call's having a well-defined external referent, that point is especially well illustrated by the "eagle call." When they see a large, broad-winged soaring hawk, vervets usually respond with the eagle call if the hawk is a martial eagle or crowned eagle, their two most dangerous avian predators. They usually do not respond if the hawk is a tawny eagle, and almost never if it's a black-chested snake eagle or white-backed vulture, which do not prey on vervets. Seen from below, black-chested snake eagles look rather similar to martial eagles in their shared pale underparts, banded tail, and black head and throat. Vervets rate as good bird-watchers, because their lives depend on it!

These examples demonstrate that vervet alarm calls are not involuntary expressions of either fear or intent. They have an external referent that may be quite exact. They are finely targeted communications that are more likely to be given honestly if the caller cares about the listener, and that may also be given dishonestly to enemies.

Skeptics go on to dispute proposed analogies between animal sounds and human speech on the further grounds that human speech is learned, but that many animals are born with the instinctive ability to utter the sounds characteristic of their species. However, young vervets appear to learn how to utter and respond to sounds appropriately, just as do human infants. The grunts of an infant vervet sound different from those of an adult. "Pronunciation" gradually improves with age until it becomes virtually adultlike around the age of two years, somewhat less than half the age for vervet puberty. That's like human children attaining adultlike pronunciation around age five; my four-year-old sons are still sometimes hard to understand. Infant vervets don't learn reliably to give the correct response to an adult's call until the age of six or seven months. Until then, an adult's snake alarm call may send the infant jumping into a bush, the correct response to an eagle but a suicidal response to a snake. Not until the age of two years does the infant consistently emit each alarm call in the correct context. Before that age, the young vervet may call "Eagle!" not only when a martial or crowned eagle goes overhead, but also when any other bird flies over, and even when a leaf flutters down from a tree. Child psychologists refer to such behavior in our own children as "overgeneralizing"—as when a child greets not just dogs but also cats and pigeons with "bow-wow."

So FAR, I've loosely applied human concepts such as "word" and "language" to vervet vocalization. Let's now compare human vocalizations and those of subhuman primates more closely. In particular, let's ask ourselves three questions. Do vervet sounds really constitute "words"? How large are animals' "vocabularies"? Do any animal vocalizations involve "grammar" and merit the term "language"?

First, as for the question of words, it's clear at least that each vervet alarm call refers to a well-defined class of external dangers. That doesn't imply, of course, that a vervet's "leopard call" designates the same animals to a vervet as the word "leopard" does to a professional zoologist—namely, members of a single animal species, defined as a collection of potentially interbreeding individuals. Scientists already know that vervets give their leopard alarm in response not just to leopards but also to two other medium-sized cat species (caracals and servals). Hence if the "leopard call" is a word at all, it wouldn't mean "leopard" but instead "medium-sized cats that are likely to attack us, hunt in a similar way, and are best avoided by running up a tree." However, many human words are used in a similar generic sense. For example, most of us other than ichthyologists and ardent fishermen apply the generic word "fish" to any cold-blooded animal with fins and a backbone that swims in the water and might be worth eating.

Instead, the real question is whether the leopard call constitutes a word ("medium-sized cat that et cetera"), a statement ("there goes a medium-sized cat"), an exclamation ("Watch out for that medium-sized cat!!"), or a proposition ("Let's run up a tree or take other appropriate action to avoid that medium-sized cat"). At present it's not clear which of those functions the leopard call fills, or whether it fills a combination of them. Similarly, I was excited when my one-year-old son Max said "Juice," which I proudly took to be one of his first words. To Max, though, the syllable "juice" was not just his academically correct identification of an external referent with certain properties, but it also served as a proposition: "Give me some juice!" Only at a later age did Max add more syllables, like "gimme juice," to distinguish propositions from pure words. Vervets show no evidence of having reached that stage.

On the second question of size of "vocabulary," even the most advanced animals seem on the basis of present knowledge to be far behind us. The average human has a daily working vocabulary of around a thousand words; my compact desk dictionary claims to contain 142,000 words; but only ten calls have been distinguished even for vervets, the most intensively studied mammal. Animals and humans surely do indeed differ in vocabulary size, yet the difference may not be as great as these numbers suggest. Remember how slow our progress has been in distinguishing vervet calls. Not until 1967 did anyone realize that these common animals had *any* calls with distinct meanings. The most experienced observers of vervets still can't separate some of their calls without machine analysis, and even with machine analysis the distinctness of some of the suspected ten calls remains unproven. Obviously, vervets (and other animals) could have many other calls whose distinctness we haven't yet recognized.

There's nothing surprising about our difficulties in distinguishing animal sounds, when one considers our difficulties in distinguishing human sounds. Children devote much of their time for the first several years of their lives to learning how to recognize and reproduce the distinctions in the utterances of adults around them. As adults, we continue to have difficulty distinguishing sounds in unfamiliar human languages. After four years of high school French between ages twelve and sixteen, my problems with understanding spoken French are embarrassing compared to the abilities of any four-year-old French child. But French is easy compared to the Iyau language of New Guinea's Lakes Plains, in which a single vowel may have eight different meanings depending on its pitch. A slight change in pitch converts the meaning of the Iyau word meaning "mother-in-law" into "snake." Naturally, it would be suicidal for an Iyau man to address his mother-in-law as "beloved snake," and Iyau children learn infallibly to hear and reproduce pitch distinctions that for years confounded even a professional linguist devoting full time to the Iyau language. Given these problems of our own with unfamiliar human languages, of course we must still be overlooking distinctions within the vervet vocabulary.

However, it's unlikely that any studies on vervets will reveal to us the limits attained by animal vocal communication, because those limits are probably reached by apes rather than by monkeys. While the sounds made by chimps and gorillas seem to our ears to be unsophisticated grunts and shrieks, so did the sounds made by vervet monkeys until they were studied carefully. Even unfamiliar human languages can sound like undifferentiated gibberish to us.

Unfortunately, vocal communication by wild chimps and other apes has never been studied by the methods applied to vervets, because of logistical problems. The width of a troop's territory is typically less than two thousand feet for vervets but several miles for chimps, making it far harder to carry out playback experiments with video cameras and hidden loudspeakers. These logistical problems can't be overcome by studying groups of wild-caught captive apes in convenient-sized zoo cages, because the captives generally constitute an artificial community of individuals caught at different African locations and thrown together in a cage. As I'll discuss later in this chapter, humans caught at different African locations, originally speaking different languages, and thrown together as slaves converse in only the crudest shadow of human language, virtually without any grammar. Similarly, wild-caught captive apes must be nearly useless for studying how sophisticated the vocal communications of wild apes are. The answer will remain unknown until someone figures out how to do for wild chimps what Cheney and Seyfarth have done for wild vervets.

Several groups of scientists have nevertheless spent years training captive gorillas, common chimps, and pygmy chimps to understand and use artificial languages based on plastic chips of different sizes and colors, or on hand signs similar to those used by deaf people, or on consoles like typewriters with each key bearing a different symbol. The animals have thereby learned the meanings of up to several hundred symbols, and a pygmy chimp has recently been observed to understand (but not to utter) a good deal of spoken English. At minimum, these studies of trained apes reveal that they possess the intellectual capabilities for mastering large vocabularies, begging the obvious question whether they have evolved such vocabularies in the wild.

It's suggestive that wild gorilla troops may be seen sitting together for a long time, grunting back and forth in seemingly undifferentiated gibberish, until suddenly all the gorillas get up at the same time and head off in the same direction. One wonders whether there really was a transaction concealed within that gibberish. Because the anatomy of apes' vocal tracts restricts their ability to produce the variety of vowels and consonants that we can, the vocabulary of wild apes is unlikely to be anywhere near as large as our own. Nevertheless, I would be surprised if wild chimp and gorilla vocabularies did *not* eclipse those reported for vervets and comprise dozens of "words," possibly including names for individual animals. In this exciting field in which new knowledge is being added rapidly, we should keep an open mind on how large the vocabulary gap is between apes and humans.

The remaining unanswered question concerns whether animal vocal communication involves anything that could be considered grammar or syntax. Humans don't just have vocabularies of thousands of words with different meanings. We also combine those words and vary their forms in ways prescribed by grammatical rules (such as rules of word order) that determine the meaning of the word combinations. Grammar thereby lets us construct a potentially infinite number of sentences from a finite number of words. To appreciate this point, consider the different meaning of the following two sentences composed of the same words and endings but with different word order:

> "Your hungry dog bit my old mother's leg."
> *or*
> "My hungry mother bit your old dog's leg."

If human language did not involve grammatical rules, those two sentences would have exactly the same meaning. Most linguists would not dignify an animal's system of vocal communication with the name of language, no matter how large its vocabulary, unless it also involved grammatical rules.

No hint of syntax has been discovered in the studies of vervets to date. Most of their grunts and alarm calls are single utterances. When a vervet gives a sequence of two or more utterances, all analyzed cases have proved to consist of the same utterance repeated, as has also been the case when one vervet has been recorded responding to another vervet's call. Capuchin monkeys and gibbons do have calls of several elements used only in certain combinations or sequences, but the meanings of these combinations remain to be deciphered (by us humans, that is).

I doubt that any student of primate vocalizations expects even wild chimps to have evolved a grammar remotely approaching the complexity of human grammar, complete with prepositions, verb tenses, and interrogative particles. However, it remains for the present an open guess whether any animal has evolved syntax. The necessary studies on the wild animals most likely to use grammar—pygmy or common chimps—simply have not yet been attempted.

In short, while the gulf between animal and human vocal communication is surely large, scientists are rapidly gaining understanding of how that gulf has been partly bridged from the animal side. Now let's trace the bridge from the human side. We have already discovered complex animal "languages"; do any truly primitive human languages still exist?

To HELP US RECOGNIZE what a primitive human language might sound like if there were any, let's remind ourselves of the ways in which normal human language differs from vervet vocalizations. One difference is the one of grammar that I just mentioned. Humans, but not vervets, possess grammar, meaning the variations in word order, prefixes, suffixes, and changes in word roots (like *they/them/their*) that modulate the sense of the roots. A second difference is that vervet vocalizations, if they constitute words at all, stand only for things that one can point to or act out. One could try to argue that vervet calls do include the equivalents of nouns ("eagle") and verbs or verb phrases ("watch out for the eagle"). Our words clearly include both nouns and verbs as distinct from each other, plus adjectives. Those three parts of speech referring to specific objects, acts, or qualities are termed "lexical items." But up to half of the words in typical human speech are purely grammatical items, with no referent that one can point to.

These grammatical words include prepositions, conjunctions, articles, and auxiliary verbs (words like "can," "may," "do," and "should"). It's much harder to understand how grammatical items could evolve than it is for lexical items. Given someone who understands no English, you can point to your nose to explain what that noun means. Apes might similarly come to agree on the meanings of grunts functioning as nouns, verbs, or adjectives. How, though, do you explain the meaning of "by," "because," "the," and "did" to someone who understands no English? How could our ancestors have stumbled on such grammatical terms?

Still another difference between human and vervet vocalizations is that ours possess a hierarchical structure, such that a modest number of items at each level creates a larger number of items at the next higher level. Our language uses many different syllables, all based on the same set of only a few dozen sounds. We assemble those syllables into thousands of words. Those words aren't merely strung haphazardly together but are organized into phrases, such as prepositional phrases. Those phrases in turn interlock to form a potentially infinite number of sentences. In contrast, vervet calls cannot be resolved into modular elements and lack even a single stage of hierarchical organization.

As children, we master all this complex structure of human language without ever learning the explicit rules producing it. Not unless we study our own language in school or learn a foreign language from books are we forced to formulate the rules. So complex is human language's structure that many of the underlying rules currently postulated by professional linguists have been proposed only in recent decades. This gulf between human language and animal vocalizations explains why most linguists never discuss how human language might have evolved from animal precursors. They instead regard that question as unanswerable and unworthy even of speculation.

BECAUSE THE EARLIEST WRITTEN LANGUAGES of five thousand years ago were as complex as those of today, human language must have achieved its modern complexity long before that. Can we at least recognize linguistic missing links by searching for primitive peoples with simple languages that might represent early stages of language evolution? After all, some tribes of hunter-gatherers retain stone tools as simple as those that characterized the whole world tens of thousands of years ago. Nineteenth-century travel books abound with tales of backward tribes who supposedly used only a few hundred words or who lacked articulated sounds, were reduced to saying "ugh," and depended on gestures for their communications. That was Darwin's first impression of the speech of Tierra del Fuego Indians. But all such tales proved to be pure myth. Darwin and other western travelers merely found it as hard to distinguish the unfamiliar sounds of non-western languages as non-westerners found English sounds, or as zoologists find the sounds of vervet monkeys.

Actually, it turns out that there is no correlation between linguistic and social complexity. Technologically primitive people don't speak primitive languages, as I discovered on my first day in the New Guinea highlands among the Foré people. Foré grammar proved deliciously complex, with postpositions like those of the Finnish language, dual as well as singular and plural forms like those of Slovenian, and verb tenses and phrase construction like no language that I had encountered previously. I already mentioned the eight vowel tones of New Guinea's Iyau people, whose sound distinctions proved imperceptibly subtle to professional linguists for years.

Thus, while some peoples in the modern world retained primitive tools, none retained primitive languages. Furthermore, Cro-Magnon archaeological sites contain lots of preserved tools but no preserved words. The absence of such linguistic missing links deprives us of what might have been our best evidence about human language origins. We are forced to try more indirect approaches.

ONE SUCH APPROACH is to ask whether some people, deprived of the opportunity to hear any of our fully evolved modern languages, ever spontaneously invented a primitive language. According to the Greek historian Herodotus, the Egyptian king Psammeticus intentionally carried out such an experiment in the hope of identifying the world's oldest language. The king assigned two newborn infants to a solitary shepherd to rear in strict silence, with instructions to listen for their first words. The shepherd duly reported that both children, after mouthing nothing but meaningless babble until the age of two, ran up to him and began constantly repeating the word "becos." Since that word meant "bread" in the Phrygian language then spoken in central Turkey, Psammeticus supposedly conceded that the Phrygians were the most ancient people.

Unfortunately, Herodotus's brief account of Psammeticus's experiment fails to convince skeptics that it was carried out rigorously as described. It illustrates why some scholars prefer to honor Herodotus as the Father of Lies, rather than as the Father of History. Certainly, solitary infants reared in social isolation, like the famous wolf boy of

Aveyron, remain virtually speechless and don't invent or discover a language. However, a variant of the Psammeticus experiment has occurred dozens of times in the modern world. In this variant, whole populations of children heard adults around them speaking a grossly simplified and variable form of language, somewhat similar to what normal children themselves speak around the age of two. The children proceeded unconsciously to evolve their own language, far advanced over vervet communication but simpler than normal human languages. The results were the new languages known as creoles. Along with their precursors, which are known as pidgins, creoles may provide us with models of missing links in the evolution of normal human language.

My first experience of a creole was with the New Guinea *lingua franca* known as either Neo-Melanesian or pidgin English. (The latter name is a confusing misnomer, since Neo-Melanesian is not a pidgin but rather a creole derived from an advanced pidgin—I'll explain the difference later—and it's only one of many independently evolved languages equally misnamed as pidgin English.) Papua New Guinea boasts about seven hundred native languages within an area similar to Sweden's, but no single one of those languages is spoken by more than 3 percent of the population. Not surprisingly, a *lingua franca* was needed and arose after the arrival of English-speaking traders and sailors in the early 1800s. Today, Neo-Melanesian serves in Papua New Guinea as the language not only of much conversation, but also of many schools, newspapers, radios, and parliamentary discussions. The advertisement on page 166 will give a sense of this newly evolved language.

When I arrived in Papua New Guinea and first heard Neo-Melanesian, I was scornful of it. It sounded like long-winded, grammarless baby talk. On speaking English according to my own notion of baby talk, I was jolted to discover that New Guineans weren't understanding me. My assumption that Neo-Melanesian words meant the same as their English cognates led to spectacular disasters, notably when I tried to apologize to a woman in her husband's presence for accidentally jostling her, only to find that Neo-Melanesian "pushim" doesn't mean "push" but instead means "have sexual intercourse with."

Neo-Melanesian proved to be as strict as English in its grammatical rules. It is a supple language that lets one express anything sayable in English. It even lets one make some distinctions that cannot be expressed in English except by means of clumsy circumlocutions. For example, the English pronoun "we" actually lumps two quite different concepts: "I plus you to whom I am speaking," and "I plus one or more other people, but not including you to whom I am speaking." In Neo-Melanesian these two separate meanings are expressed by the words "yumi" and "mipela," respectively. After I have been using Neo-Melanesian for months and then meet an English speaker who starts talking about "we," I often find myself wondering, "Am I included or not in your 'we'?"

Neo-Melanesian's deceptive simplicity and actual suppleness stem partly from its vocabulary, partly from its grammar. Its vocabulary is based on a modest number of core words whose meaning varies with context and becomes extended metaphorically. For instance, while Neo-Melanesian "gras" can mean English "grass" (whence "gras bilong solwara [salt water]" means "seaweed"), it also can mean "hair," whence "man i no gat gras long head bilong em" becomes "bald man."

The derivation of Neo-Melanesian "banis bilong susu" as the word for "bra" further illustrates how supple is the core vocabulary. "Banis," meaning "fence," comes from that English word as spoken by New Guineans who have difficulty pronouncing our consonant *f* and our double consonants like *nc*. "Susu," taken over from Malay as the word for "milk," is extended to mean "breast" as well. That sense in turn provides the expressions for "nipple" ("ai [eye] bilong susu"), "prepubertal girl" ("i no gat susu bilong em"), "adolescent girl" ("susu i sanap [stand up]"), and "aging woman" ("susu i pundaun pinis [fall down finish]"). Combining these two roots, "banis bilong susu" denotes a bra as the fence to keep the breasts in, just as "banis pik" denotes pigpen as the fence to keep pigs in.

Neo-Melanesian grammar appears deceptively simple because of what it lacks or else expresses by circumlocutions. These lacks include such seemingly standard grammatical items as plural and case forms of nouns, inflectional endings of verbs, the passive voice of verbs, and most prepositions and verb tenses. Yet Neo-Melanesian has passed far beyond baby talk and vervet sounds in many other respects, including its conjunctions and auxiliary verbs and pronouns, and its ways of expressing verb moods and aspects. It is a normal complex language in its hierarchical organization of phonemes, syllables, and words. It lends itself so well to hierarchical organization of phrases and sentences that election speeches by New Guinea politicians rival the German prose of Thomas Mann in their convoluted structure.

AT FIRST, I ignorantly assumed that Neo-Melanesian was a delightful aberration among the world's languages. It had obviously arisen just in the two centuries since English ships started visiting New Guinea, but I supposed that it had somehow developed from baby talk that colonists spoke to natives they believed incapable of learning English. As it turns out, though, dozens of other languages resemble Neo-Melanesian in structure. They have arisen independently around the globe, with vocabularies variously derived largely from English, French, Dutch, Spanish, Portuguese, Malay, or Arabic. They appeared especially in plantation, fort, and trading-post situations, where populations speaking different languages came into contact and needed to communicate, but where social circumstances impeded the usual solution of each group's learning the other's language. Many cases throughout the tropical Americas and Australia, and on tropical islands of the Caribbean, Pacific, and Indian Ocean, involved European colonists importing workers who came from afar and spoke many different tongues. Other European colonists set up forts or trading posts in already densely populated areas of China, Indonesia, or Africa.

Strong social barriers between the dominant colonists and the imported workers or local populations made the former unwilling, the latter unable, to learn the other's language. Usually the colonists scorned the local people, but in China the scorn was mutual: when English traders set up a post at Canton in 1664, the Chinese would no more abase themselves by learning the foreign devils' language or teaching them Chinese than would the English learn from or teach the heathen Chinese. Even if those social barriers had not existed, the workers would have had few opportunities to learn the colonists' tongue, because workers so greatly outnumbered colonists. Conversely, the colonists would also have found it difficult to learn "the" workers' tongue, because so many different languages were often represented among the workers.

Out of the temporary linguistic chaos that followed the plantations' or forts' founding, simplified but stabilized new languages emerged. Consider the evolution of Neo-Melanesian as an example. After English ships began to visit Melanesian islands just east of New Guinea around 1820, the English also took islanders to work on the sugar plantations of Queensland and Samoa, where workers of many language groups were thrown together. From this Babel sprang somehow the Neo-Melanesian language, whose vocabulary is 80 percent English, 15 percent Tolai (the Melanesian group that furnished many of the workers), and the rest Malay and other languages.

LINGUISTS DISTINGUISH two stages in the emergence of the new languages: initially, the crude languages termed pidgins, then later the more complex ones referred to as creoles. Pidgins arise as a second language for colonists and workers who speak differing native (first) languages and need to communicate with each other. Each group (colonists or workers) retains its native language for use within its own group; each group uses the pidgin to communicate with the other group, and in addition workers on a polyglot plantation may use pidgin to communicate with other groups of workers.

Compared to normal languages, pidgins are greatly impoverished in their sounds, vocabulary, and syntax. A pidgin's sounds are generally only those common to the two or more native languages thrown together. For example, many New Guineans find it hard to pronounce our consonants *f* and *v,* but I and other native English speakers find it hard to pronounce the vowel tones and nasalized sounds rampant in many New Guinea languages. Such sounds became largely excluded from New Guinea pidgins and then from the Neo-Melanesian creole that developed from them. Words of early-stage pidgins consist largely of nouns, verbs, and adjectives, with few or no articles, auxiliary verbs, conjunctions, or prepositions. As for grammar, early-stage pidgin discourse typically consists of short strings of words with little phrase construction, no regularity in word order, no subordinate clauses, and no inflectional endings on words. Along with that impoverishment, variability of speech within and between individuals is a hallmark of early-stage pidgins, which approximate an anarchic linguistic free-for-all.

Pidgins that are used only casually by adults who otherwise retain their own separate native languages persist at this rudimentary level.

For example, a pidgin known as Russonorsk grew up to facilitate barter between Russian and Norwegian fishermen who encountered each other in the Arctic. That *lingua franca* persisted throughout the nineteenth century but never developed further, as it was used only to transact simple business during brief visits. Both those groups of fishermen spent most of their time speaking Russian or Norwegian with their compatriots. In New Guinea, on the other hand, the pidgin gradually became more regular and complex over many generations because it was used intensively on a daily basis, but most children of New Guinea workers continued to learn their parents' native languages as their first languages until after World War II.

However, pidgins evolve rapidly into creoles whenever a generation of the groups contributing to a pidgin begins to adopt the pidgin itself as its native language. (I'll discuss later what particular members of the generation do the adopting, and why they do so.) That generation then finds itself using pidgin for all social purposes, not just for discussing plantation tasks or bartering. Compared to pidgins, creoles have a larger vocabulary, much more complex grammar, and consistency within and between individuals. Creoles can express virtually any thought expressible in a normal language, whereas trying to say anything even slightly complex is a desperate struggle in pidgin. Somehow, without any equivalent of the Académie Française to lay down explicit rules, a pidgin expands and stabilizes to become a uniform and fuller language.

This process of creolization is a natural experiment in language evolution that has unfolded independently dozens of times in the modern world. The sites for the experiment have ranged from mainland South America through Africa to Pacific islands; the laborers, from Africans through Portuguese and Chinese to New Guineans; the dominant colonists, from English to Spaniards to other Africans and Portuguese; and the century, from at least the seventeenth to the twentieth. What is striking is that the linguistic outcomes of all these independent natural experiments share so many similarities, both in what they lack and in what they possess. On the negative side, creoles are simpler than normal languages in mostly lacking conjugations of verbs for tense and person, declensions of nouns for case and number, most prepositions, and agreement of words for gender. On the positive side, creoles are advanced over pidgins in many respects: consistent word order; singular and plural pronouns for the first, second, and third persons; relative clauses; indications of the anterior tense (describing actions occurring before the time under discussion, whether or not that time is the present); and particles or auxiliary verbs preceding the main verb and indicating negation, anterior tense, conditional mood, and continuing as opposed to completed actions. Furthermore, most creoles agree in placing a sentence's subject, verb, and object in that particular order, and also agree in the order of particles or auxiliaries preceding the main verb.

The factors responsible for this remarkable convergence are still controversial among linguists. It's as if you drew a dozen cards fifty times from well-shuffled decks and almost always ended up with no hearts or diamonds, but with one queen, a jack, and two aces. The interpretation I find most convincing is that of linguist Derek Bickerton, who views many of the similarities among creoles as resulting from our possessing a genetic blueprint for language.

Bickerton derived his view from his studies of creolization in Hawaii, where sugar planters imported workers from China, the Philippines, Japan, Korea, Portugal, and Puerto Rico in the late nineteenth century. Out of that linguistic chaos, and following Hawaii's annexation by the United States in 1898, a pidgin based on English developed into a full-fledged creole. The immigrant workers themselves retained their original native languages. They also learned pidgin that they heard, but they did not improve on it, despite its gross deficiencies as a medium of communication. That, however, posed a big problem for the immigrants' Hawaii-born children. Even if the kids were lucky enough to hear a normal language at home because both mother and father were from the same ethnic group, that normal language was useless for communicating with kids and adults from other ethnic groups. Many children were less fortunate and heard nothing but pidgin even at home, when mother and father came from different ethnic groups. Nor did the children have adequate opportunities to learn English because of the social barriers isolating them and their worker parents from the English-speaking plantation owners. Presented with an inconsistent and impoverished model of human language in the form of pidgin, Hawaiian laborers' children spontaneously "expanded" pidgin into a consistent and complex creole within a generation.

In the mid-1970s Bickerton was still able to trace the history of this creolization by interviewing working-class people born in Hawaii between 1900 and 1920. Like all of us, those people had soaked up language skills in their early years but then became fixed in their ways, so that their speech in their old age continued to reflect the language spoken around them in their youth. (My children too will soon be wondering why their father persists in saying "icebox" rather than "refrigerator," decades after the iceboxes of my parents' own childhood disappeared.) The old adults of various ages whom Bickerton interviewed in the 1970s gave him virtually frozen snapshots of various stages in Hawaii's pidgin-to-creole transition, depending on the subjects' birth year. In that way, Bickerton was able to conclude that creolization had begun by 1900, was complete by 1920, and was accomplished by children in the process of their acquiring the ability to speak.

In effect, the Hawaiian children lived out a modified version of the Psammeticus experiment. Unlike the Psammeticus children, the Hawaiian children did hear adults speaking and were able to learn words. Unlike normal children, however, the Hawaiian children heard little grammar, and what they did hear was inconsistent and rudimentary. Instead, they created their own grammar. That they did indeed create it, rather than somehow borrowing grammar from the language of Chinese laborers or English plantation owners, is clear from the many features of Hawaiian creole that differ from English and from the workers' languages. The same is true for Neo-Melanesian: its vocabulary is largely English, but its grammar includes many features absent from English.

I DON'T WANT to exaggerate the grammatical similarities among creoles by implying that they're all essentially the same. Creoles do vary depending on the social history surrounding creolization—especially on the initial ratio between the numbers of plantation owners (or colonists) and workers, how quickly and to what extent that ratio changed, and for how many generations the early-stage pidgin could gradually borrow more complexity from existing languages. But many similarities remain, particularly among those creoles quickly arising from early-stage pidgins. How did each creole's children come so quickly to agree on a grammar, and why did the children of different creoles tend to reinvent the same grammatical features again and again?

It wasn't because they did it in the easiest or sole way possible to devise a language. For instance, creoles use prepositions (short words preceding nouns), as do English and some other languages, but there are other languages that dispense with prepositions in favor of post-positions following nouns, or else noun case endings. Again, creoles happen to resemble English in placing subject, verb, and object in that order, but borrowing from English can't be the explanation, because creoles derived from languages with a different word order still use the subject-verb-object order.

These similarities among creoles seem likely to stem from a genetic blueprint that the human brain possesses for learning language during childhood. Such a blueprint has been widely assumed ever since the linguist Noam Chomsky argued that the structure of human language is far too complex for a child to learn within just a few years, in the absence of any hard-wired instructions. For example, at age two my twin sons were just beginning to use single words. As I write this paragraph a bare twenty months later, still several months before their fourth birthday, they have already mastered most rules of basic English grammar that people who immigrate to English-speaking countries as adults often fail to master after decades. Even before the age of two, my children had learned to make sense of the initially incomprehensible babble of adult sound coming at them, to recognize groupings of syllables into words, and to realize which groupings constituted underlying words despite variations of pronunciation within and between adult speakers.

Such difficulties convinced Chomsky that children learning their first language would face an impossible task unless much of language's structure was already preprogrammed into them. Chomsky concluded that we are born with a "universal grammar" already wired into our brains to give us a spectrum of grammatical models encompassing the range of grammars in actual languages. This prewired universal grammar would be like a set of switches, each with various alternative positions. The switch positions would then become fixed to match the grammar of the local language that the growing child hears.

However, Bickerton goes further than Chomsky and concludes that we are preprogrammed not just to a universal grammar with adjustable switches, but to a particular set of switch settings: the settings that surface again and again in creole grammars. The preprogrammed settings can be overridden if they turn out to conflict with what a child hears in the local language around it. But if a child hears no local switch settings at all because it grows up amidst the structureless anarchy of a pidgin language, the creole settings can persist.

If Bickerton is correct that we really are preprogrammed at birth with creole settings that can be overridden by later experience, then one would expect children to learn creolelike features of their local languages earlier and more easily than features conflicting with creole grammar. This reasoning might explain the notorious difficulty of English-speaking children in learning how to express negatives: they insist on creolelike double negatives such as "Nobody don't have this." The same reasoning could explain the difficulties of English-speaking children with word order in questions.

To pursue the latter example, English happens to be among the languages that uses the creole word order of subject, verb, and object for statements: for instance, "I want juice." Many languages, including creoles, preserve this word order in questions, which are merely distinguished by an altered tone of voice ("You want juice?"). However, the English language does not treat questions in this way. Instead, our questions deviate from creole word order by inverting the subject and verb ("Where are you?," not "Where you are?"), or by placing the subject between an auxiliary verb (such as "do") and the main verb ("Do you want juice?"). My wife and I have been barraging our sons from early infancy onward with grammatically correct English questions as well as statements. My sons quickly picked up the correct order for statements, but both of them are still persisting in the incorrect creolelike order for questions, despite the hundreds of correct counterexamples that my wife and I model for them every day. Today's samples from Max and Joshua include "Where it is?," "What that letter is?," "What the handle can do?," and "What you did with it?" It's as if they're not yet ready to accept the evidence of their ears, because they're still convinced that their preprogrammed creolelike rules are correct.

Now LET'S PULL TOGETHER all these animal and human studies to try to form a coherent picture of how our ancestors progressed from grunts to Shakespeare's sonnets. A well-studied early stage is represented by vervet monkeys, with at least ten different calls that are under voluntary control, are used for communication, and have external referents. The calls may function as words, explanations, propositions, or as all of those things simultaneously. Scientists' difficulties in identifying those ten calls have been such that surely more await identification, but we still don't know how large the vervet vocabulary really is. We also don't know how far other animals may have progressed beyond vervets, because the vocal communications of the species most likely to have eclipsed vervets, the common and pygmy chimp, have yet to be studied carefully in the wild. At least in the laboratory, chimps can master hundreds of symbols that we teach them, suggesting that they have the necessary intellectual equipment to master symbols of their own.

The single words of young toddlers, like "juice" as uttered by my son Max, constitute a next stage beyond animal grunts. Like vervet grunts, Max's "juice" may have functioned as some combination of a word, an explanation, and a proposition. But Max has made an enormous advance on vervets by assembling his "juice" word from the smaller units of vowels and consonants, thereby scaling the lowest level of modular linguistic organization. A few dozen such phonetic units can be reshuffled to produce a very large number of words, such as the 142,000 words in my English desk dictionary. That principle of modular organization lets us recognize far more distinctions than can vervets. For example, they name only six types of animals, whereas we name nearly two million.

A further step toward Shakespeare is exemplified by two-year-old children, who in all human societies proceed spontaneously from a one-word to a two-word stage and then to a multiword stage. But those multiword utterances are still mere word strings with little grammar, and their words are still nouns, verbs, and adjectives with concrete referents. As Bickerton points out, those word strings are rather like the pidgin languages that human adults spontaneously reinvent when necessary. They also resemble the strings of symbols produced by captive apes whom we have instructed in the use of those symbols.

From pidgins to creoles, or from the word strings of two-year-olds to the complete sentences of four-year-olds, is another giant step. In that step are added words lacking external referents and serving purely grammatical functions; elements of grammar such as word order, prefixes and suffixes, and word root variation; and more levels of hierarchical organization to produce phrases and sentences. Perhaps that step is what triggered the Great Leap Forward discussed earlier in this book. Nevertheless, creole languages reinvented in modern times still give us clues to how these advances arose, through the creoles' circumlocutions to express prepositions and other grammatical elements.

If you compare the Neo-Melanesian advertisement below with a Shakespeare sonnet, you might conclude that a huge gap still remains. In fact, I'd argue that, with an advertisement like "Kam insait long stua bilong mipela ..." we have come 99.9 percent of the way from vervet calls to Shakespeare. Creoles already constitute expressive complex languages. For example, Indonesian, which arose as a creole to become the language of conversation and government for the world's fifth most populous country, is also a vehicle for serious literature.

Thus, animal communication and human language once seemed to be separated by an unbridgeable gulf. Now we have identified not only parts of bridges starting from both opposite shores, but also a series of islands and bridge segments spaced across the gulf. We are beginning to understand in broad outline how the most unique and important attribute that distinguishes us from animals arose from animal precursors.

NEO-MELANESIAN, IN ONE EASY LESSON

Try to understand this Neo-Melanesian advertisement for a department store:

> Kam insait long stua bilong mipela—stua bilong salim olgeta samting—mipela i-ken helpim yu long kisim wanem samting yu laikim bikpela na liklik long gutpela prais. I-gat gutpela kain kago long baiim na i-gat stap long helpim yu na lukautim yu long taim yu kam insait long dispela stua.

If some of the words look strangely familiar but don't quite make sense, read the ad aloud to yourself, concentrate on the sounds, and ignore the strange spelling. As the next step, here is the same ad rewritten with English spelling:

> Come inside long store belong me-fellow—store belong sellim altogether something—me-fellow can helpim you long catchim what-name something you likim, big-fellow na liklik, long good-fellow price. He-got good-fellow kind cargo long buyim, na he-got staff long helpem you na lookoutim you long time you come inside long this-fellow store.

A few explanations should help you make sense of the remaining strangenesses. Almost all the words in this sample of Neo-Melanesian are derived from English, except for the word "liklik" for "little," derived from a New Guinea language (Tolai). Neo-Melanesian has only two pure prepositions: "bilong," meaning "of" or "in order to," and "long," meaning almost any other English preposition. The English consonant *f* becomes *p* in Neo-Melanesian, as in "stap" for "staff," and "pela" for "fellow." The suffix "-pela" is added to monosyllabic adjectives (hence "gutpela" for "good," "bikpela" for "big"), and also makes the singular pronouns "me" and "you" into plural ones (for "we" and "you" plural). "Na" means "and." Hence the ad means:

> Come into our store—a store for selling everything—we can help you get whatever you want, big and small, at a good price. There are good types of goods for sale, and staff to help you and look after you when you visit the store.

CHAPTER 9

Animal Origins
of Art

GEORGIA O'KEEFFE'S DRAWINGS WERE SLOW TO WIN RECOGNITION for her, but Siri's drawings brought her acclaim as soon as other knowledgeable artists saw them. "They had a kind of flair and decisiveness and originality"—that was the first reaction of the famous abstract expressionist painter Willem de Kooning. Jerome Witkin, an authority on abstract expressionism who teaches art at Syracuse University, was even more effusive: "These drawings are very lyrical, very, very beautiful. They are so positive and affirmative and tense, the energy is so compact and controlled, it's just incredible. . . . This drawing is so graceful, so delicate. . . . This drawing indicates a grasp of the essential mark that makes the emotion."

Witkin applauded Siri's balance of positive and negative space, and her placement and orientation of images. Having seen the drawings but knowing nothing about who made them, he guessed correctly that the artist was female and interested in Asian calligraphy. But Witkin didn't guess that Siri was eight feet tall and weighed four tons. She was an Asian elephant who drew by holding a pencil in her trunk.

De Kooning's response to being told Siri's identity was "That's a damned talented elephant." Actually, Siri was not extraordinary by elephant standards. Wild elephants often use their trunks to make drawing motions in the dust, while captive elephants often spontaneously scratch marks on the ground with a stick or stone. Hanging in many doctors' and lawyers' offices are paintings by an elephant named Carol, dozens of whose works were sold at prices of up to $500.

Supposedly, art is the noblest distinctively human attribute—one that sets us apart from animals at least as sharply as does spoken language, by differing in basic ways from anything that any animal does. Art ranks as even nobler than language, since language is really "just" a highly sophisticated advance on animal communication systems, serves an obvious biological function in helping us to survive, and obviously developed from the sounds made by other primates. In contrast, art serves no such transparent function, and its origins are considered a sublime mystery. But it's clear that elephant art could have implications for our own. At minimum, it's a similar physical activity resulting in products that even experts couldn't distinguish from human products accepted as constituting art. Of course, there are also huge differences between Siri's art and ours, not least of which is that Siri wasn't trying to communicate her message to other elephants. Nevertheless, we can't just dismiss her art as a quirk of one individual beast.

In this chapter I'll go beyond elephants to examine artlike activities of some other animals. I believe that the comparisons will help us understand the original functions of human art. Although we usually consider art to be the antithesis of science, there may really be a science of art.

To APPRECIATE that our art must have some animal precursors, recall that it's only about seven million years since we branched off from our closest living relatives, the chimpanzees. Seven million years sounds like a lot on the scale of a human lifetime, but it is barely 1 percent of the history of complex life on Earth. We still share over 98 percent of our DNA with chimps. Hence art and those other features that we consider uniquely human must be due to just a tiny fraction of our genes. They must have arisen only a few moments ago on the evolutionary time clock.

Modern studies of animal behavior have been shrinking the list of features once considered uniquely human, such that most differences between us and so-called animals now appear to be only matters of degree. For example, I described in the preceding chapter how vervet monkeys have a rudimentary language. You may not have considered vampire bats allied with us in nobility, but they prove to practice reciprocal altruism regularly (toward other vampire bats, of course). Among our darker qualities, murder has now been documented in innumerable animal species, genocide in wolves and chimps, rape in ducks and orangutans, and organized warfare and slave raids in ants.

As absolute distinctions between us and animals, these discoveries leave us few characteristics besides art, which we managed to dispense with for the first 6,960,000 of the 7 million years since we diverged from chimps. Perhaps the earliest art forms were wood carving and body painting, but we wouldn't know it because they wouldn't have remained preserved. The first preserved hints of human art consist of some flower remains around Neanderthal skeletons, and some scratches on animal bones at Neanderthal campsites. However, the interpretation that they were arranged or scratched intentionally is in doubt. Not until the Cro-Magnons, beginning around forty thousand years ago, do we have our first unequivocal evidence for art, surviving in the form of the famous cave paintings at Lascaux, statues, necklaces, and flutes and other musical instruments.

If we're going to claim that true art is unique to humans, then in what ways do we claim that it differs from superficially similar productions of animals, like bird songs? Three supposed distinctions are often put forward: that human art is nonutilitarian, that it's just for aesthetic pleasure, and that it's transmitted by learning rather than through our genes. Let's scrutinize these claims more closely.

First, as Oscar Wilde said, "All art is quite useless." The implicit meaning a biologist sees behind this quip is that art is nonutilitarian in a narrow sense employed within the fields of animal behavior and evolutionary biology. That is, human art doesn't help us to survive or to pass on our genes, which are the readily discernible functions of most animal behaviors. Of course, most human art is utilitarian in the broader sense that the artist thereby communicates something to fellow humans, but transmitting one's thoughts to the next generation isn't the same thing as transmitting one's genes. In contrast, bird song serves the obvious functions of wooing a mate, defending a territory, and thereby transmitting genes.

Regarding the second claim, that human art is instead motivated by aesthetic pleasure, a dictionary defines art as "the making or doing of things that have form or beauty." While we can't ask mocking-birds and nightingales if they similarly enjoy the form or beauty of their songs, it's suspicious that they sing mainly during the breeding season. Hence they're probably not singing just for aesthetic pleasure.

As for human art's third claimed distinction, each human group has a distinctive art style, and the knowledge of how to make and enjoy that particular style is learned, not inherited. For example, it's easy to distinguish typical songs being sung today in Tokyo and in Paris. But those stylistic differences aren't hard-wired in our genes, as are the differences between Parisians and Japanese in their eyes. Parisians and Japanese can and often do visit each other's cities and learn each other's songs. In contrast, many species of birds (the so-called nonpasserine birds) inherit the knowledge of how to produce and respond to the particular song of their species. Each of those birds would produce the right song even if it had never heard it, and even if it had heard only the songs of other species. It's as if a French baby adopted by Japanese parents, flown in infancy to Tokyo and educated there, began spontaneously to sing the "Marseillaise."

At this point, we may seem to be light-years removed from ele-phant art. Elephants aren't even closely related to us evolutionarily. Much more relevant to us are the artworks that were produced by two captive chimpanzees named Congo and Betsy, a gorilla called Sophie, an orangutan named Alexander, and a monkey named Pablo. These primates variously mastered the media of brush or finger painting and pencil, chalk, or crayon drawing. Congo did up to thirty-three paintings in one day, apparently for his own satisfaction, as he didn't show his work to other chimps and threw a tantrum when his pencil was taken away. For human artists, the ultimate proof of success is a one-man show, but Congo and Betsy were honored by a two-chimp show of their paintings in 1957 at London's Institute of Contemporary Art. The following year, Congo had a one-chimp show at London's Royal Festival Hall. What's more, al-most all of the paintings on exhibit at those chimp shows sold (to human buyers); plenty of human artists can't make that boast. Still other ape paintings were surreptitiously entered into exhibits by human artists and were enthusiastically acclaimed by unsuspecting art critics for their dynamism, rhythm, and sense of balance.

Equally unsuspecting were child psychologists who were given paintings by chimps from the Baltimore Zoo and were asked to diagnose the painters' problems. The psychologists guessed that a painting by a three-year-old male chimp was instead by an aggressive seven- or eight-year-old boy with paranoid tendencies. Two paintings by the same one-year-old female chimp were attributed to different ten-year-old girls, one painting indicating a belligerent girl of the schizoid type, the other a paranoid girl with strong father identification. It's a tribute to the psychologists' insight that they intuited the artists' sex correctly in each case; they were only wrong about the artists' species.

These paintings by our closest relatives do start to blur the distinction between human art and animal activities. Like human paintings, the ape paintings served no narrow utilitarian function of transmitting genes, and were instead just produced for satisfaction. One could object that the ape artists, like the elephant Siri, made their pictures just for their own satisfaction, while most human artists aim to communicate to other humans. The apes didn't even keep their paintings to enjoy but just discarded them. But that objection doesn't strike me as compelling, since the simplest human art (doodling) is also regularly discarded, and since one of the best pieces of art I own is a wood statue carved by a New Guinea villager who discarded it under his house after carving it. Even some human art that later became famous was created by artists for their private satisfaction: the composer Charles Ives published little of his music, and Franz Kafka not only didn't publish his three great novels but even forbade his executor to do so. (Fortunately, the executor disobeyed, thereby forcing Kafka's novels to take on a communicative function posthumously.)

However, there's a more serious objection against claiming a parallel between ape art and human art. Ape painting is just an unnatural activity of captive animals. One might insist that, since it's not a natural behavior, it couldn't illuminate the animal origins of art. Hence let's turn now to an undeniably natural and illuminating behavior: bowerbirds' building of bowers, the most elaborate structures built and decorated by any animal species other than humans.

*   *   *

IF I HADN'T already heard of bowers, I'd have mistaken the first one I saw for something man-made, as did nineteenth-century explorers in New Guinea. I had set out that morning from a New Guinea village, with its circular huts, neat rows of flowers, people wearing decorative beads, and little bows and arrows carried by children in imitation of their fathers' larger ones. Suddenly, in the jungle, I came across a beautifully woven circular hut eight feet in diameter and four feet high, with a doorway large enough for a child to enter and sit inside. In front of the hut was a lawn of green moss, clean of debris except for hundreds of natural objects of various colors that had obviously been placed there intentionally as decorations. They mainly consisted of flowers and fruits and leaves, but also some butterfly wings and fungi. Objects of similar color were grouped together, such as red fruits next to a group of red leaves. The largest decorations were a tall pile of black fungi facing the door, with another pile of orange fungi a few yards further from the door. All blue objects were grouped inside the hut, red ones outside, and yellow, purple, black, and a few green ones in other locations.

That hut was not a child's playground. It had instead been built and decorated by an otherwise unimpressive jay-size bird called a bowerbird, a member of a family of eighteen species confined to New Guinea and Australia. Bowers are erected by males for the sole purpose of seducing females, who then bear the sole responsibility for building the nest and rearing the young. Males are polygamous, try to mate with as many females as possible, and provide the female with nothing except sperm. Females, often in groups, cruise around the bowers and inspect all those in the vicinity before selecting one at which to mate. Human equivalents of such scenes are played out every night on Sunset Strip, a few miles from my home in Los Angeles.

Female bowerbirds select their bedmate by the quality of his bower, its number of decorations, and its conformity to local rules, which vary among species and populations of bowerbirds. Some populations prefer blue decorations, others red or green or gray, while some replace the hut with one or two towers, or a two-walled avenue, or a four-walled box. There are populations that paint their bowers with crushed leaves or else with oils that they excrete. These local differences in rules appear not to be hard-wired into the birds' genes. Instead, they are learned through younger birds' observing older birds during the many years that it takes a bowerbird to reach adulthood. Males learn the locally correct way to decorate, while females learn those same rules for the purpose of choosing a male.

At first, this system strikes us as absurd. After all, what a female bowerbird is trying to do is to pick a good mate. The evolutionary winner in such a mate-selection contest is that female bowerbird who picks that male bowerbird who makes it possible for her to leave the largest number of surviving offspring. What good does it do her to pick the guy with the blue fruits?

All animals, including us, face similar problems of mate selection. Consider those species (such as most European and North American songbirds) whose males carve out mutually exclusive territories that each male will share with his mate. The territory contains the nest site and food resources for the female to use in rearing her young. Hence a part of the female's task is to assess the quality of each male's territory. Or suppose that the male himself will assist in feeding and protecting the young, and in hunting cooperatively with the female. Then the female and male must assess each other's parenting and hunting skills and the quality of their relationship. All these things are hard enough to assess, but it's even harder for the female to assess a male when he provides nothing but sperm and genes, as is the case with male bowerbirds. How on earth is an animal to assess a prospective mate's genes, and what have blue fruits to do with good genes?

Animals don't have the time to produce ten offspring with each of many prospective mates, and to compare the outcomes (the eventual number of surviving offspring). Instead, they have to resort to shortcuts by relying on mating signals such as songs or ritualized displays. It's now a hotly contested problem in animal behavior to understand why, or even whether, those mating signals serve as veiled indicators of good genes. We have only to reflect on our own difficulties in selecting mates and in assessing the true wealth, parenting skills, and genetic quality of our various prospective partners.

In this light, reflect what it means when a female bowerbird finds a male with a good bower. She knows at once that that male is strong, since the bower he assembled weighs hundreds of times his own weight, and since he had to drag some individual decorations half his weight from dozens of yards distant. She knows that the male has the mechanical dexterity needed to weave hundreds of sticks into a hut, tower, or walls. He must have a good brain, to carry out the complex design correctly. He must have good vision and memory to search out the required hundreds of decorations in the jungle. He must be good at coping with life, to have survived to the age of perfecting all those skills. And he must be dominant over other males: since males spend much of their leisure time trying to wreck and steal from each other's bowers, only the best males end up with intact bowers and many decorations.

Thus, bower building provides a comprehensive test of male genes. It's as if women put each of their suitors in sequence through a weight-lifting contest, sewing contest, chess tournament, eye test, and boxing tournament, and finally went to bed with the winner. By comparison with bowerbirds, our efforts to identify mates with good genes are pathetic. We grasp at external bagatelles like facial features and earlobe lengths, or sex appeal and Porsche ownership, which tell nothing about intrinsic genetic worth. Think of all the human suffering caused by the sad truth that beautiful, sexy women or handsome, Porsche-owning men often prove to have miserable genes for other traits. It's no wonder that so many marriages end in divorce, as we belatedly realize how badly we chose and how flimsy our criteria were.

How did bowerbirds evolve to use art so cleverly for such important purposes? Most male birds woo females by advertising their colorful bodies, songs, displays, or offerings of food as dim indicators of good genes. Males of two groups of birds of paradise in New Guinea go one step further by clearing areas on the jungle floor, as bowerbirds do, to enhance their displays and show off their fancy plumage. Males of one of those birds of paradise have gone still further by decorating their cleared areas with objects useful to a nesting female: pieces of snakeskin to line her nest, pieces of chalk or mammal feces to eat for their minerals, and fruits to eat for their calories. Finally, bowerbirds have learned that decorative objects useless in themselves may nevertheless be useful indicators of good genes, if the objects are ones that were difficult to acquire and keep.

We can easily relate to that concept. Just think of all those ads showing handsome men presenting diamond rings to seemingly fertile young women. You can't eat a diamond ring, but a woman knows that the gift of such a ring tells far more about the resources that her suitor commands (and might devote to her offspring and herself) than a gift of a box of chocolates would tell. Yes, chocolates provide a few useful calories, but they're quickly gone and any jerk can afford to buy them. In contrast, the man who can afford that inedible diamond ring has money to support the woman and her kids, and also has whatever genes (for intelligence, persistence, energy, etc.) that it took to acquire or hold on to the money.

Thus, in the course of bowerbird evolution the female's attention has been lured from ornaments that are permanent parts of the male's body to ornaments that the male gathers. Whereas sexual selection in most species has produced differences between males and females in their bodily ornaments, in bowerbirds it has shifted toward causing males to emphasize collected ornaments separate from their bodies. From this perspective, bowerbirds are rather human. We, too, rarely court (or at least rarely initiate courtship) by displaying the beauties of our unadorned naked bodies. Instead, we swathe ourselves in colored cloths, spray or daub ourselves with perfumes and paints and powders, and augment our beauty with decorations ranging from jewels to sports cars. The parallel between bowerbirds and humans may be even closer if, as friends of mine who are into sports cars assure me, duller young men tend to decorate themselves with fancier sports cars.

Now LET'S REEXAMINE, in the light of bowerbirds, those three criteria supposedly separating human art from any animal production. Both bower styles and our art styles are learned rather than inherited, so there's no difference by the third criterion. As for the second criterion (doing it for aesthetic pleasure), it's unanswerable. We can't ask bowerbirds whether they get pleasure out of their art, and I suspect that many humans who claim to do so are just putting on cultural affectations. That leaves only the first criterion: Oscar Wilde's assertion that art is useless, in a narrow biological sense. His statement is definitely untrue of bower art, which serves a sexual function. But it's absurd to pretend any longer that our own art lacks biological functions. Instead, there are several ways in which art helps us to survive and to pass on our genes.

First, art often brings direct sexual benefits to its owner. It's not just a joke that a man bent on seduction invites a woman to view his etchings. In real life, dance and music and poetry are common preludes to sex.

Second and much more important, art brings indirect benefits to its owner. Art is a quick indicator of status, which—in human as in animal societies—is a key to acquiring food, land, and sex partners. Yes, bowerbirds get the credit for discovering the principle that ornaments separate from one's body are more flexible status symbols than ornaments that one has to grow. But we still get credit for running away with that principle. Cro-Magnons decorated their bodies with bracelets, pendants, and ocher; New Guinea villagers today decorate theirs with shells, fur, and bird-of-paradise plumes. In addition to these art forms for bodily adornment, both Cro-Magnons and New Guinea villagers produced larger art (carvings and paintings) of world quality. We know that New Guinea art signals superiority and wealth, because birds of paradise are hard to hunt, beautiful statues take talent to make, and both are very expensive to buy. These badges of distinction are essential for marital sex in New Guinea: brides are bought, and part of the price consists of luxury art. Elsewhere as well, art is often viewed as a signal of talent, money, or both.

In a world where art is a coin of sex, it's only a small further step for some artists to be able to convert art into food. There are whole societies that support themselves by making art for trade to food-producing groups. For example, the Siassi islanders, who lived on tiny islets with little room for gardens, survived by carving beautiful bowls that other tribes coveted for bride payments and paid for in food.

The same principles hold even more strongly in the modern world. Where we once signaled our status with bird feathers on our bodies and giant clam shells in our huts, we now do it with diamonds on our bodies and Picassos on our walls. Where Siassi islanders sold a carved bowl for the equivalent of twenty dollars, Richard Strauss built himself a villa with the proceeds from his opera *Salome* and earned a fortune from *Der Rosenkavalier*. Nowadays we read increasingly often of art sold at auction for tens of millions of dollars, and of art theft. In short, precisely because it serves as a signal of good genes and ample resources, art can be cashed in for still more genes and resources.

So far, I've considered only the benefits that art brings to individuals. But art also helps define human groups. Humans have always formed competing groups whose survival is essential if the individuals in that group are to pass on their genes. Human history largely consists of the details of groups' killing, enslaving, or expelling other groups. The winner takes the loser's land, sometimes also the loser's women, and thus the loser's opportunity to perpetuate genes. But group cohesion depends on the group's distinctive culture—especially its language, religion, and art (including stories and dances). Hence art is a significant force behind group survival. Even if you have better genes than most of your fellow tribesmen, that will do you no good should your whole tribe (including you) get annihilated by some other tribe.

By now, you're probably protesting that I've gone completely overboard in ascribing utility to art. What about all of us who just enjoy art, without converting it to status or sex? What about all the artists who remain celibate? Aren't there easier ways to seduce a sex partner than to take piano lessons for ten years? Isn't private satisfaction a (the?) main reason for our art, just as for Siri and Congo?

Of course. Such expansion of behaviors far beyond their original role is typical of animal species whose foraging efficiency gives them much leisure time, and who have brought their survival problems under control. Bowerbirds and birds of paradise have much leisure time, because they're big and feed on wild fruit trees that they can kick smaller birds out of. We have much leisure time because we use tools to obtain food. Animals with leisure time can channel it into more lavish signals to outdo the next guy. Those behaviors may then come to serve other purposes, such as representing information (a suggested function of Cro-Magnon cave paintings of hunted animals), relieving boredom (a real problem for captive apes and elephants), channeling neurotic energy (a problem for us as well as for them), and just providing pleasure. To maintain that art is useful isn't to deny that art provides pleasure. Indeed, if we weren't programmed to enjoy art, it couldn't serve most of its useful functions for us.

Perhaps we can now answer the question why art as we know it characterizes us but no other animal. Since chimps paint in captivity, why don't they do so in the wild? As an answer, I suggest that wild chimps still have their day filled with problems of finding food, surviving, and fending off rival chimp groups. If wild chimps had more leisure time plus the means to manufacture paints, they would be painting. The proof of my theory is that it actually happened: we're still 98 percent chimps in our genes.

CHAPTER 10

Agriculture's
Mixed Blessings

To SCIENCE WE OWE DRAMATIC CHANGES IN OUR SMUG SELF-IMAGE. Astronomy taught us that our Earth is not the center of the universe but merely one of nine planets circling one of billions of stars. From biology we learned that humans were not specially created by God but evolved along with tens of millions of other species. Now archaeology is demolishing another sacred belief: that human history over the last million years has been a long tale of progress.

In particular, recent discoveries suggest that the adoption of agriculture (plus animal husbandry), supposedly our most decisive step toward a better life, was actually a milestone for the worse as well as for the better. With agriculture came not only greatly increased food production and food storage, but also the gross social and sexual inequality, the disease and despotism, that curse modern human existence. Thus, among the human cultural hallmarks being discussed in Part Three of this book, agriculture represents in its mixed blessings a halfway station between our noble traits already discussed (art and language) and our unmitigated vices still to be discussed (drug abuse, genocide, and environmental destructiveness).

At first, the evidence for progress and against this revisionist interpretation will strike twentieth-century Americans and Europeans as irrefutable. We are better off in almost every respect than people of the Middle Ages, who in turn had it easier than Ice Age cavemen, who were still better off than apes. If you're inclined to be cynical, just count our advantages. We enjoy the most abundant and varied food, the best tools and material goods, the longest and healthiest lives in human history. Most of us are safe from starvation and predators. We obtain most of our energy from oil and machines, not just from our sweat. What neo-Luddite among us would really trade today's life for that of a medieval peasant, caveman, or ape?

For most of our history, all humans had to practice a primitive life-style termed "hunting and gathering": they hunted wild animals and gathered wild plant food. That hunter-gatherer life-style is often characterized by anthropologists as "nasty, brutish, and short." Since no food is grown and little is stored, there is (according to this view) no respite from the time-consuming struggle that starts anew each day to find wild foods and avoid starving. Our escape from this misery was launched only after the end of the last Ice Age, when people began independently in different parts of the world to domesticate plants and animals. The agricultural revolution gradually spread until today it is nearly universal and few tribes of hunter-gatherers survive.

From the progressivist perspective on which I was brought up, the question "Why did almost all our hunter-gatherer ancestors adopt agriculture?" is silly. Of course they adopted it because agriculture is an efficient way to get more food for less work. Our planted crops yield far more tons per acre than do wild roots and berries. Just imagine savage hunters, exhausted from searching for nuts and chasing wild animals, suddenly gazing for the first time at a fruit-laden orchard or a pasture full of sheep. How many milliseconds do you think it took those hunters to appreciate the advantages of agriculture?

The progressivist party line goes further and credits agriculture with giving rise to art, the noblest flowering of the human spirit. Since crops can be stored, and since it takes less time to grow food in gardens than to find it in the jungle, agriculture gave us free time that hunter-gatherers never had. But free time is essential for creating art and enjoying it. Thus ultimately it was agriculture that, as its greatest gift, enabled us to build the Parthenon and compose the *B Minor Mass*.

AMONG OUR MAJOR CULTURAL HALLMARKS, agriculture is especially recent, having begun to emerge only ten thousand years ago. None of our primate relatives practices anything remotely resembling agriculture. For the most similar animal precedents, we must turn to ants, which invented not only plant domestication but also animal domestication.

Plant domestication is practiced by a group of several dozen related species of New World ants. All those ants cultivate specialized species of yeasts or fungi in gardens within the ants' nest. Rather than relying on natural soil, each gardener-ant species gathers its own particular type of compost: some ants grow their crop on caterpillar feces, others on insect corpses or dead plant material, and still others (the so-called leaf-cutter ants) on fresh leaves, stems, and flowers. For example, leaf-cutter ants clip off leaves, slice them into pieces, scrape off foreign fungi and bacteria, and take the pieces into underground nests. There the leaf fragments are crushed into moist pellets of a pastelike consistency, manured with ant saliva and feces, and seeded with the ants' preferred species of fungus, which serves as the ants' main or sole food. In an operation the equivalent of weeding a garden, the ants continually remove any spores or threads of other fungus species that they may find growing on their leaf paste. When a queen ant goes off to found a new colony, she carries with her a starting culture of the precious fungus, just as human pioneers take along seeds to plant.

As for animal domestication, ants obtain a concentrated sugary secretion termed honeydew from diverse insects, ranging from aphids, mealybugs, and scale insects to caterpillars, treehoppers, and spittle insects. In return for the honeydew, the ants protect their "cows" from predators and parasites. Some aphids have evolved into virtually the insect equivalent of domestic cattle: they lack offensive structures of their own, excrete honeydew from their anuses, and have a specialized anal anatomy designed to hold the droplet in place while an ant drinks it. To milk their cows and stimulate honeydew flow, ants stroke the aphids with their antennae. Some ants care for their aphids in the ants' nest during the cold winter, then in the spring carry the aphids at the correct stage of development to the correct part of the correct food plant. When aphids eventually develop wings and disperse in search of new habitat, some lucky ones are discovered by ants and "adopted."

Obviously, we didn't inherit plant and animal domestication directly from ants but reinvented it. Actually, "reevolved" is a better term than "reinvented," since our early steps toward agriculture did not consist of conscious experimentation toward an articulated goal. Instead, agriculture grew from human behaviors, and from responses or changes in plants and animals, leading without plan toward domestication. For example, animal domestication arose partly from people's keeping captive wild animals as pets, partly from wild animals' learning to profit from the proximity of people (e.g., wolves following human hunters to catch crippled prey). Similarly, early stages of plant domestication included people's harvesting wild plants and discarding seeds, which were thereby accidentally "planted." The inevitable result was unconscious selection of those plant and animal species and individuals most useful to humans. Eventually, conscious selection and care followed.

Now LET'S RETURN to the progressivist view of this agricultural revolution of ours. As I explained at the outset of this chapter, we're accustomed to assuming that the transition from the hunter-gatherer life-style to agriculture brought us health, longevity, security, leisure, and great art. While the case for this view *seems* overwhelming, it is hard to prove. How do you actually show that lives of people ten thousand years ago got better when they abandoned hunting for farming? Until recently, archaeologists couldn't test this question directly. Instead, they had to resort to indirect tests, whose results (surprisingly) failed to support the view of agriculture as an unmixed blessing.

Here is one example of such an indirect test. If agriculture had been visibly such a great idea, you'd expect it to have spread quickly, once it arose in some source area. In fact, the archaeological record shows that agriculture advanced across Europe at a snail's pace: barely one thousand yards per year! From its origins in the Near East around 8000 B.C., agriculture crept northwestward to reach Greece around 6000 B.C., Britain and Scandinavia only 2,500 years later.

That's hardly what you can call a wave of enthusiasm. As recently as the nineteenth century, all the Indians of California, now the fruit basket of America, remained hunter-gatherers, even though they knew of agriculture through trade with farming Indians in Arizona. Were California Indians really blind to their self-interest? Or could it instead be that they were smart enough to see, hidden beyond agriculture's glittering façade, the drawbacks that ensnared the rest of us?

Another indirect test of the progressivist view is to study whether surviving twentieth-century hunter-gatherers really are worse off than farmers. Scattered throughout the world, mainly in areas unsuitable for agriculture, several dozen groups of so-called "primitive people," like the Kalahari Desert Bushmen, continued to live as hunter-gatherers in recent years. Astonishingly, it turns out that these hunters generally have leisure time, sleep a lot, and work no harder than their farming neighbors. For instance, the average time devoted each week to obtaining food has been reported to be only twelve to nineteen hours for Bushmen; how many readers of this book can boast of such a short work week? As one Bushman replied when asked why he had not emulated neighboring tribes by adopting agriculture, "Why should we plant, when there are so many mongongo nuts in the world?"

Of course, one's belly isn't filled just by finding food; the food also has to be processed for eating, and that can take a lot of time for things like mongongo nuts. Hence it would be a mistake to swing to the opposite extreme from the progressivist view and to regard hunter-gatherers as living the life of leisure, as some anthropologists have done. However, it would also be a mistake to view them as working much harder than farmers. Compared to my physician and lawyer friends today, and to my shopkeeper grandparents in the early twentieth century, hunter-gatherers really do have more free time.

While farmers concentrate on high-carbohydrate crops like rice and potatoes, the mixture of wild plants and animals in the diets of surviving hunter-gatherers provides more protein and a better balance of other nutrients. The Bushmen's average daily food intake is 2,140 calories and 93 grams of protein, considerably greater than the RDA (Recommended Daily Allowance) for people of their small size but vigorous activity. Hunter-gatherers are healthy, suffer from little disease, enjoy a very diverse diet, and do not experience the periodic famines that befall farmers, dependent on few crops. It is almost inconceivable for Bushmen, who utilize eighty-five edible wild plants, to die of starvation, as did about a million Irish farmers and their families during the 1840s when a blight attacked potatoes, their staple crop.

Thus, the lives of at least the surviving modern hunter-gatherers aren't "nasty, brutish, and short," even though farmers have pushed them into the world's worst real estate. Hunters of the past, who still occupied fertile lands, could hardly have been worse off than modern hunters. But all those modern hunter societies have been affected by farming societies for thousands of years and don't tell us about the condition of hunters before the agricultural revolution. The progressivist view is really making a claim about the distant past: that the lives of people in each part of the world got better when they switched from hunting to farming. Archaeologists can date that switch by distinguishing remains of wild plants and animals from remains of domestic ones in prehistoric garbage dumps. How can one deduce the health of the prehistoric garbage makers, and thereby test directly for agriculture's supposed blessings?

THAT QUESTION has become answerable only in recent years, through the newly emerging science of "paleopathology": looking for signs of disease (the science of pathology) in remains of ancient peoples (from the Greek root *paleo* = ancient, as in paleontology). In some lucky situations, the paleopathologist has almost as much material to study as does a pathologist. For example, archaeologists in the deserts of Chile found well-preserved mummies whose medical condition at time of death could be determined by an autopsy, just as one would do on a fresh corpse in a hospital today. Feces of long-dead Indians who lived in dry caves in Nevada remained sufficiently well preserved to be examined for hookworm and other parasites.

Usually, though, the only human remains available for paleopathologists to study are skeletons, but they still permit a surprising number of deductions about health. To begin with, a skeleton identifies its owner's sex, and his/her weight and approximate age at time of death. Thus, with enough skeletons, one can construct mortality tables like those used by life-insurance companies to calculate expected life span and risk of death at any given age. Paleopathologists can also calculate growth rates by measuring bones of people of different ages, can examine teeth for cavities (signs of a high-carbohydrate diet) or enamel defects (signs of a poor diet in childhood), and can recognize scars that many diseases such as anemia, tuberculosis, leprosy, and osteoarthritis leave on bones.

One straightforward example of what paleopathologists have learned from skeletons concerns historical changes in height. Many modern cases illustrate how improved childhood nutrition leads to taller adults: for instance, we stoop to pass through doorways of medieval castles built for a shorter, malnourished population. Paleopathologists studying ancient skeletons from Greece and Turkey found a striking parallel. The average height of hunter-gatherers in that region toward the end of the Ice Age was a generous five feet ten inches for men, five feet six inches for women. With the adoption of agriculture, height crashed, reaching by 4000 B.C. a low value of only five feet three for men, five feet one for women. By classical times, heights were very slowly on the rise again, but modern Greeks and Turks have still not regained the heights of their healthy hunter-gatherer ancestors.

Another example of paleopathologists at work is the study of thousands of American Indian skeletons excavated from burial mounds in the Illinois and Ohio river valleys. Corn, first domesticated in Central America thousands of years ago, became the basis of intensive farming in those valleys around A.D. 1000. Until then, Indian hunter-gatherers had skeletons "so healthy it is somewhat discouraging to work with them," as one paleopathologist complained. With the arrival of corn, Indian skeletons suddenly became interesting to study. The number of cavities in an average adult's mouth jumped from fewer than one to nearly seven, and tooth loss and abscesses became rampant. Enamel defects in children's milk teeth imply that pregnant and nursing mothers were severely undernourished. Anemia quadrupled in frequency; tuberculosis became established as an epidemic disease; half the population suffered from yaws or syphilis; and two-thirds suffered from osteoarthritis and other degenerative diseases. Mortality rates at every age increased, with the result that only 1 percent of the population survived past age fifty, as compared to 5 percent in the golden days before corn. Almost one-fifth of the whole population died between the ages of one and four, probably because weaned toddlers succumbed to malnutrition and infectious diseases. Thus corn, usually considered among the New World's blessings, actually proved to be a public-health disaster. Similar conclusions about the transition from hunting to farming emerge from studies of skeletons elsewhere in the world.

There are at least three sets of reasons to explain these findings that agriculture was bad for health. First, hunter-gatherers enjoyed a varied diet with adequate amounts of protein, vitamins, and minerals, while farmers obtained most of their food from starchy crops. In effect, the farmers gained cheap calories at the cost of poor nutrition. Today just three high-carbohydrate plants—wheat, rice, and corn—provide more than 50 percent of the calories consumed by the human species.

Second, because of that dependence on one or a few crops, farmers ran a greater risk of starvation if one food crop failed than did hunters. The Irish potato famine is merely one of many examples.

Finally, most of today's leading infectious diseases and parasites of mankind could not become established until after the transition to agriculture. These killers persist only in societies of crowded, malnourished, sedentary people constantly reinfected by each other and by their own sewage. The cholera bacterium, for example, does not survive for long outside the human body. It spreads from one victim to the next through contamination of drinking water with feces of cholera patients. Measles dies out in small populations once it has either killed or immunized most potential hosts; only in populations numbering at least a few hundred thousand people can it maintain itself indefinitely. Such crowd epidemics could not persist in small, scattered bands of hunters who often shifted camp. Tuberculosis, leprosy, and cholera had to await the rise of farming, while smallpox, bubonic plague, and measles appeared only in the past few thousand years with the rise of even denser populations in cities.

BESIDES MALNUTRITION, starvation, and epidemic diseases, farming brought another curse to humanity: class divisions. Hunter-gatherers have little or no stored food, and no concentrated food sources like orchards or herds of cows. Instead, they live off the wild plants and animals that they obtain each day. Everybody except for infants, the sick, and the old join in the search for food. Thus there can be no kings, no full-time professionals, no class of social parasites who grow fat on food seized from others.

Only in a farming population could contrasts between the disease-ridden masses and a healthy, nonproducing elite develop. Skeletons from Greek tombs at Mycenae around 1500 B.C. suggest that royals enjoyed a better diet than commoners, since the royal skeletons were two or three inches taller and had better teeth (on the average, 1 instead of 6 cavities or missing teeth). Among mummies from Chilean cemeteries around A.D. 1000, the elite were distinguished not only by ornaments and gold hair clips, but also by a fourfold lower rate of bone lesions stemming from infectious diseases.

These signs of health differentials within local communities of farmers in the past appear on a global scale in the modern world. To most American and European readers, the argument that humanity could on the average be better off as hunter-gatherers than we are today sounds ridiculous, because most people in industrial societies today enjoy better health than most hunter-gatherers. However, Americans and Europeans are an elite in today's world, dependent on oil and other materials imported from countries with large peasant populations and much lower health standards. If you could choose between being a middle-class American, a Bushman hunter, and a peasant farmer in Ethiopia, the first would undoubtedly be the healthiest choice, but the third might be the least healthy.

While giving rise to class divisions for the first time, farming may also have exacerbated sexual inequality already in existence. With the advent of agriculture, women often became beasts of burden, were drained by more frequent pregnancies (see below), and thus suffered poorer health. For example, among the Chilean mummies from A.D. 1000, women exceeded men in osteoarthritis and in bone lesions from infectious disease. In New Guinea farming communities today I often see women staggering under loads of vegetables and firewood while the men walk empty-handed. In one case I offered to pay some villagers to carry supplies from an airstrip to my mountain camp, and a group of men, women, and children volunteered. The heaviest item was a 110-pound bag of rice, which I lashed to a pole and assigned to a team of four men to shoulder the poles together. When I eventually caught up with the villagers, the men were carrying light loads, while one small woman weighing less than the bag of rice was bent under it, supporting its weight by a cord across her temples.

As for the claim that agriculture laid the foundations of art by providing us with leisure time, modern hunter-gatherers have on the average at least as much free time as do farmers. I grant that some people in industrial and farming societies enjoy more leisure than do hunter-gatherers, at the expense of many others who support them and have far less leisure. Farming undoubtedly made it possible to sustain full-time craftsmen and artists, without whom we would not have such large-scale art projects as the Sistine Chapel and Cologne Cathedral. However, the whole emphasis on leisure time as a critical factor in explaining artistic differences among human societies seems to me misguided. It's not lack of time that prevents us today from surpassing the beauty of the Parthenon. While postagricultural technological advances did make new art forms possible and art preservation easier, great paintings and sculptures on a smaller scale than that of Cologne Cathedral were already being produced by Cro-Magnon hunter-gatherers fifteen thousand years ago. Great art was still being produced in modern times by hunter-gatherers such as Eskimos and Pacific Northwest Indians. In addition, when we count up the specialists whom society became able to support after the advent of agriculture, we should recall not only Michelangelo and Shakespeare but also standing armies of professional killers.

WITH THE ADVENT OF AGRICULTURE an elite became healthier, but many people became worse off. Instead of the progressivist party line that we chose agriculture because it was good for us, a cynic might ask how we got trapped by agriculture despite its being such a mixed blessing.

The answer boils down to the adage "Might makes right." Farming could support far more people than hunting, whether or not it also brought on the average more food per mouth. (Population densities of hunter-gatherers are typically one person or less per square mile, while densities of farmers average at least ten times higher.) Partly, this is because an acre of field planted entirely in edible crops produces far more tons of food, hence lets one feed far more mouths, than an acre of forest with scattered edible wild plants. Partly, too, it's because nomadic hunter-gatherers have to keep their children spaced at four-year intervals by infanticide and other means, since a mother must carry her toddler until it's old enough to keep up with the adults. Because sedentary farmers don't have that problem, a woman can and does bear a child every two years. Perhaps the main reason we find it so hard to shake off the traditional view that farming was unequivocally good for us is that there's no doubt that it meant more tons of food per acre. We forget that it also resulted in more mouths to feed. and that health and quality of life depend on the amount of food per mouth.

As population densities of hunter-gatherers slowly rose at the end of the Ice Age, bands had to "choose," whether consciously or unconsciously, between feeding more mouths by taking the first steps toward agriculture, or else finding ways to limit growth. Some bands adopted the former solution, unable to anticipate the evils of farming, and seduced by the transient abundance they enjoyed until population growth caught up with increased food production. Such bands outbred and then drove off or killed the bands that chose to remain hunter-gatherers, because ten malnourished farmers can still outfight one healthy hunter. It's not that hunter-gatherers abandoned their lifestyle, but that those sensible enough not to abandon it were forced out of all areas except ones that farmers didn't want. Modern hunter-gatherers persist mainly in scattered areas useless for agriculture, such as the Arctic and deserts.

At this point it's ironic to recall the common complaint that archaeology is an expensive luxury, concerned with the remote past, and offering no lessons of present relevance. Archaeologists studying the rise of farming have reconstructed for us a stage at which we made one of the most crucial decisions in human history. Forced to choose between limiting population growth and trying to increase food production, we opted for the latter and ended up with starvation, warfare, and tyranny. The same choice faces us today, with the difference that we now can learn from the past.

HUNTER-GATHERERS PRACTICED the most successful and long-persistent life-style in the career of our species. In contrast, we are still struggling with the problems into which we descended with agriculture, and it is unclear whether we can solve them. Suppose that an archaeologist who had visited us from Outer Space were trying to explain human history to his fellow spacelings. The visitor might illustrate the results of his digs by a twenty-four-hour clock on which one hour of clock time represents a hundred thousand years of real past time. If the history of the human race began at midnight, then we would now be almost at the end of our first day. We lived as hunter-gatherers for nearly the whole of that day, from midnight through dawn, noon, and sunset. Finally, at 11:54 P.M. we adopted agriculture. In retrospect, the decision was inevitable, and there is now no question of turning back. But as our second midnight approaches, will the present plight of African peasants gradually spread to engulf all of us? Or will we somehow achieve those seductive blessings that we imagine behind agriculture's glittering façade, and that have so far eluded us except in mixed form?

CHAPTER 11

Why Do We Smoke, Drink, and Use Dangerous Drugs?

CHERNOBYL—FORMALDEHYDE IN DRYWALL—LEAD POISONING—smog—the *Valdez* oil spill—Love Canal—asbestos—Agent Orange ... Hardly a month goes by without our learning of yet another way in which we and our children have been exposed to toxic chemicals through the negligence of others. The public's outrage, sense of help-lessness, and demand for change are growing. Why, then, do we do to ourselves what we cannot stand for others to do to us? How do we explain the paradox that many people intentionally consume, inject, or inhale toxic chemicals, such as alcohol, cocaine, and the chemicals in tobacco smoke? Why are various forms of this willful self-damage native to many contemporary societies, from primitive tribes to high-tech urbanites, and extending back into the past as far as we have written records? How did drug abuse become a hallmark virtually unique to the human species?

The problem isn't so much in understanding why we continue to take toxic chemicals once we've started. In part, that's because our drugs of abuse are addictive. Instead, the greater mystery is what impels us to start at all. Evidence for the damaging or lethal effects of alcohol, cocaine, and tobacco is by now overwhelming and familiar. Only the existence of some strong countervailing motives could explain why people consume these poisons voluntarily, even eagerly. It's as if unconscious programs were driving us to do something we know to be dangerous. What could those programs be?

Naturally, there is no single explanation: different motives carry different weight with different people and in different societies. For instance, some people drink to overcome their inhibitions or to join friends, others to deaden their feelings or drown their sorrows, still others because they like the taste of alcoholic beverages. Naturally, too, differences among human populations and social classes in their options for achieving satisfying lives largely account for geographic and class differences in chemical abuse. It's not surprising that self-destructive alcoholism is a bigger problem in high-unemployment areas of Ireland than in southeast England, or that cocaine and heroin addiction is commoner in Harlem than in affluent suburbs. One could be tempted to dismiss drug abuse as a human hallmark with obvious social and cultural causes, and in no need of a search for animal precedents.

However, none of the motives that I've just mentioned goes to the heart of the paradox of our actively seeking what we know to be harmful. In this chapter I'll propose one other contributing motive, which does address that paradox. It relates our chemical self-assaults to a wide range of seemingly self-destructive traits in animals, and to a general theory of animal signaling. It unifies a wide range of phenomena in our culture, from smoking and alcoholism to drug abuse. It has potential cross-cultural validity, for it may explain not just phenomena of the western world but also some otherwise mystifying customs elsewhere, such as kerosene drinking by Indonesian kung fu experts. I'll also reach into the past and apply the theory to the seemingly bizarre practice of ceremonial enemas in ancient Mayan civilization.

LET ME BEGIN by relating how I arrived at this idea. One day, I was abruptly struck by the puzzle that companies manufacturing toxic chemicals for human use advertise their use explicitly. This business practice would seem a sure route to bankruptcy. Yet, while we don't tolerate ads for cocaine, ads for tobacco and alcohol are so widespread that we cease to regard their existence as puzzling. It hit me only after I had been living with New Guinea hunters in the jungle for many months, far from any advertising.

Day after day, my New Guinea friends had been asking me about western customs, and I had come to realize through their astonished responses how senseless many of our customs are. Then the months of fieldwork ended with one of those sudden transitions that modern transportation has made possible. On June 25 I was still in the jungle, watching a brilliant-colored male bird of paradise flap awkwardly across a clearing, dragging its three-foot-long tail behind it. On June 26 I was sitting in a Boeing 747 jet, reading the magazines and catching up on the wonders of western civilization.

I leafed through the first magazine. It fell open to a page with a photo of a tough-looking man on horseback chasing cows, and the name of a brand of cigarette in large letters below. The American in me knew what the photo was about. But part of me was still in the jungle, looking at that photo naïvely. Perhaps my reaction won't seem so strange to you if you try to imagine yourself completely unfamiliar with western society, seeing the ad for the first time, and trying to fathom the connection between chasing cows and smoking (or not smoking) cigarettes.

The naïve part of me, fresh out of the jungle, thought: such a brilliant antismoking ad! It's well known that smoking impairs athletic ability and causes cancer and early death. Cowboys are widely regarded as athletic and admirable. This ad must be a devastating new appeal by the antismoking forces, telling us that if we smoke that particular brand of cigarette, we won't be fit to be cowboys. What an effective message to our youth!

But then it became obvious that the ad had been put there by the cigarette company itself, which somehow hoped that readers would draw exactly the opposite message from the ad. How on earth did the company let its public-relations department talk it into such a disastrous miscalculation? Surely that ad would dissuade any person concerned about his/her strength and self-image from starting to smoke.

Still half immersed in the jungle, I turned to another page. There I saw a photo of a whiskey bottle on a table, a man sipping presumably the bottle's contents from a glass, and an obviously fertile young woman gazing at him admiringly as if she were on the verge of sexual surrender. How can that be? I asked myself. Everyone knows that alcohol interferes with sexual function, tends to make men impotent, makes one likely to stumble, impairs judgment, and predisposes to cirrhosis of the liver and other debilitating conditions. In the immortal words of the porter in Shakespeare's *Macbeth*, "It [alcohol] provokes the desire, but it takes away the performance." A man with such performance handicaps should conceal them at all costs from a woman he aims to seduce. Why is the man in the photo intentionally displaying those handicaps? Do whiskey manufacturers think that pictures of this impaired individual will help sell their product? One could expect that Mothers Against Drunk Driving would be the ones producing such ads, and that the whiskey companies would be suing to prevent publication.

Page after page of ads flaunted the use of cigarettes or strong alcohol, and hinted at their benefits. There were even pictures of young people smoking in the presence of attractive members of the opposite sex, as if to imply that smoking too brought sexual opportunities. Yet any nonsmoker who has ever been kissed by (or tried to kiss) a smoker knows how severely the smoker's bad breath compromises his or her sex appeal. The ads paradoxically implied not just sexual benefits but also platonic friendships, business opportunities, vigor, health, and happiness, when the direct conclusion to be drawn from the ads was actually the reverse.

As the days passed and I reimmersed myself in western civilization, I gradually stopped noticing its apparently self-defeating ads. I retreated into analyzing my field data and wondering instead about an entirely different paradox, involving bird evolution. But that paradox was what led me finally to understand one rationale behind cigarette and whiskey ads.

THE NEW PARADOX concerned why that male bird of paradise I had been watching on June 25 had evolved the impediment of a tail three feet long. Males of other bird of paradise species evolved other bizarre impediments, such as long plumes growing out of their eyebrows, the habit of hanging upside-down, and brilliant colors and loud calls likely to attract hawks. All those features must impair male survival, yet they also serve as the advertisements by which male birds of paradise woo female birds of paradise. Like many other biologists, I found myself wondering why male birds of paradise use such handicaps as advertisements, and why females find the handicaps attractive.

At that point I recalled a remarkable paper published in 1975 by an Israeli biologist, Amotz Zahavi. In that paper Zahavi proposed a novel general theory, still hotly debated by biologists, about the role of costly or self-destructive signals in animal behavior. For example, he attempted to explain how deleterious male traits might attract a female precisely because they constitute handicaps. On reflection, I decided that Zahavi's hypothesis might apply to the birds of paradise I studied. Suddenly I realized, with growing excitement, that his theory perhaps could also be extended to explain the paradox of our use of toxic chemicals, and our touting it in ads.

Zahavi's theory as he proposed it concerned the broad problem of animal communication. All animals need to devise quick, easily understood signals for conveying messages to their mates, potential mates, offspring, parents, rivals, and would-be predators. For example, consider a gazelle that notices a lion stalking it. It would be in the gazelle's interests to give a signal that the lion would interpret to mean, "I am a superior fast gazelle! You'll never succeed in catching me, so don't waste your time and energy on trying." Even if that gazelle really is able to outrun a lion, giving a signal that dissuades the lion from trying would save time and energy for the gazelle too.

But what signal will unequivocally tell the lion that it's hopeless? The gazelle can't take the time to run a demonstration hundred-yard dash in front of every lion that shows up. Perhaps gazelles could agree on some quick arbitrary signal that lions learn to understand: e.g., pawing the ground with the left hind foot means, "I claim that I'm fast!" However, such a purely arbitrary signal opens the door to cheating: any gazelle can easily give the signal regardless of its speed. Lions will then catch on that many slow gazelles giving the signal are lying, and lions will learn to ignore the signal. It's in the interests both of lions and of fast gazelles that the signal be believable. What type of signal could convince a lion of the gazelle's honesty?

The same dilemma arises in the problem of sexual selection and mate choice that I discussed in earlier chapters. This is especially a problem of how females pick males, since females invest more in reproduction, have more to lose, and have to be choosier. Ideally, a female should pick a male for his good genes to pass on to her offspring. Since genes themselves are hard to assess, a female should look for quick indicators of good genes in a male, and a superior male should provide such indicators. In practice, male traits such as plumage, songs, and displays usually serve as indicators. Why do males "choose" to advertise with those particular indicators, why should females trust a male's honesty and find those indicators attractive, and why do they imply good genes?

I've described the problem as if a gazelle or courting male voluntarily picks out some indicator from among many possible ones, and as if a lion or a female decides on reflection whether it's really a valid indicator of speed or good genes. In practice, of course, those "choices" are the result of evolution and become specified by genes. Those females who select males on the basis of indicators that really denote good male genes, and those males that use unambiguous indicators of good genes for self-advertisement, tend to leave the most offspring, as do those gazelles and lions that spare themselves unnecessary chases.

As it turns out, many of the advertising signals evolved by animals pose a paradox similar to that posed by cigarette ads. The indicators often seem not to suggest speed or good genes but instead to constitute handicaps, expenses, or sources of risk. For example, a gazelle's signal to a lion that it sees approaching consists of a peculiar behavior termed "stotting." Instead of running away as fast as possible, the gazelle runs slowly while repeatedly jumping high into the air with stiff-legged leaps. Why on earth should the gazelle indulge in this seemingly self-destructive display, which wastes time and energy and gives the lion a chance to catch up? Or think of the males of many animal species that sport large structures, such as a peacock's tail or a bird of paradise's plumes, that make movement difficult. Males of many more species have bright colors, loud songs, or conspicuous displays that attract predators. Why should a male advertise such an impediment, and why should a female like it? These paradoxes remain an important unsolved problem in animal behavior today.

Zahavi's theory goes to the heart of this paradox. According to his theory, those deleterious structures and behaviors constitute valid indicators that the signaling animal is being honest in its claim of superiority, precisely *because* those traits themselves impose handicaps. A signal that entails no cost lends itself to cheating, since even a slow or inferior animal can afford to give the signal. Only costly or deleterious signals are guarantees of honesty. For example, a slow gazelle that stotted at an approaching lion would seal its fate, whereas a fast gazelle could still outrun the lion after stotting. By stotting, the gazelle boasts to the lion, "I'm so fast that I can escape you even after giving you this head start." The lion thereby has grounds for believing in the gazelle's honesty, and both the lion and the gazelle profit by not wasting time and energy on a chase whose outcome is certain.

Similarly, as applied to males displaying toward females, Zahavi's theory reasons that any male that has managed to survive despite the handicap of a big tail or conspicuous song must have terrific genes in other respects. He has proved that he must be *especially* good at escaping predators, finding food, and resisting disease. The bigger the handicap, the more rigorous the test that he has passed. The female who selects such a male is like the medieval damsel testing her knight suitors by watching them slay dragons. When she sees a one-armed knight who can still slay a dragon, she knows that she has finally found a knight with great genes. And that knight, by flaunting his handicap, is actually flaunting his superiority.

It seems to me that Zahavi's theory applies to many costly or dangerous human behaviors aimed at achieving status in general or at sexual benefits in particular. For instance, men who woo women with costly gifts and other displays of wealth are in effect saying, "I have plenty of money to support you and children, and you can believe my boast because you see how much money I'm spending now without blanching." People who show off expensive jewels, sports cars, or works of art gain status because the signal can't be faked; everyone else knows what those ostentatious objects cost. American Indians of the Pacific Northwest used to seek status by competing to give away as much wealth as possible in ceremonies known as potlatch rituals. In the days before modern medicine, tattooing was not only painful but dangerous because of the risk of infection; hence tattooed people in effect were advertising two facets of their strength, resistance to disease plus tolerance of pain. Men on the Pacific island of Malekula have traditionally showed off by the insanely dangerous practice, now imitated elsewhere by bungee jumpers, of building a high tower and jumping off it headfirst, after tying one end of some stout vines to each ankle and the other end to the top of the tower. The length of the vines is calculated to stop the braggart's plunge while his head is still a few feet above the ground.

Survival guarantees that the jumper is courageous, carefully calculating, and a good builder.

Zahavi's theory can also be extended to human abuse of chemicals.
Especially in adolescence and early adulthood, the age when drug
abuse is most likely to begin, we are devoting much energy to as-
serting our status. I suggest that we share the same unconscious
instinct that leads birds to indulge in dangerous displays. Ten thousand years ago, we "displayed" by challenging a lion or a tribal
enemy. Today, we do it in other ways, such as by fast driving or by
consuming dangerous drugs.

The messages of our old and new displays nevertheless remain the
same: I'm strong and superior. Even to take drugs only once or twice,
I must be strong enough to get past the burning, choking sensation of
my first puff on a cigarette, or to get past the misery of my first hang-
over. To do it chronically and remain alive and healthy, I must be
superior (so I imagine). It's a message to our rivals, our peers, our prospective mates—and to ourselves. The smoker's kiss may taste awful,
and the drinker may be impotent in bed, but he or she still hopes to
impress peers or attract mates by the implicit message of superiority.

Alas, the message may be valid for birds, but for us it's a false one.
Like so many animal instincts in us, this one has become maladaptive
in modern human society. If you can still walk after drinking a bottle
of whiskey, it may prove that you have high levels of liver alcohol
dehydrogenase, but it implies no superiority in other respects. If you
haven't developed lung cancer after chronically smoking several
packs of cigarettes daily, you may have a gene for resistance to lung
cancer, but that gene doesn't convey intelligence, business acumen, or
the ability to create happiness for your spouse and children.

It's true that animals with only brief lives and courtships have no
alternative except to develop quick indicators, since prospective mates
don't have enough time to measure each other's real quality. But we,
with our long lives and courtships and business associations, have
ample time to scrutinize each other's worth. We needn't rely on
superficial, misleading indicators. Drug abuse is a classic instance of
a once-useful instinct—the reliance on handicap signals—that has
turned foul in us. It's that old instinct to which the tobacco and
whiskey companies are directing their clever, obscene ads. If we
legalized cocaine, the drug lords too would soon have ads appealing
to the same instinct. You can easily picture it: the photo of the cowboy

on his horse, or the suave man and the attractive maiden, above the tastefully displayed packet of white powder.

Now, let's test my theory by jumping from western industrialized society to the other side of the world. Drug abuse did not begin with the industrial revolution. Tobacco was a native American Indian crop, native alcoholic beverages are widespread in the world, and cocaine and opium came to us from other societies. The oldest preserved code of laws, that of the Babylonian king Hammurabi (1792–50 B.C.), already contained a section regulating drinking houses. Hence my theory, if it's valid, should apply to other societies as well. As an instance of its cross-cultural explanatory power, I'll cite a practice you may not have heard of: kung fu kerosene drinking.

I learned of this practice when I was working in Indonesia with a wonderful young biologist named Ardy Irwanto. Ardy and I had come to like and admire each other, and to look out for each other's well-being. At one point, when we reached a troubled area and I expressed concern about dangerous people we might encounter, Ardy assured me, "No problem, Jared. I have kung fu grade eight." He explained that he practiced the Oriental martial art of kung fu and had reached a high level of proficiency, such that he could single-handedly fight off a group of eight attackers. To illustrate, Ardy showed me a scar in his back stemming from an attack by eight ruffians. One had knifed him, whereupon Ardy broke the arms of two and the skull of a third and the remainder fled. I had nothing to fear in Ardy's company, he told me.

One evening at our campsite, Ardy walked with his drinking cup up to our jerrycans. As usual, we had two cans: a blue one for water, and a red one for kerosene for our pressure lamp. To my horror, I watched Ardy pour from the red jerrycan and raise the cup to his lips. Remembering an awful moment during a mountaineering expedition when I had taken a sip of kerosene by mistake and spent all the next day coughing it back up, I screamed to Ardy to stop. But he raised his hand and said calmly, "No problem, Jared. I have kung fu grade eight."

Ardy explained that kung fu gave him strength, which he and his fellow kung fu masters tested each month by drinking a cup of kerosene. Without kung fu, of course, kerosene would make a weaker person sick: heaven forbid that I, Jared, for instance, should try it. But it did him, Ardy, no harm, because he had kung fu. He calmly retired to his tent to sip his kerosene and emerged the next morning, happy and healthy as usual.

I can't believe that kerosene did Ardy no harm. I wish that he could have found a less damaging way to make periodic tests of his preparedness. But for him and his kung fu associates, it served as an indicator of their strength and their advanced level of kung fu. Only a really robust person could get through that test. Kerosene drinking illustrates the handicap theory of toxic chemical use, in a form as startlingly repellent to us as our cigarettes and alcohol were to a horrified Ardy.

As MY LAST EXAMPLE, I'll generalize my theory further by extending its applications to the past—in this case, to the civilization of Mayan Indians that flourished in Central America one or two thousand years ago. Archaeologists have been fascinated by Mayan success at creating an advanced society in the middle of tropical rain forest. Many Mayan achievements, such as their calendar, writing, astronomical knowledge, and agricultural practices, are now understood to varying degrees. But archaeologists were long puzzled by slender tubes of unknown purpose that they kept finding in Mayan excavations.

The tubes' function finally became clear with the discovery of painted vases showing scenes of the tubes' use: to administer intoxicating enemas. The vases depict a high-status figure, evidently a priest or a prince, receiving a ceremonial enema in the presence of other people. The enema tube is shown as connected to a bag of a frothy, beerlike beverage—probably containing either alcohol or hallucinogens or both, as suggested by practices of other Indian groups. Many Central and South American Indian tribes formerly practiced similar ritual enemas when first encountered by European explorers, and some still do so today. The substances known to be administered range from alcohol (made by fermenting agave sap or a tree bark) to tobacco, peyote, LSD derivatives, and mushroom-derived hallucinogens. Thus, the ritual enema is similar to our consumption of intoxicants by mouth, but there are four reasons why an enema constitutes a more effective and valid indicator of strength than does drinking.

First, it is possible to relapse into solitary drinking and thus to lose all opportunity for signaling one's high status to others. However, it is more difficult for a solitary person to administer the same beverage to himself or herself unassisted as an enema. An enema encourages one to enlist associates, and thereby automatically creates an occasion for self-advertisement. Second, more strength is required to handle alcohol as an enema than as a drink, since the alcohol goes directly into the intestine and thence to the bloodstream, and it isn't first diluted with food in the stomach. Third, drugs absorbed from the small intestine after ingestion by mouth pass first to the liver, where many drugs are detoxified before they can reach the brain and other sensitive organs. But drugs absorbed from the rectum after an enema bypass the liver. Finally, nausea may limit one's intake of drinks but not of enemas. Hence an enema seems to me a more convincing advertisement of superiority than are our whiskey ads. I recommend this concept to an ambitious public-relations firm competing for the account of one of the large distilleries.

LET'S NOW STEP BACK and summarize the perspective on chemical abuse that I've suggested. Although frequent self-destruction by chemicals may be unique to humans, I see it as fitting into a broad pattern of animal behavior and thus as having innumerable animal precedents. All animals have had to evolve signals for quickly communicating messages to other animals. If the signals were ones that any individual animal could master or acquire, they would lend themselves to rampant cheating and hence to disbelief. To be valid and believable, a signal must be one that guarantees the honesty of the signaler, by entailing a cost, risk, or burden that only superior individuals can afford. Many animal signals that would otherwise strike us as counterproductive—such as stotting by gazelles, or costly structures and risky displays with which males court females—can be understood in this light.

It seems to me that this perspective has contributed to the evolution not only of human art but also of human chemical abuse. Both art and chemical abuse are widespread human hallmarks characteristic of most known human societies. Both beg explanation, since it's not immediately obvious why they promote our survival through natural selection, or why they help us acquire mates through sexual selection. I argued earlier that art often serves as a valid indicator of an individual's superiority or status, since art requires skill to create and requires status or wealth to acquire. But those individuals perceived by their fellows as enjoying status thereby acquire enhanced access to resources and mates. I'm arguing now that humans seek status through many other costly displays besides art, and that some of those displays (like jumping from towers, fast driving, and chemical abuse) are surprisingly dangerous. The former costly displays advertise status or wealth; the latter dangerous ones advertise that the displaying individual can master even such risks and hence must be superior.

I don't claim, though, that this perspective affords a total understanding of art or chemical abuse. As I mentioned in connection with art, complex behaviors acquire a life of their own, go far beyond their original purpose (if there ever was just a single purpose), and may even originally have served multiple functions. Just as art is now motivated far more by pleasure than by need for advertisement, chemical abuse too is now clearly much more than an advertisement. It's also a way to get past inhibitions, drown sorrows, or just enjoy a good-tasting drink.

I also don't deny that, even from an evolutionary perspective, there remains a basic difference between human abuse of chemicals and its animal precedents. Stotting, long tails, and all the animal precedents that I described involve costs, but those behaviors persist because the costs are outweighed by the benefits. A stotting gazelle loses a possible head start in a chase, but gains by decreasing the likelihood that a lion will embark on a serious chase at all. A long-tailed male bird is encumbered in finding food or escaping predators, but those survival disadvantages imposed by natural selection are more than compensated by mating advantages gained through sexual selection. The net balance is more rather than fewer offspring to pass on the male's genes. Hence these animal traits only appear to be self-destructive; they are actually self-promoting.

In the case of our chemical abuse, though, the costs outweigh the benefits. Drug addicts and drunkards not only lead shorter lives, but they lose rather than gain attractiveness in the eyes of potential mates and lose the ability to care for children. These traits don't persist because of hidden advantages outweighing costs; they persist mainly because they are chemically addicting. Overall, they are self-destructive, not self-promoting, behaviors. While gazelles may occasionally miscalculate in stotting, they don't commit suicide through addiction to the excitement of stotting. In that respect, our self-destructive abuse of chemicals diverged from its animal precursors to become truly a human hallmark.

CHAPTER 12

Alone
in a Crowded
Universe

THE NEXT TIME YOU'RE OUTDOORS ON A CLEAR NIGHT AWAY FROM CITY lights, look up at the sky and get a sense of the myriads of stars. Next, find a pair of binoculars, train them on the Milky Way, and appreciate how many more stars escaped your naked eye. Then look at a photo of the Andromeda nebula as seen through a powerful telescope to realize how enormous is the number of stars that escaped your binoculars as well.

Once all those numbers have sunk in, you'll finally be ready to ask: How could humans possibly be unique in the universe? How many civilizations of intelligent beings like ourselves must be out there, looking back at us? How long before we are in communication with them, before we visit them, or before we are visited?

On Earth, we certainly are unique. No other species possesses language, art, or agriculture of a complexity remotely approaching ours. No other species abuses drugs. But we've seen in the last four chapters that, for each of those human hallmarks, there are many animal precedents or even precursors. Similarly, human intelligence arose directly from chimpanzee intelligence, which is impressive by the standards of other animals though still far below ours. Isn't it likely that some other species on some other planets have also developed such widespread animal precursors to the level of our own art, language, and intelligence?

Alas, most human hallmarks lack effects detectable at a distance of many light-years. If there were creatures enjoying art or addicted to drugs on planets orbiting even the nearest stars, we'd never know it. But fortunately there are two signs of intelligent beings elsewhere that might be detectable on Earth: space probes and radio signals. We ourselves are already becoming effective at sending out both, so surely other intelligent creatures have mastered the necessary skills. Where, then, are the expected flying saucers?

This seems to me one of the greatest puzzles in all of science. Given the billions of stars, and given the abilities that we know did develop in our own species, we ought to be detecting flying saucers or at least radio signals. There's no question about there being billions of stars. What is there about the human species, then, that could explain the missing saucers? Could we really be unique not only on Earth, but also in the accessible universe? In this chapter I'll argue that we can obtain a fresh perspective on our uniqueness by looking carefully at some other unique creatures here on Earth.

FOR A LONG TIME, people have asked themselves such questions. Already around 400 B.C. the philosopher Metrodorus wrote, "To consider the Earth the only populated world in an infinite space is as absurd as to assert that in an entire field sown with millet only one grain will grow." Not until 1960, however, did scientists make a serious first attempt to find the answer, by listening (unsuccessfully) for radio transmissions from two nearby stars. In 1974 astronomers at the giant Arecibo radio telescope tried to establish an interstellar dialogue, by beaming a powerful radio signal to the star cluster M13 in the constellation Hercules. The signal described to Hercules denizens what we earthlings look like, how many of us there are, and where the Earth is located in our solar system. Two years later the search for extraterrestrial life provided the main motivation behind the Viking missions to Mars, whose cost of about a billion dollars dwarfed all the U.S. National Science Foundation's expenditures (since its inception) for classifying the life known to exist on Earth.

More recently the U.S. government has decided to spend a further hundred million dollars to detect radio signals from any intelligent beings who might exist outside our solar system. Several spacecraft that we launched are now heading out of our solar system, carrying sound tapes and photographic records of our civilization to inform spacelings who might be encountered.

It's easy to understand why lay people as well as biologists would consider the detection of extraterrestrial life as possibly the most exciting scientific discovery ever made. Just imagine how it would affect our self-image to find that the universe holds other intelligent creatures, with complex societies, languages, and learned cultural traditions, and capable of communicating with us. Among those of us who believe in an afterlife and an ethically concerned deity, most would agree that an afterlife awaits humans but not beetles (or even chimpanzees). Creationists believe that our species had a separate origin through divine creation. Suppose, though, that we should detect on another planet a society of seven-legged creatures more intelligent and ethical than we are, and able to converse with us, but having radio receivers and transmitters in place of eyes and mouths. Shall we believe that those creatures (but still not chimps) share the afterlife with us, and that they too were divinely created?

Many scientists have tried to calculate the odds of there being intelligent creatures out there somewhere. Those calculations have spawned a whole new field of science termed exobiology—the sole scientific field whose subject matter has not yet been shown to exist. Let's now consider the numbers that encourage exobiologists to believe in their subject matter.

Exobiologists calculate the number of advanced technical civilizations in the universe by an equation known as the Green Bank formula, which multiplies a string of estimated numbers. Some of those numbers can be estimated with considerable confidence. There are billions of galaxies, each with billions of stars. Astronomers conclude that many stars probably have one or more planets each, and that many of those planets probably have an environment suitable for life. Biologists conclude that, where conditions suitable for life exist, life will probably evolve eventually. Multiplying all of those probabilities or numbers together, we conclude that there are likely to be billions of billions of planets supporting living creatures.

Now let's estimate what fraction of those planetary biotas include intelligent beings with advanced technical civilizations, which we'll define operationally as civilizations capable of interstellar radio communication. (This is a less demanding definition than flying-saucer capability, since our own development suggests that interstellar radio communication will precede interstellar probes.) Two arguments suggest that that fraction may be considerable. First, the sole planet where we are certain that life evolved—our own—did evolve an advanced technical civilization. We have already launched interplanetary probes. We have made progress with techniques for freezing and thawing life and for making life from DNA—techniques relevant to preserving life as we know it for the long duration of an interstellar trip. Technical progress in recent decades has been so rapid that manned interstellar probes surely will be feasible within a few centuries at the very most, since some of our unmanned interplanetary probes are already on their way out of our solar system.

However, this first argument suggesting that many other planetary biotas have evolved advanced technical civilizations isn't a compelling one. To use the language of statisticians, it suffers from the obvious flaw of very small sample size (how can you generalize from one case?) and very high ascertainment bias (we picked out that one case precisely because it evolved our own advanced technical civilization).

A second, stronger argument is that life on Earth is characterized by what biologists term convergent evolution. That is, seemingly whatever ecological niche or physiological adaptation you consider, many groups of creatures have converged by evolving independently to exploit that niche, or to acquire that adaptation. An obvious example is the independent evolution of flight by birds, bats, pterodactyls, and insects. Other spectacular cases are the independent evolution of eyes, and even of devices for electrocuting prey, by many animals. Within the past two decades, biochemists have recognized convergent evolution at the molecular level, such as the repeated evolution of similar protein-splitting enzymes. So common is convergent evolution of anatomy, physiology, biochemistry, and behavior that whenever biologists observe two species to be similar in some respect, one of the first questions they now ask is: did that similarity result from common ancestry or from convergence?

There's nothing surprising about the seeming ubiquity of convergent evolution. If you expose millions of species for millions of years to similar selective forces, of course you can expect similar solutions to emerge time and time again. We know that there has been much convergence among species on Earth, but by the same reasoning there should also be much convergence between Earth's species and those elsewhere. Hence, although radio communication is one of those things that happens to have evolved here only once so far, considerations of convergent evolution lead us to expect its evolution on some other planets as well. As the *Encyclopedia Britannica* puts it, "It is difficult to imagine life evolving on another planet without progressing towards intelligence."

But that conclusion brings us back to the puzzle I mentioned earlier. If many or most stars have planetary systems, and if many of those systems include at least one planet with conditions suitable for life, and if life is likely eventually to evolve where suitable conditions exist, and if about one percent of planets with life include an advanced technical civilization—then one estimates that our own galaxy alone contains about a million planets supporting advanced civilizations. But within only a few dozen light-years of us are several hundred stars, some (most?) of which surely have planets like ours, supporting life. Then where are all the flying saucers that we would expect? Where are the intelligent beings that should be visiting us, or at least directing radio signals at us? The silence is deafening.

Something must be wrong with the astronomers' calculations. They know what they're talking about when they estimate the number of planetary systems, and the fraction of those likely to be supporting life. I find these estimates plausible. Instead, the problem is likely to lie in the argument, based on convergent evolution, that a significant fraction of biotas will evolve advanced technical civilizations. Hence let's scrutinize more closely the inevitability of convergent evolution.

WOODPECKERS PROVIDE a good test case, because woodpecking is a terrific life-style that offers much more food than do flying saucers or radios. The "woodpecker niche" is based on digging holes in live wood and on prying off pieces of bark. That means dependable food sources all year round in the form of sap, insects living under bark, and insects burrowing into wood. It also means an excellent place for a nest, since a hole in a tree affords protection from wind, rain, predators, and temperature fluctuations. Other bird species besides woodpeckers can pull off the easier feat of digging nest holes in dead wood, but there are many fewer dead trees than live trees available.

These considerations mean that if we're counting on convergent evolution of radio communication, we can surely count on convergent evolution among many species to exploit the woodpecker niche. Not surprisingly, woodpeckers are very successful birds. There are nearly two hundred species, many of them common. They come in all sizes, from tiny birds the size of kinglets up to crow-sized. They are widespread over most of the world, except on oceanic islands too remote for them to reach by flying.

How hard is it to evolve to become a woodpecker? Two considerations might seem to suggest "Not very hard." Woodpeckers are not an extremely distinctive old group without close relatives, like egg-laying mammals. Instead, ornithologists have agreed for a long time that their closest relatives are honey guides, toucans, and barbets, to which woodpeckers are fairly similar except in their special adaptations for woodpecking. Woodpeckers have numerous such adaptations, but none is remotely as extraordinary as building radios, and all are readily seen as extensions of adaptations possessed by other birds. The adaptations fall into four groups.

First and most obvious are the adaptations for drilling in live wood. These include a chisel-like bill, nostrils protected with feathers to keep out sawdust, a thick skull, strong head and neck muscles, and a hinge between the base of the bill and the front of the skull to help spread the shock of pounding. These features for drilling in live wood can be traced to features of other birds much more easily than our radios can be traced to any primitive radios of chimpanzees. Many other birds, such as parrots, peck or bite holes in dead wood. Within the woodpecker family there is a gradation of drilling ability—from wrynecks, which can't excavate at all, to the many woodpeckers that drill in softer wood, to hardwood specialists like sapsuckers.

Another set of adaptations are those for perching vertically on bark, such as a stiff tail to press against bark as a brace, strong muscles for manipulating the tail, short legs, and long, curved toes. The evolution of these adaptations can be traced even more easily than can the adaptations for woodpecking. Even within the woodpecker family, wrynecks and piculets do not have stiff tails for use as braces. Many birds outside the woodpecker family, including creepers and pygmy parrots, do have stiff tails that they evolved to prop themselves on bark.

The third adaptation is an extremely long and extensible tongue, fully as long as our own tongues in some woodpeckers. Once a woodpecker has broken into the tunnel system of wood-dwelling insects at one point, the bird uses its tongue to lick out many branches of the system without having to drill a new hole for each branch. Woodpeckers' tongues have many animal precedents, including the similarly long insect-catching tongues of frogs, anteaters, and aardvarks.

Finally, woodpeckers have tough skins to withstand insect bites plus the stresses from pounding and from strong muscles. Anyone who has skinned and stuffed birds knows that some birds have much tougher skins than others. Taxidermists groan when given a pigeon, whose paper-thin skin tears almost as soon as you look at it, but smile when given a woodpecker, hawk, or parrot.

Thus, while woodpeckers have many adaptations for woodpecking, most of those adaptations have also evolved convergently in other birds or animals, and the unique skull adaptations can at least be traced to precursors. You might therefore expect the whole package of woodpecking to have evolved repeatedly, with the result that there would now be many groups of large animals capable of excavating into live wood for food or nest sites. But all modern woodpeckers are more closely related to each other than to any nonwoodpecker, proving that woodpecking evolved only once. Even on remote landmasses that woodpeckers never reached, like Australia, New Guinea, New Zealand, nothing else has evolved to exploit the splendid opportunities made available by the woodpecker life-style. Some birds and mammals on those landmasses do excavate dead wood or bark, but they are only feeble excuses for woodpeckers, and none can excavate in live wood. If woodpeckers hadn't evolved that one time in the Americas or Old World, a terrific niche would be flagrantly vacant over the whole Earth.

I HAVE DWELT on woodpeckers to illustrate that convergence is not universal, and that not all splendid opportunities are seized. I could have illustrated the same point with other, equally flagrant examples. The most ubiquitous opportunity available to animals is to consume

plants, much of whose mass consists of cellulose. Yet no higher
animal has managed to evolve a cellulose-digesting enzyme. Those
animal herbivores that digest cellulose (such as cows) instead have to
rely on microbes housed within their intestines. To take another
example that I discussed in an earlier chapter, growing your own
food would seem to offer obvious advantages for animals, but the
only animals to master the trick before the dawn of human agriculture ten thousand years ago were leaf-cutter ants and a few other
insects, which cultivate fungi or domesticate aphid "cows."

Thus, it has proved extraordinarily difficult to evolve even such
obviously valuable adaptations as woodpecking, digesting cellulose
efficiently, or growing one's own food. Radios do much less for one's
food needs and would seem far less likely to evolve. Are our radios
a fluke, unlikely to have been duplicated on any other planet?

Consider what biology might have taught us about the inevitability
of radio evolution on Earth. If radio building were like woodpecking,
some species might have evolved certain elements of the package or
evolved them in inefficient form, although only one species managed
to evolve the complete package. For instance, we might have found
today that turkeys build radio transmitters but no receivers, while
kangaroos build receivers but no transmitters. The fossil record might
have shown dozens of now-extinct animals experimenting over the
last half-billion years with metallurgy and increasingly complex electronic circuits, leading to electric toasters in the Triassic, battery-
operated rat traps in the Oligocene, and finally radios in the Holocene.
Fossils might have revealed five-watt transmitters built by trilobites,
two-hundred-watt transmitters amidst bones of the last dinosaurs,
and five-hundred-watt transmitters in use by sabertooths, until humans finally upped the power output enough to be the first to broadcast into space.

But none of that happened. Neither fossils nor living animals—not
even our closest living relatives, the common and pygmy
chimpanzees—had even the most remote precursors of radios. It's
instructive to consider the experience of the human line itself. Neither australopithecines nor early *Homo sapiens* developed radios. As
recently as 150 years ago, modern *Homo sapiens* didn't even have the
concepts that would lead to radios. The first practical experiments
didn't begin until around 1888; it's still less than one hundred years
since Marconi built the first transmitter capable of broadcasting one

mere mile; and we still aren't sending signals targeted at other stars, though the 1974 Arecibo experiment was our first attempt.

I mentioned early in this chapter that the existence of radios on the one planet known to us seemed at first to suggest a high probability of radios' evolving on other planets. In fact, closer scrutiny of Earth's history supports exactly the opposite conclusion: radios had a vanishingly low probability of evolving here. Only one of the billions of species that have existed on Earth showed any proclivities toward radios, and even it failed to do so for the first 69,999/70,000ths of its seven-million-year history. A visitor from Outer Space who had come to Earth as recently as the year 1800 would have written off any prospects of radios' being built here.

You might object that I'm being too stringent in looking for early precursors of radios themselves, when I should instead look just for the two qualities necessary to make radios: intelligence and mechanical dexterity. But the situation there is little more encouraging. Based on the very recent evolutionary experience of our own species, we arrogantly assume intelligence and dexterity to be the best way of taking over the world, and to have evolved inevitably. Think again about that quote I cited earlier from the *Encyclopedia Britannica*: "It is difficult to imagine life evolving on another planet without progressing towards intelligence." Earth history again supports exactly the opposite conclusion. In reality, vanishingly few animals on Earth have bothered with much of either intelligence or dexterity. No animal has acquired remotely as much of either as have we; those that have acquired a little of one (smart dolphins, dexterous spiders) have acquired none of the other; and the only other species to acquire a little of both (common and pygmy chimpanzees) have been rather unsuccessful. Earth's really successful species have instead been dumb and clumsy rats and beetles, which found better routes to their current dominance.

WE HAVE ONLY STILL to consider the last missing variable in the Green Bank formula for calculating the likely number of civilizations capable of interstellar radio communication. That variable is the lifetime of such a civilization. The intelligence and dexterity required to build radios are useful for other purposes that have been our species' hallmarks for much longer than have radios: purposes such as mass killing devices and means of environmental destruction. We have become so potent at both that we are gradually stewing in our civilization's juices. We may not enjoy the luxury of an end by slow stewing. Half-a-dozen countries now possess the means for bringing us all to a quick end, and still other countries are eagerly seeking to acquire those means. The wisdom of some past leaders of bomb-possessing nations, or of some present leaders of bomb-seeking nations, doesn't encourage us to believe that there will be radios on Earth for much longer.

It was an extremely unlikely fluke that we developed radios at all, and more of a fluke that we developed them before we developed the technology that could end us in a slow stew or fast bang. While Earth's history thus offers little hope that radio civilizations exist elsewhere, it also suggests that any that might exist are short-lived. Other intelligent civilizations that rose elsewhere probably reversed their own progress overnight, just as we now risk doing.

We're very lucky that that's so. I find it mind-boggling that the astronomers now eager to spend a hundred million dollars on the search for extraterrestrial life have never thought seriously about the most obvious question: what would happen if we found it, or if it found us. The astronomers tacitly assume that we and the little green monsters would welcome each other and settle down to fascinating conversations. Here again, our own experience on Earth offers useful guidance. We've already discovered two species that are very intelligent but technically less advanced than we are—the common chimpanzee and pygmy chimpanzee. Has our response been to sit down and try to communicate with them? Of course not. Instead we shoot them, dissect them, cut off their hands for trophies, put them on exhibit in cages, inject them with AIDS virus as a medical experiment, and destroy or take over their habitats. That response was predictable, because human explorers who discovered technically less advanced humans also regularly responded by shooting them, decimating their populations with new diseases, and destroying or taking over their habitats.

Any advanced extraterrestrials who discovered us would surely treat us in the same way. Think again of those astronomers who beamed radio signals into Space from Arecibo, describing Earth's location and its inhabitants. In its suicidal folly that act rivaled the folly of the last Inca emperor, Atahualpa, who described to his gold-crazy Spanish captors the wealth of his capital and provided them with guides for the journey. If there really are any radio civilizations within listening distance of us, then for heaven's sake let's turn off our own transmitters and try to escape detection, or we're doomed.

Fortunately for us, the silence from Outer Space is deafening. Yes, out there are billions of galaxies with billions of stars. Out there must be some transmitters as well, but not many, and they won't last long. Probably there are no others in our galaxy, and surely none within hundreds of light-years of us. What woodpeckers teach us about flying saucers is that we're unlikely ever to see one. For practical purposes, we're unique and alone in a crowded universe. Thank God!

PART FOUR

WORLD
CONQUERORS

Part Three discussed some of our cultural hallmarks and their animal precedents or precursors. Those cultural hallmarks—especially language, agriculture, and advanced technology—have been the causes of our rise. They are what permitted us to expand over the globe and become world conquerors.

That expansion, though, didn't just consist of our conquering areas previously unoccupied by the human species. It also involved the expansion of particular human populations that conquered, expelled, or killed other populations. We became conquerors of each other, as well as of the world. Thus, our expansion has been marked by yet another human hallmark that has animal precursors and that we've taken far beyond its animal limits—namely, our propensity to kill other members of our species en masse. Along with our environmental destructiveness, it now poses one of the two potential reasons that we may fall.

To appreciate our rise to the status of world conquerors, recall that most animal species are distributed over only a small part of the Earth's surface. For example, Hamilton's frog is confined to one forest patch of 37 acres plus one rock pile covering 720 square yards in New Zealand. The most widespread wild land mammal other than humans used to be the lion, which as of ten thousand years ago occupied most of Africa, much of Eurasia, and North plus northern South America. Even at that time of its greatest extent, though, the lion didn't reach Southeast Asia, Australia, southern South America, polar regions, or islands.

Humans used to have a typically mammalian restricted distribution, in warm, nonforested areas of Africa. As recently as fifty thousand years ago, we were still confined to tropical and mild-temperature areas of Africa and Eurasia. Then we expanded in turn to Australia and New Guinea (around fifty thousand years ago), cold parts of Europe (by thirty thousand years ago), Siberia (by twenty thousand years ago), North and South America (around eleven thousand years ago), and Polynesia (between thirty-six hundred and one thousand years ago). Today we occupy or at least visit not only all lands but also the surface of all the oceans, and we are starting to probe into Space and the oceans' abysses.

In the process of this world conquest, our species underwent a basic change in the relation among its populations. Most animal species with sufficiently wide geographic ranges fall into populations in contact with neighboring populations but not with distant ones. In this respect, too, humans used to be just another species of big mammal. Until relatively recently, most people spent their entire lives within a few dozen miles of their birthplaces, and had no way of learning even of the existence of people living at much greater distances. Relations between neighboring tribes were marked by an uneasy shifting balance between trade and xenophobic hostility.

This fragmentation promoted, and was reinforced by, the tendency for each human population to develop its own language and culture. Initially, the massive expansion of our species' geographic range involved a massive increase in our linguistic and cultural diversity. Among the "new" parts of our range occupied only within the last fifty thousand years, New Guinea and North and South America alone came to account for about half of the modern world's languages. But much of that long heritage of cultural diversity has been erased in the last five thousand years by the expansion of centralized political states. Freedom of travel—a modern invention—is now accelerating that homogenization of our language and culture. However, in a few areas of the world, notably New Guinea, Stone Age technology and our traditional xenophobic outlook persisted into the twentieth century, giving us a last glimpse of what the rest of the world used to be like.

The outcome of conflicts between expanding human groups has been heavily influenced by group differences in our cultural hallmarks. Especially decisive have been differences in military and maritime technology, in political organization, and in agriculture. Groups with more advanced agriculture thereby acquired the military advantage of larger population numbers, ability to support a permanent military caste, and resistance to infectious diseases against which sparser populations had evolved no defense.

All those cultural differences used to be ascribed to genetic superiority of conquering "advanced" peoples over conquered "primitive" ones. However, no evidence of such genetic superiority has been forthcoming. The likelihood of genetics' playing such a role is refuted by the ease with which the most dissimilar human groups have mastered each other's cultural techniques, given adequate learning opportunities. New Guineans born of Stone Age parents now pilot airplanes, while Amundsen and his Norwegian crew mastered Eskimo dog-sledding methods to reach the South Pole.

Instead, one has to ask why some people acquired the cultural advantages that let them conquer other people, despite lack of any evident genetic advantages. For example, was it purely by chance that Bantu peoples originating from equatorial Africa displaced Khoisan people over most of southern Africa, rather than vice versa? While we can't expect to identify ultimate environmental factors behind small-scale conquests, chance is likely to play less of a role and ultimate factors to be more compelling if we focus on large-scale population shifts over long times. Hence two chapters will examine two of the largest-scale shifts in recent history: the modern expansion of Europeans over the New World and Australia; and the perennial puzzle of how Indo-European languages managed earlier to overrun so much of Eurasia from an initially restricted homeland. We'll see clearly in the former case, and more speculatively in the latter, how each human society's culture and competitive position have been shaped by its biological and geographical heritage, especially by the plant and animal species available to it for domestication.

Competition among members of the same species isn't unique to humans. Among all animal species as well, the closest competitors are inevitably members of the same species, because they share the closest ecological similarity. What varies greatly among species is the form that competitive strife takes. In the most inconspicuous form, rival animals compete merely by consuming food potentially available to each other and exhibit no overt aggression. Mild escalation involves ritualized displays, or chasing. As a last resort, now documented in many species, rival animals kill each other.

The competing units also vary greatly among animal species. In most songbirds, such as American or European robins, individual males or else male-female pairs face off. Among lions and common chimpanzees, small groups of males who may be brothers fight, sometimes unto death. Packs of wolves or hyenas do battle, while ant colonies engage in large-scale wars with other colonies. Although for some species these contests may end in deaths, there is no animal species whose survival as a species is even remotely threatened by such deaths.

Humans compete with each other for territory, as do members of most animal species. Because we live in groups, much of our competition has taken the form of wars between adjacent groups, on the model of the wars between ant colonies rather than the small-scale contests between robins. As with adjacent groups of wolves and common chimps, relations of adjacent human tribes were traditionally marked by xenophobic hostility, intermittently relaxed to permit exchanges of mates (and, in our species, of goods as well). Xenophobia comes especially naturally to our species, because so much of our behavior is culturally rather than genetically specified, and because cultural differences among human populations are so marked. Those features make it easy for us, unlike wolves and chimps, to recognize members of other groups at a glance by their clothes or hair style.

What makes human xenophobia much more lethal than chimp xenophobia is of course our recent development of weapons for mass killing at a distance. While Jane Goodall described males of one group of common chimps gradually killing off individuals of the neighboring group and usurping their territory, those chimps had no means to kill chimps of a remote group, or to exterminate all chimps (including themselves). Thus, xenophobic murder has innumerable animal precursors, but only we have developed it to the point of threatening to bring about our fall as a species. Threatening our own existence has now joined art and language as a human hallmark. This section of the book will conclude by surveying the history of human genocide, to make clear the ugly tradition from which Dachau's ovens and modern nuclear warfare spring.

CHAPTER 13

The Last
First Contacts

On August 4, 1938, a biological exploring expedition from the American Museum of Natural History made a discovery that hastened toward its end a long phase of human history. That was the date on which the advance patrol of the Third Archbold Expedition (named after its leader, Richard Archbold) became the first outsiders to enter the Grand Valley of the Balim River, in the supposedly uninhabited interior of west New Guinea. To everyone's astonishment, the Grand Valley proved to be densely populated—by fifty thousand Papuans, living in the Stone Age, previously unknown to the rest of humanity and themselves unaware of others' existence. In search of undiscovered birds and mammals, Archbold had found an undiscovered human society.

To appreciate the significance of Archbold's finding, we need to understand the phenomenon of "first contact." As I mentioned earlier, most animal species occupy a geographic range confined to a small fraction of the Earth's surface. Of those species (like lions and grizzly bears) occurring on several continents, it's not the case that individuals from one continent visit one another. Instead, each continent, and usually each small part of a continent, has its own distinctive population, in contact with close neighbors but not with distant members of the same species. (Migratory songbirds constitute only apparently a glaring exception. Yes, they do commute seasonally between continents, but along a traditional path, and both the summer breeding range and the winter nonbreeding range of a given population tend to be quite circumscribed).

This geographic fidelity of animals is reflected in their geographic variability: populations of the same species in different geographic areas tend to evolve into different-looking subspecies, because most breeding remains within the same population. For example, no gorilla of the East African lowland species has ever turned up in West Africa or vice versa, though the eastern and western subspecies look sufficiently different for biologists to recognize a wanderer if there were any.

In these respects, we humans have been typical animals throughout most of our evolutionary history. Like other animals, each human population is genetically molded to its area's climate and diseases. But human populations are also impeded from freely mixing by linguistic and cultural barriers far stronger than in other animals. An anthropologist can identify from a person's naked appearance roughly where that person originated, and a linguist or student of dress styles can pinpoint origins much more closely. That's testimony to how sedentary human populations have been.

While we think of ourselves as travelers, we were quite the opposite throughout several million years of human evolution. Every human group was ignorant of the world beyond its own lands and those of its immediate neighbors. Only in recent millennia did changes in political organization and technology permit some people routinely to travel afar, to encounter distant peoples, and to learn firsthand about places and peoples that they had not personally visited. This process accelerated with Columbus's voyage of 1492, until today there remain only a few tribes in New Guinea and South America still awaiting first contact with remote outsiders. The Archbold Expedition's entry into the Grand Valley will be remembered as one of the last first contacts of a large human population. It was thus a landmark in the process by which humanity became transformed from thousands of tiny societies, collectively occupying only a fraction of the globe, to world conquerors with world knowledge.

How could such a numerous people as the Grand Valley's fifty thousand Papuans remain completely unknown to outsiders until 1938? How could those Papuans in turn remain completely ignorant of the outside world? How did first contact change human societies? I'll argue that this world before first contact—a world that is finally ending within our own generation—holds a key to the origins of human cultural diversity. As world conquerors, our species now numbers over five billion, compared to the mere ten million people who existed before the advent of agriculture. Ironically, though, our cultural diversity has plunged even as our numbers have soared.

To ANYONE who has not been to New Guinea, the long concealment of fifty thousand people there seems incomprehensible. After all, the Grand Valley lies only 115 miles from either New Guinea's north coast or south coast. Europeans discovered New Guinea in 1526, Dutch missionaries took up residence in 1852, and European colonial governments were established in 1884. Why did it take another fifty-four years longer to find the Grand Valley?

The answers—terrain, food, and porters—become obvious as soon as one sets foot in New Guinea and tries to walk away from an established trail. Swamps in the lowlands, endless series of knife-edge ridges in the mountains, and jungle that covers everything reduce one's progress to a few miles per day under the best conditions. On my 1983 expedition into the Kumawa Mountains, it took me and a team of twelve New Guineans two weeks to penetrate seven miles inland. But we had it easy, compared to the British Ornithologists' Union Jubilee Expedition. On January 4, 1910, they landed on New Guinea's coast and set off for the snow-capped mountains that they could see only a hundred miles inland. On February 12, 1911, they finally gave up and turned back, having covered less than half the distance (forty-five miles) in those thirteen months.

Compounding those terrain problems is the impossibility of living off the land, because of New Guinea's lack of big game animals. In lowland jungle the staple food plant of New Guineans is a tree called the sago palm, whose pith yields a substance with the consistency of rubber and the flavor of vomit. However, not even New Guineans can find enough wild foods to survive in the mountains. This problem was illustrated by a horrible sight on which the British explorer

Alexander Wollaston stumbled while descending a New Guinea jungle trail: the bodies of thirty recently dead New Guineans and two dying children, who had starved while trying to return from the lowlands to their mountain gardens without carrying enough provisions.

The paucity of wild foods in the jungle compels explorers going through uninhabited areas, or unable to count on obtaining food from native gardens, to bring their own rations. A porter can carry forty pounds, the weight of the food necessary to feed himself for about fourteen days. Thus, until the advent of planes made airdrops possible, all New Guinea expeditions that penetrated more than seven days' walk from the coast (fourteen days' round trip) did so by having teams of porters going back and forth, building up food depots inland. Here's a typical plan: fifty porters start from the coast with seven hundred man-days of food, deposit two hundred man-days' food five days inland, and return in another five days to the coast, having consumed the remaining five hundred man-days' food (fifty men times ten days) in the process. Then fifteen porters march to that first depot, pick up the cached two hundred man-days' food, deposit fifty man-days' food a further five days' march inland, and return to the first depot (reprovisioned in the meantime), having consumed the remaining hundred fifty man-days' food in the process. Then...

The expedition that came closest to discovering the Grand Valley before Archbold, the 1921–22 Kremer Expedition, used eight hundred porters, two hundred tons of food, and ten months of relaying to get four explorers inland to just beyond the Grand Valley. Unfortunately for Kremer, his route happened to pass a few miles west of the valley, whose existence he did not suspect because of intervening ridges and jungle.

Apart from these physical difficulties, the interior of New Guinea seemed to hold no attractions for missionaries or colonial governments, because it was believed to be virtually uninhabited. European explorers landing on the coast or rivers discovered many tribes in the lowlands living off sago and fish, but few people eking out an existence in the steep foothills. From either the north or south coast, the snow-capped Central Cordillera that forms New Guinea's backbone presents steep faces. It was assumed that the northern and southern faces met in a ridge. What remained invisible from the coasts was the existence of broad intermontane valleys, hidden behind those faces and suitable for agriculture.

For east New Guinea, the myth of an empty interior was shattered on May 26, 1930, when two Australian miners, Michael Leahy and Michael Dwyer, crossed a ridge of the Bismarck Mountains in search of gold, looked down at night on the valley beyond, and were alarmed to see countless dots of light: the cooking fires of thousands of people. For west New Guinea, the myth ended with Archbold's second survey flight on June 23, 1938. After hours of flying over jungle with few signs of humans, Archbold was astonished to spot the Grand Valley, looking like Holland: a cleared landscape devoid of jungle, neatly divided into small fields outlined by irrigation ditches, and with scattered hamlets. It took six more weeks before Archbold could establish camps at the nearest lake and river where his seaplane could land, and before patrols from those camps could reach the Grand Valley to make first contact with its inhabitants.

THAT'S WHY THE OUTSIDE WORLD didn't know of the Grand Valley till 1938. Why didn't the valley's inhabitants, now referred to as the Dani people, know of the outside world?

Part of the reason, of course, is the same logistic problems that faced the Kremer Expedition on its march inland, but in reverse. Yet those problems would be minor in areas of the world with gentler terrain and more wild foods than New Guinea, and thus they don't explain why all other human societies in the world also used to live in relative isolation. Instead, at this point we have to remind ourselves of a modern perspective that we take for granted. Our perspective didn't apply to New Guinea until very recently, and it didn't apply anywhere in the world ten thousand years ago.

Recall that the whole globe is now divided into political states, whose citizens enjoy more or less freedom to travel within the boundaries of their state and to visit other states. Anyone with the time, money, and desire can visit almost any country except for a few xenophobic holdouts, like North Korea. As a result, people and goods have diffused around the globe, and many items such as Coca-Cola are now available on every continent. I recall with embarrassment my visit in 1976 to a Pacific island called Rennell, whose isolated location, vertical sea cliffs without beaches, and fissured coral landscape had preserved its Polynesian culture unchanged until recently. Setting out at dawn from the coast, I plodded through jungle with not a trace of humans. When in the late afternoon I finally heard a woman's voice ahead and glimpsed a small hut, my head whirled with fantasies of the beautiful, unspoilt, grass-skirted, bare-breasted Polynesian maiden who awaited me at this remote site on this remote island. It was bad enough that the lady proved to be fat and with her husband. What humiliated my self-image as intrepid explorer was the University of Wisconsin sweatshirt that she wore.

In contrast, for all but the last ten thousand years of human history, unfettered travel was impossible, and diffusion of sweatshirts was very limited. Each village or band constituted a political unit, living in a perpetually shifting state of wars, truces, alliances, and trade with neighboring groups. Hence New Guinea highlanders spent their entire lives within ten miles of their birthplaces. They might occasionally enter lands bordering their village lands by stealth during a war raid, or by permission during a truce, but they had no social framework for travel beyond immediately neighboring lands. The notion of tolerating unrelated strangers was as unthinkable as the notion that any such stranger would dare appear.

Even today, the legacy of this no-trespassing mentality persists in many parts of the world. Whenever I go bird-watching in New Guinea, I take pains to stop at the nearest village to request permission to bird-watch on that village's land or rivers. On two occasions when I neglected that precaution (or asked permission at the wrong village) and proceeded to boat up a river, I found the river barred on my return by canoes of stone-throwing villagers, furious that I had violated their territory. When I was living among Elopi tribespeople in west New Guinea and wanted to cross the territory of the neighboring Fayu tribe to reach a nearby mountain, the Elopis explained to me matter-of-factly that the Fayus would kill me if I tried. From a New Guinea perspective, it seemed so perfectly natural and self-explanatory. Of course the Fayus will kill any trespasser; you surely don't think they're so stupid that they'd admit strangers to their territory? Strangers would just hunt their game animals, molest their women, introduce diseases, and reconnoiter the terrain in order to stage a raid later.

While most precontact peoples had trade relations with their neighbors, many thought they were the only humans in existence.

Perhaps the smoke of fires on the horizon, or an empty canoe floating past down a river did prove the existence of other people. But to venture out of one's territory to meet those humans, even if they lived only a few miles away, was equivalent to suicide. As one New Guinea highlander recalled his life before the first arrival of whites in 1930, "We had not seen far places. We knew only this side of the mountains. And we thought that we were the only living people."

Such isolation bred great genetic diversity. Each valley in New Guinea has not only its own language and culture, but also its own genetic abnormalities and local diseases. The first valley where I worked was the home of the Foré people, famous to science for their unique affliction with a fatal viral disease called "kuru," or laughing sickness, which accounted for over half of all deaths (especially among women) and left men outnumbering women three to one in some Foré villages. At Karimui, sixty miles to the west of the Foré area, kuru is completely unknown, and the people are instead afflicted with the world's highest incidence of leprosy. Still other tribes are unique in their high frequency of deaf mutes, or of male pseudo-hermaphrodites lacking penises, or of premature aging, or of delayed puberty.

Today we can picture areas of the globe that we haven't visited, from films and TV. We can read about them in books. English dictionaries exist for all the world's major languages, and most villages speaking minor languages still contain individuals who understand one of the world's major languages. For example, missionary linguists have studied literally hundreds of New Guinea and South American Indian languages in recent decades, and I have found some inhabitant speaking either Indonesian or Neo-Melanesian in every New Guinea village that I have visited, no matter how remote. Hence linguistic barriers no longer impede the worldwide flow of information. Almost every village in the world today has thereby obtained fairly direct accounts of the outside world and has yielded fairly direct accounts of itself.

In contrast, pre-contact peoples had no way to picture the outside world, or to learn about it directly. Information instead arrived via long chains of languages, with accuracy lost at each step—as in the children's game called "telephone," in which one child in a circle whispers a message to the next child, who in turn whispers it to his/her neighbor, until by the time the message is whispered back to the first child its meaning has become changed beyond recognition. As a result, New Guinea highlanders had no concept of the ocean a hundred miles distant, and knew nothing about the white men who had been prowling their coasts for several centuries. When highlanders tried to figure out why the first arriving white men wore trousers and belts, one theory was that the clothes served to conceal an enormously long penis coiled around the waist. Some Dani believed that a neighboring group of New Guineans munched grass and had their hands joined behind their back.

First-contact patrols had a traumatic effect that is difficult for those of us living in the modern world to imagine. Highlanders "discovered" by Michael Leahy in the 1930s, and interviewed fifty years later, still recalled perfectly where they were and what they were doing at that moment of first contact. Perhaps the closest parallel, to modern Americans and Europeans, is our recollection of one or two of the most important political events of our lives. Most Americans of my age recall that moment on December 7, 1941, when they heard of the Japanese attack on Pearl Harbor. We knew at once that our lives would be very different for years to come, as a result of the news. Yet even the impact of Pearl Harbor and of the resulting war on American society was minor, compared to the impact of a first-contact patrol on New Guinea highlanders. On that day, their world changed forever.

The patrols revolutionized the highlanders' material culture by bringing steel axes and matches, whose superiority over stone axes and fire drills was immediately obvious. The missionaries and government administrators who followed the patrols suppressed ingrained cultural practices like cannibalism, polygyny, homosexuality, and war. Other practices were discarded spontaneously by tribespeople themselves, in favor of new practices that they saw. But there was also a more profoundly unsettling revolution, in the highlanders' view of what comprised the universe. They and their neighbors were no longer the sole humans, with the sole way of life.

A book by Bob Connolly and Robin Anderson entitled *First Contact* poignantly relates that moment in the eastern highlands, as recalled in their old age by New Guineans and whites who met there as young adults or children in the 1930s. Terrified highlanders took the whites for returning ghosts, until the New Guineans dug up and scrutinized the whites' buried feces, sent terrified young girls to have sex with the intruders, and discovered that whites defecated and were men like themselves. Leahy wrote in his diaries that highlanders smelled bad, while at the same time the highlanders were finding the whites' smell strange and frightening. Leahy's obsession with gold was as bizarre to the highlanders as their obsession with their own form of wealth and currency—cowry shells—was to him. For the survivors of those Grand Valley Dani and Archbold Expedition members who met in 1938, such an account of first contact has yet to be written.

I SAID AT THE OUTSET that Archbold's entry into the Grand Valley was not only a watershed for the Dani, but also part of a watershed in human history. What difference did it make that all human groups used to live in relative isolation, awaiting first contact, while only a few such groups remain today? We can infer the answer by comparing those areas of the world where isolation ended long ago with those other areas where it persisted into modern times. We can also study the rapid changes that followed historical first contacts. These comparisons suggest that contact between distant peoples gradually obliterated much of the human cultural diversity that had arisen during millennia of isolation.

Take artistic diversity as one obvious example. Styles of sculpture, music, and dance used to vary greatly from village to village within New Guinea. Some villagers along the Sepik River and in the Asmat swamps produced carvings that are now world-famous because of their quality. But New Guinea villagers have been increasingly coerced or seduced into abandoning their artistic traditions. When I visited an isolated tribelet of 578 people at Bomai in 1965, the missionary controlling the only store had just manipulated the people into burning all their art. Centuries of unique cultural development ("heathen artifacts," as the missionary put it) had thus been destroyed in one morning. On my first visit to remote New Guinea villages in 1964, I heard log drums and traditional songs; on my visits in the 1980s, I heard guitars, rock music, and battery-operated boom boxes. Anyone who has seen the Asmat carvings at New York's Metropolitan Museum of Art, or heard log drums played in antiphonal duet at breathtaking speed, can appreciate the enormous tragedy of post-contact loss of art.

There has been massive loss of languages as well. For example, Europe today has only about fifty languages, most of them belonging to a single language family (Indo-European). In contrast, New Guinea, with less than one-tenth of Europe's area and less than one-hundredth of its population, has about a thousand languages, many of them unrelated to any other known language in New Guinea or elsewhere! The average New Guinea language is spoken by a few thousand people living within a radius of ten miles. When I traveled sixty miles from Okapa to Karimui in New Guinea's eastern highlands, I passed through six languages, starting with Foré (a language with postpositions, like Finnish) and ending with Tudawhe (a language with alternative tones and nasalized vowels, like Chinese).

New Guinea shows linguists what the world used to be like, with each isolated tribe having its own language, until agriculture's rise permitted a few groups to expand and spread their tongue over large areas. It was only about six thousand years ago that the Indo-European expansion began, leading to the extermination of all prior western European languages except Basque. The Bantu expansion within the last few millennia similarly exterminated most other languages of tropical and sub-Saharan Africa, just as the Austronesian expansion did in Indonesia and the Philippines. In the New World alone, hundreds of American Indian languages have become extinct in recent centuries.

Isn't language loss a good thing, because fewer languages mean easier communication among the world's people? Perhaps, but it's a bad thing in other respects. Languages differ in structure and vocabulary, in how they express causation and feelings and personal responsibility, hence in how they shape our thoughts. There's no single-purpose "best" language: instead, different languages are better suited for different purposes. For instance, it may not have been an accident that Plato and Aristotle wrote in Greek, while Kant wrote in German. The grammatical particles of those two languages, plus their ease in forming compound words, may have helped make them the preeminent languages of western philosophy. Another example, familiar to all of us who studied Latin, is that highly inflected languages (ones in which word endings suffice to indicate sentence structure) can use variations of word order to convey nuances impossible with English. Our English word order is severely constrained by having to serve as the main clue to sentence structure. If English becomes a world language, that won't be because English was necessarily the best language for diplomacy.

The range of cultural practices in New Guinea also eclipses that within equivalent areas elsewhere in the modern world, because isolated tribes were able to live out social experiments that others would find utterly unacceptable. Forms of self-mutilation and cannibalism varied from tribe to tribe. At the time of first contact, some tribes went naked, others concealed their genitals and practiced extreme sexual prudery, and still others (including the Grand Valley Dani) flagrantly advertised the penis and testes with various props. Child-rearing practices ranged from extreme permissiveness (including freedom for Foré babies to grab hot objects and burn themselves), through punishment of misbehavior by rubbing a Baham child's face with stinging nettles, to extreme repression resulting in Kukukuku child suicide. Barua men pursued institutionalized bisexuality by living with the young boys in a large communal homosexual house, while each man had a separate small heterosexual house for his wife and daughters and infant sons. Tudawhes instead had two-story houses in which women, infants, unmarried girls, and pigs lived in the lower story, while men and unmarried boys lived in the upper story accessed by a separate ladder from the ground.

We wouldn't mourn the shrinking cultural diversity of the modern world if it only meant the end of self-mutilation and child suicide. But the societies whose cultural practices have now become dominant were selected just for economic and military success. Those qualities aren't necessarily the ones that foster happiness or promote long-term human survival. Our consumerism and our environmental exploitation serve us well at present but bode ill for the future. Features of American society that already rate as disasters in anyone's book include our treatment of old people, adolescent turmoil, abuse of psychotropic chemicals, and gross inequality. For each of these problem areas, there are (or were before first contact) many New Guinea societies that found far better solutions to the same issue.

UNFORTUNATELY, ALTERNATIVE MODELS of human society are rapidly disappearing, and the time has passed when humans could try out new models in isolation. Surely there are no remaining uncontacted populations anywhere as large as the one encountered by Archbold's patrol on that August day of 1938. When I worked on New Guinea's Rouffaer River in 1979, missionaries nearby had just found a tribe of four hundred nomads, who reported another uncontacted band five days' travel upstream. Small bands have also been turning up in remote parts of Peru and Brazil. But at some point within this last decade of the twentieth century, we can expect the last first contact, and the end of the last separate experiment at designing human society.

While that last first contact won't mean the end of human cultural diversity, much of which is proving capable of surviving television and travel, it certainly does mean a drastic reduction. That loss is to be mourned, for the reasons that I've just been discussing. But our xenophobia was tolerable only as long as our means to kill each other were too limited to bring about our fall as a species. When I try to think of reasons why nuclear weapons won't inexorably combine with our genocidal tendencies to break the records we've already set for genocide in the first half of the twentieth century, our accelerating cultural homogenization is one of the chief grounds for hope that I can identify. Loss of cultural diversity may be the price that we have to pay for survival.

CHAPTER 14

Accidental Conquerors

Sₒₘₑ ₒf ₜₕₑ ₘₒₛₜ ₒbᵥᵢₒᵤₛ fₑₐₜᵤᵣₑₛ ₒf ₒᵤᵣ dₐᵢₗᵧ ₗᵢᵥₑₛ ₚₒₛₑ ₜₕₑ hardest questions for scientists. If you look around you in most locations in the United States or Australia, most of the people you see will be of European ancestry. At the same locations five hundred years ago, everyone without exception would have been an American Indian in the United States, a native (aboriginal) Australian in Australia. Why is it that Europeans came to replace most of the native population of North America and Australia, instead of Indians or native Australians coming to replace most of the original population of Europe?

This question can be rephrased to ask: Why was the ancient rate of technological and political development fastest in Eurasia, slower in the Americas (and in Africa south of the Sahara), and slowest in Australia? For example, in 1492 much of the population of Eurasia used iron tools, had writing and agriculture, had large centralized states with oceangoing ships, and was on the verge of industrialization. The Americas had agriculture, only a few large centralized states, writing in only one area, and no oceangoing ships or iron tools, and they were technologically and politically a few thousand years behind Eurasia. Australia lacked agriculture, writing, states, and ships, was still in a pre-first-contact condition, and used stone tools comparable to ones made over ten thousand years earlier in Eurasia. It was those technological and political differences—not the biological differences determining the outcome of competition among animal populations—that permitted Europeans to expand to other continents.

Nineteenth-century Europeans had a simple, racist answer to such questions. They concluded that they acquired their cultural head start through being inherently more intelligent, and that they therefore had a manifest destiny to conquer, displace, or kill "inferior" peoples. The trouble with this answer is that it is not just loathsome and arrogant, but also wrong. It's obvious that people differ enormously in the knowledge they acquire, depending on their circumstances as they grow up. But no convincing evidence of genetic differences in mental ability among peoples has been found, despite much effort.

Because of this legacy of racist explanations, the whole subject of human differences in level of civilization still reeks of racism. Yet there are obvious reasons why the subject begs to be properly explained. Those technological differences led to great tragedies in the past five hundred years, and their legacies of colonialism and conquest still powerfully shape our world today. Until we can come up with a convincing alternative explanation, the suspicion that racist genetic theories might be true will linger.

In this chapter I'll argue that continental differences in level of civilization arose from geography's effect on the development of our cultural hallmarks, not from human genetics. Continents differed in the resources on which civilization depended—especially in the wild animal and plant species that proved useful for domestication. Continents also differed in the ease with which domesticated species could spread from one area to another. Even today, Americans and Europeans are painfully aware how distant geographical features, like the Persian Gulf or the Isthmus of Panama, affect our lives. But geography and biogeography have been molding human lives even more profoundly for hundreds of thousands of years.

Why do I emphasize plant and animal species? As the biologist J. B. S. Haldane remarked, "Civilization is based, not only on men, but on plants and animals." Agriculture and herding, though they also brought the disadvantages discussed in Chapter 10, still made it possible to feed far more people per square mile of land than could live on the wild foods available in that same area. Storable food surpluses grown by some individuals permitted other individuals to devote themselves to metallurgy, manufacturing, writing—and to serving in full-time professional armies. Domestic animals provided not only meat and milk to feed people, but also wool and hides to clothe people, and power to transport people and goods. Animals also provided power to pull plows and carts, and thus to increase agricultural productivity greatly over that previously attainable by human muscle power alone.

As a result, the world's human population rose from about ten million around 10,000 B.C., when we were all still hunter-gatherers, to over five billion today. Dense populations were prerequisite to the rise of centralized states. Dense populations also promoted the evolution of infectious diseases, to which exposed populations then evolved some resistance but other populations didn't. All these factors determined who colonized and conquered whom. Europeans' conquest of America and Australia was due not to their better genes but to their worse germs (especially smallpox), more advanced technology (including weapons and ships), information storage through writing, and political organization—all stemming ultimately from continental differences in geography.

Let's start with the differences in domestic animals. By around 4000 B.C. west Eurasia already had its "Big Five" domestic livestock that continue to dominate today: sheep, goats, pigs, cows, and horses. East Asians domesticated four other cattle species that locally replace cows: yaks, water buffalo, gaur, and banteng. As already mentioned, these animals provided food, power, and clothing, while the horse was also of incalculable military value. (It was the tank, the truck, and the jeep of warfare until the nineteenth century.) Why didn't American Indians reap similar benefits by domesticating the corresponding native American mammal species: mountain sheep, mountain goats, peccaries, bison, and tapirs? Why didn't Indians mounted on tapirs, and native Australians mounted on kangaroos, invade and terrorize Eurasia?

The answer is that, even today, it has proved possible to domesticate only a tiny fraction of the world's wild mammal species. This becomes clear when one considers all the attempts that failed. Innumerable species reached the necessary first step of being kept captive as tame pets. In New Guinea villages I routinely find tamed possums and kangaroos, while I saw tamed monkeys and weasels in Amazonian Indian villages. Ancient Egyptians tamed gazelles, antelopes, cranes, and even hyenas and possibly giraffes. Romans were terrorized by the tamed African elephants with which Hannibal crossed the Alps (*not* Asian elephants, the tame elephant species in circuses today).

But all these incipient efforts at domestication failed. Domestication requires not just capturing individual wild animals and taming them, but getting them to breed in captivity and modifying them through selective breeding so as to be more useful to us. Since the domestication of horses around 4000 B.C. and reindeer a few thousand years later, no large European mammal has been added to our repertoire of successful domesticates. Thus, our few modern species of domestic mammals were quickly winnowed from hundreds of others that had been tried and abandoned.

Why have efforts at domesticating most animal species failed? It turns out that a wild animal must possess a whole suite of unusual characteristics for domestication to succeed. First, in most cases it must be a social species living in herds. A herd's subordinate individuals have instinctive submissive behaviors that they display toward dominant individuals, and that they can transfer toward humans. Asian mouflon sheep (the ancestors of domestic sheep) have such behaviors but North American bighorn sheep do not—a crucial difference that prevented Indians from domesticating the latter. Except for cats and ferrets, solitary territorial species have not been domesticated.

Second, species such as gazelles and many deer and antelopes, which instantly take flight at signs of danger instead of standing their ground, prove too nervous to manage. Our failure to domesticate deer is especially striking, since there are few other wild animals with which humans have been so closely associated for tens of thousands of years. Although deer have always been intensively hunted and often tamed, reindeer alone among the world's forty-one deer species were successfully domesticated. Territorial behavior, flight reflexes,

or both eliminated the other forty species as candidates. Only reindeer had the necessary tolerance of intruders and gregarious, non-
territorial behavior.

Finally, as zoos often discover to their dismay, captive animals that
are docile and healthy may nevertheless refuse to breed in cages. You
yourself wouldn't want to carry out a lengthy courtship and copulate
under the watchful eyes of others; many animals don't want to either.
This problem of getting captive animals to breed has derailed per-
sistent attempts to domesticate some potentially very valuable animals. For example, the finest wool in the world comes from the
vicuña, a small camel species native to the Andes. But neither the
Incas nor modern ranchers have ever been able to domesticate it, and
wool must still be obtained by capturing wild vicuñas. Princes from
ancient Assyrian kings to nineteenth-century Indian maharajahs have
tamed cheetahs, the world's swiftest land mammals, for hunting. But
every prince's cheetah had to be captured from the wild, and not even
zoos were able to breed them until 1960.

Collectively, these reasons help explain why Eurasians succeeded
in domesticating the Big Five but not other closely related species,
and why American Indians did not domesticate bison, peccaries,
tapirs, and mountain sheep or goats. The military value of the
horse is especially interesting in illustrating what seemingly slight
differences make one species uniquely prized, another useless.
Horses belong to the order of mammals termed Perissodactyla,
which consists of the hoofed mammals with an odd number of
toes: horses, tapirs, and rhinoceroses. Of the seventeen living species
of Perissodactyla, all four tapirs and all five rhinos, plus five of the
eight wild horse species, have never been domesticated. Africans or
Indians mounted on rhinos or tapirs would have trampled any European invaders, but it never happened.

A sixth wild horse relative, the wild ass of Africa, gave rise to
domestic donkeys, which proved splendid as pack animals but useless
as military chargers. The seventh wild horse relative, the onager of
western Asia, may have been used to pull wagons for some centuries
after 3000 B.C. But all accounts of the onager blast its vile disposition
with adjectives like "bad-tempered," "irascible," "unapproachable,"
"unchangeable," and "inherently intractable." The vicious beasts had
to be kept muzzled to prevent them from biting their attendants.
When domesticated horses reached the Middle East around 2300 B.C.,

onagers were finally kicked onto the scrap heap of failed domesticates.

Horses revolutionized warfare in a way that no other animal, not
even elephants or camels, ever rivaled. Soon after their domestication,
they may have enabled herdsmen speaking the first Indo-European
languages to begin the expansion that would eventually stamp their
languages on much of the world. A few millennia later, hitched to
battle chariots, horses became the unstoppable Sherman tanks of
ancient war. After the invention of saddles and stirrups, they enabled
Attila the Hun to devastate the Roman Empire, Genghis Khan to
conquer an empire from Russia to China, and military kingdoms to
arise in West Africa. A few dozen horses helped Cortés and Pizarro,
leading only a few hundred Spaniards each, to overthrow the two
most populous and advanced New World states, the Aztec and Inca
empires. With futile Polish cavalry charges against Hitler's invading
armies in September 1939, the military importance of this most universally prized of all domestic animals finally came to an end after six
thousand years.

Ironically, relatives of the horses that Cortés and Pizarro rode had
formerly been native to the New World. Had those horses survived,
Montezuma and Atahualpa might have shattered the conquistadores
with cavalry charges of their own. But, in a cruel twist of fate,
America's horses had become extinct long before that, along with 80
or 90 percent of the other large animal species of the Americas and
Australia. It happened around the time that the first human settlers—
ancestors of modern Indians and native Australians—reached those
continents. The Americas lost not only their horses but also other
potentially domesticable species like large camels, ground sloths, and
elephants. Australia and North America ended up with no domesticable mammal species at all, unless Indian dogs were derived from
North American wolves. South America was left with only the guinea
pig (used for food), alpaca (used for wool), and llama (used as a pack
animal, but too small to carry a rider).

As a result, domestic mammals made no contribution to the protein needs of native Australians and Americans except in the Andes,
where their contribution was still much slighter than in the Old
World. No native American or Australian mammal ever pulled a
plough, cart, or war chariot, gave milk, or bore a rider. The civilizations of the New World limped forward on human muscle power

alone, while those of the Old World ran on the power of animal muscle, wind, and water.

Scientists still debate whether the prehistoric extinctions of most large American and Australian mammals were due to climatic factors or were caused by the first human settlers themselves. Whichever was the case, the extinctions may have virtually ensured that the descendants of those first settlers would be conquered over ten thousand years later by people from Eurasia and Africa, the continents that retained most of their large mammal species.

Do similar arguments apply to plants? Some parallels jump out immediately. As true of animals, only a tiny fraction of all wild plant species have proved suitable for domestication. For example, plant species in which a single hermaphroditic individual can pollinate itself (like wheat) were domesticated earlier and more easily than cross-pollinated species (like rye). The reason is that self-pollinating varieties are easier to select and then maintain as true strains, since they're not continually mixing with their wild relatives. As another example, although acorns of many oak species were a major food source in prehistoric Europe and North America, no oak has ever been domesticated, perhaps because squirrels remained much better than humans at selecting and planting acorns. For every domesticated plant that we still use today, many others were tried in the past and discarded. (What living American has eaten sumpweed, which Indians in the eastern United States domesticated for its seeds by around 2000 B.C.?).

Such considerations help explain the slow rate of human technological development in Australia. That continent's relative poverty in wild plants appropriate for domestication, as in appropriate wild animals, undoubtedly contributed to the failure of aboriginal Australians to develop agriculture. But it's not so obvious why agriculture in the Americas lagged behind that in the Old World. After all, many food plants now of worldwide importance were domesticated in the New World: corn, potatoes, tomatoes, and squash, to name just a few. The answer to this puzzle requires closer scrutiny of corn, the New World's most important crop.

Corn is a cereal—i.e., a grass with edible starchy seeds, like barley kernels or wheat grains. Cereals still provide most of the calories consumed by the human race. While all civilizations have depended on cereals, different native cereals have been domesticated by different civilizations: e.g., wheat, barley, oats, and rye in the Near East and Europe; rice, foxtail millet, and broomcorn millet in China and southeast Asia; sorghum, pearl millet, and finger millet in sub-Saharan Africa; but only corn in the New World. Soon after Columbus discovered America, corn was taken back to Europe by early explorers and spread around the globe, so that it now exceeds all other crops except wheat in world acreage planted. Why, then, didn't corn enable American Indian civilizations to develop as fast as the Old World civilizations fed by wheat and other cereals?

It turns out that corn was a much bigger pain in the neck to domesticate and grow, and gave an inferior product. Those will be fighting words to all of you who, like me, love hot buttered corn-on-the-cob. Throughout my childhood, I looked forward to late summer as the season to stop at roadside stands and pick out the best-looking fresh ears. Corn is the most important crop in the United States today, worth $22 billion to us and $50 billion to the world. But before you charge me with slander, please hear me out on the differences between corn and other cereals.

The Old World had over a dozen wild grasses that were easy to domesticate and grow. Their large seeds, favored by the Near East's highly seasonal climate, made their value obvious to incipient farmers. They were easy to harvest en masse with a sickle, easy to grind, easy to prepare for cooking, and easy to sow. Another subtle advantage was first recognized by University of Wisconsin botanist Hugh Iltis: we didn't have to figure out for ourselves that they could be stored, since wild rodents in the Near East already made caches of up to sixty pounds of those wild grass seeds.

The Old World grains were already productive in the wild: one can still harvest up to seven hundred pounds of grain per acre from wild wheat growing naturally on hillsides in the Near East. In a few weeks a family could harvest enough to feed itself for a year. Hence even before wheat and barley were domesticated, there were sedentary villages in Palestine that had already invented sickles, mortars and pestles, and storage pits, and that were supporting themselves on wild grains.

Domestication of wheat and barley wasn't a conscious act. It wasn't the case that several hunter-gatherers sat down one day, mourned the extinction of big game animals, discussed which particular wheat plants were best, planted the seeds of those plants, and thereby became farmers the next year. Instead, the process we call plant domestication—the changes in wild plants under cultivation—was an unintended by-product of people's preferring some wild plants over others, and hence accidentally spreading seeds of the preferred plants. In the case of wild cereals, people naturally preferred to harvest those with big seeds, those whose seeds were easy to remove from the seed coverings, and those with firm nonshattering stalks that held all the seeds together. It took only a few mutations, favored by this unconscious human selection, to produce the large-seeded, nonshattering cereal varieties that we refer to as domesticated rather than wild.

By around 8000 B.C., wheat and barley remains from archaeological digs at ancient Near Eastern village sites are beginning to show these changes. The development of bread wheats, other domestic varieties, and intentional sowing soon followed. Gradually, fewer remains of wild foods are found at the sites. By 6000 B.C., crop cultivation had been integrated with animal herding into a complete food-production system in the Near East. For better or worse, people were no longer hunter-gatherers but farmers and herders, en route to being civilized.

Now contrast these relatively straightforward Old World developments with what happened in the New World. Because the parts of the Americas where farming began lacked the Near East's highly seasonal climate, they lacked large-seeded grasses that were already productive in the wild. North American and Mexican Indians did start to domesticate three small-seeded wild grasses—maygrass, little barley, and a wild millet—but these were displaced by the arrival of corn and then of European cereals. Instead, the ancestor of corn was a Mexican wild grass that did have the advantage of big seeds but in other respects hardly seemed like a promising food plant: annual teosinte.

Teosinte ears look so different from corn ears that scientists argued about teosinte's precise role in corn's ancestry till recently, and even now some scientists remain unconvinced. No other crop underwent such drastic changes on domestication as did teosinte. It has only six to twelve kernels per ear, and they are inedible, because they're enclosed in stone-hard cases. One can chew teosinte stalks like sugar cane, as Mexican farmers still do. But no one uses its seeds today, and there is no indication that anyone did prehistorically either.

Hugh Iltis identified the key step in teosinte's becoming useful: a permanent sex change! In teosinte the side branches end in tassels composed of male flowers; in corn they end in female structures, the ears. Although that sounds like a drastic difference, it's really a simple hormonally controlled change that could have been started by a fungus, virus, or change in climate. Once some flowers on the tassel had changed sex to female, they would have produced edible naked grains likely to catch the attention of hungry hunter-gatherers. The tassel's central branch would then have been the beginning of a corncob. Early Mexican archaeological sites have yielded remains of tiny ears, barely an inch and a half long and much like the tiny ears of our Tom Thumb corn variety.

With that abrupt sex change, teosinte (alias corn) was now finally on the road to domestication. However, in contrast to the case with Near Eastern cereals, thousands of years of development still lay ahead before high-yield corns capable of sustaining villages or cities resulted. The final product was still much more difficult for Indian farmers to manage than were the cereals of Old World farmers. Corn ears had to be harvested individually by hand, rather than en masse with a sickle; the cobs had to be shucked; the kernels didn't fall off but had to be scraped or bitten off; and sowing the seeds involved planting them individually, rather than scattering them en masse. And the result was still poorer nutritionally than Old World cereals: lower protein content, deficiencies of nutritionally important amino acids, deficiency of the vitamin niacin (tending to cause the disease pellagra), and need for alkali treatment of the grain to partially overcome these deficiencies.

In short, characteristics of the New World's staple food crop made its potential value much harder to discern in the wild plant, harder to develop by domestication, and harder to extract even after domestication. Much of the lag between New World and Old World civilization may have been due to those peculiarities of one plant.

So FAR, I've discussed geography's biogeographic role in providing the local wild animal and plant species suitable for domestication. But there's another major role of geography that deserves mention. Each civilization has depended not only on its own food plants domesticated locally, but also on other food plants that arrived after having been first domesticated elsewhere. The predominantly north-south axis of the New World made such diffusion of food plants difficult; the predominantly east-west axis of the Old World made it easy (see Figure 6).

Today, we take plant diffusion so much for granted that we seldom stop to think where our foods originated. A typical American or European meal might consist of chicken (of Southeast Asian origin) with corn (from Mexico) or potatoes (from the southern Andes), seasoned with pepper (from India), accompanied by a piece of bread (from Near Eastern wheat) and butter (from Near Eastern cattle), and washed down by a cup of coffee (from Ethiopia). But this diffusion of valued plants and animals didn't begin just in modern times: it has been happening for thousands of years.

Plants and animals spread quickly and easily within a climate zone to which they're already adapted. To spread out of this zone, they have to develop new varieties with different climate tolerances. A glance at the map of the Old World in Figure 6 shows how species could shift long distances without encountering a change of climate. Many of these shifts proved enormously important in launching farming or herding in new areas, or enriching it in old areas. Species

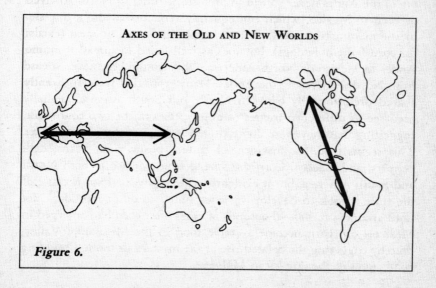

AXES OF THE OLD AND NEW WORLDS

Figure 6.

moved between China, India, the Near East, and Europe without ever leaving temperate latitudes of the northern hemisphere. Ironically, the U.S. patriotic song "America the Beautiful" invokes America's own spacious skies, its amber waves of grain. In reality, the most spacious skies of the northern hemisphere are in the Old World, where amber waves of related grains came to stretch for seven thousand miles from the English Channel to the China Sea.

The ancient Romans were already growing wheat and barley from the Near East, peaches and citrus fruits from China, cucumbers and sesame from India, and hemp and onions from central Asia, along with oats and poppies originating locally in Europe. Horses that spread from the Near East to West Africa revolutionized military tactics there, while sheep and cattle spread down from the highlands of East Africa to launch herding in southern Africa among the Hottentots, who lacked locally domesticated animals of their own. African sorghum and cotton reached India by around 2000 B.C., while bananas and yams from tropical Southeast Asia crossed the Indian Ocean to enrich agriculture in tropical Africa.

In the New World, however, the temperate zone of North America is isolated from the temperate zone of the Andes and southern South America by thousands of miles of tropics, in which temperate-zone species can't survive. As a result, the llama, alpaca, and guinea pig of the Andes never spread in prehistoric times to North America or even to Mexico, which consequently remained without any domestic mammals to carry packs or to produce wool or meat (except for corn-fed edible dogs). Potatoes as well failed to spread from the Andes to Mexico or North America, while sunflowers never spread from North America to the Andes. Many crops that were apparently shared prehistorically between North and South America actually occurred as different varieties or even species in the two continents, suggesting that they were domesticated independently in both areas. This seems true, for instance, of cotton, beans, lima beans, chili peppers, and tobacco. Corn did spread from Mexico to both North and South America, but it evidently wasn't easy, perhaps because of the time it took to develop varieties suited to other latitudes. Not until around A.D. 900—thousands of years after corn had emerged in Mexico—did corn become a staple food in the Mississippi Valley, thereby triggering the belated rise of the mysterious mound-building civilization of the American Midwest.

If the Old and New Worlds had each been rotated ninety degrees about their axes, the spread of crops and domestic animals would have been slower in the Old World, faster in the New World. The rates of rise of civilization would have been correspondingly different. Who knows whether that difference would have sufficed to let Montezuma or Atahualpa invade Europe, despite their lack of horses?

I'VE ARGUED, then, that continental differences in the rates of rise of civilization weren't an accident caused by a few individual geniuses. They weren't produced by the biological differences determining the outcome of competition among animal populations—e.g., some populations being able to run faster or digest food more efficiently than others. They also weren't the result of average differences among whole peoples in inventiveness; there is no evidence for such differences anyway. Instead, they were determined by biogeography's effect on cultural development. If Europe and Australia had exchanged their human populations twelve thousand years ago, it would have been the former native Australians, transplanted to Europe, who eventually invaded America and Australia from Europe.

Geography sets ground rules for the evolution, both biological and cultural, of all species, including our own. Geography's role in determining our modern political history is even more obvious than its role in determining the rate at which we domesticate plants and animals. From this perspective, it's almost funny to read that half of all American schoolchildren don't know where Panama is, but not at all funny when politicians display comparable ignorance. Among the many notorious examples of disasters brought on by politicians ignorant of geography, two must suffice: the unnatural boundaries drawn on the map of Africa by nineteenth-century European colonial powers, thereby undermining the stability of some modern African states that inherited those borders; and the borders of Eastern Europe drawn at the Treaty of Versailles in 1919 by politicians who knew little of that region, thereby helping fuel World War II.

Geography used to be a required subject in schools and colleges until a few decades ago, when it began to be dropped from many curricula. The mistaken belief arose then that geography consisted of little more than memorizing the names of capital cities. But twenty weeks of geography in the seventh grade isn't enough to teach our future politicians about the effects that maps really have on us. The fax machines and satellite communications that span the globe can't erase the differences among us bred by differences in location. In the long run, and on a broad scale, where we live has contributed heavily to making us who we are.

CHAPTER 15

Horses, Hittites, and History

Y KSI, KAKSI, KOLME, NELJÄ, VIISI."

I watched the little girl counting out five marbles, one by one. Her act was familiar, but her words were strange. Almost anywhere else in Europe, I would have heard words like our English "one two three": "uno due tre" in Italy, "ein zwei drei" in Germany, "odin dva tri" in Russia. But I was vacationing in Finland, and Finnish is one of Europe's few non-Indo-European languages.

Today, most European languages and many Asian languages as far east as India are very similar to each other (see table of vocabulary on page 250). No matter how we complain while memorizing French word lists in school, these so-called "Indo-European" languages resemble English and each other, and differ from all the world's other languages, in vocabulary and grammar. Only 140 of the modern world's 5,000 tongues belong to this language family, but their importance is far out of proportion to their numbers. Thanks to the global expansion of Europeans since 1492—especially of people from England, Spain, Portugal, France, and Russia—nearly half the

Indo-European vs. non-Indo-European vocabulary

INDO-EUROPEAN LANGUAGES

	one	two	three	mother	brother	sister
English	one	two	three	mother	brother	sister
German	ein	zwei	drei	Mutter	Bruder	Schwester
French	un	deux	trois	mère	frère	soeur
Latin	unus	duo	tres	mater	frater	soror
Russian	odin	dva	tri	mat'	brat	sestra
Old Irish	oen	do	tri	mathir	brathir	siur
Tocharian	sas	wu	trey	macer	procer	ser
Lithuanian	vienas	du	trys	motina	brolis	seser
Sanskrit	eka	duva	trayas	matar	bhratar	svasar
PIE*	oynos	dwo	treyes	mater	bhrater	suesor

NON-INDO-EUROPEAN LANGUAGES

Finnish	yksi	kaksi	kolme	äiti	veli	sisar
Foré*	ka	tara	kakaga	nano	naganto	nanona

* PIE stands for Proto-Indo-European, the reconstructed mother tongue of the first Indo-Europeans. Foré is a language of the New Guinea Highlands. Note that most words are very similar among the Indo-European languages and totally different among the non-Indo-European languages.

world's present population of five billion now speak an Indo-European language as their native tongue.

To us it may seem perfectly natural, and in no need of further explanation, that most European languages resemble each other. Not until we go to parts of the world with great linguistic diversity do we realize how weird Europe's homogeneity is, and how it cries out for explanation. For example, in areas of the New Guinea highlands where I work and where first contact with the outside world began only in the twentieth century, languages as different as Chinese is from English replace each other over short distances. Eurasia as well must have been diverse in its pre-first-contact condition, and gradually became less so until finally some people speaking the mother tongue of the Indo-European language family steamrollered almost all other European languages out of existence.

Of all the processes by which the modern world lost its earlier linguistic diversity, the Indo-European expansion has been the most important. Its first stage, which long ago carried Indo-European languages over Europe and much of Asia, was followed by a second stage that began in 1492 and carried them to all other continents.

When and where did the steamroller start, and what gave it its power? Why wasn't Europe overrun instead by speakers of a language related to, say, Finnish or Assyrian?

While the Indo-European problem is the most famous problem of historical linguistics, it's a problem of archaeology and history as well. In the case of those Europeans who carried out the second stage of the Indo-European expansion, beginning in 1492, we know not only their vocabularies and grammars but also the ports from which they set out, the dates of their sailings, the names of their leaders, and why they succeeded in conquering. But the quest to understand the first stage is a search for an elusive people whose language and society lie veiled in the preliterate past, even though they became world conquerors and founded today's dominant societies. That quest is also a great detective story, whose solution depends on a language discovered behind a secret wall in a Buddhist monastery, and on an Italian language inexplicably preserved on the linen wrappings of an Egyptian mummy.

When you first think about it, you might be excused for dismissing the Indo-European problem as obviously insoluble. Since the Indo-European mother tongue arose before the origin of writing, isn't it almost by definition impossible to study? Even if we found the skeletons or pottery of the first Indo-Europeans, how would we recognize them? The skeletons and pottery of modern Hungarians, living in the center of Europe, are as typically European as goulash is typically Hungarian. A future archaeologist excavating a Hungarian city would never guess that Hungarians speak a non-Indo-European language, if no examples of writing itself were recovered. Even if we could somehow locate the place and time of the first Indo-Europeans, how could we hope to deduce what advantage let their language triumph?

Remarkably, it turns out that linguists have been able to extract answers to these questions from the languages themselves. I'll first explain why we are so confident that language distributions today reflect a steamroller in the past. Then I'll try to figure out when and where the mother tongue was spoken, and how it managed to take over so much of the world.

How can we infer that the modern Indo-European languages replaced other, now-vanished languages? I'm not talking about the observed second-stage replacements of the past five hundred years, which saw English and Spanish dislodge most native tongues of the Americas and Australia. Those modern expansions were obviously due to the advantages Europeans gained from guns, germs, iron, and political organization. Instead, I'm talking about the inferred first-stage replacement that saw Indo-European dislodge older languages of Europe and western Asia, and that must have happened before writing reached those areas.

The map in Figure 7 shows the distribution of Indo-European language branches surviving in 1492, just before Spanish began to leap across the Atlantic with Columbus. Three branches are especially familiar to most Europeans and Americans: Germanic (including English and German), Italic (including French and Spanish), and Slavic (including Russian), each branch with 12 to 16 surviving languages and 300 to 500 million speakers. The largest branch, however, is Indo-Iranian, with 90 languages and nearly 700 million speakers from Iran to India (including Romany, the language of Gypsies). Relatively tiny surviving branches are Greek, Albanian, Armenian, Baltic (consisting of Lithuanian and Latvian), and Celtic (including Welsh and Gaelic), each with only 2 to 10 million speakers. In addition, at least two Indo-European branches, Anatolian and Tocharian, vanished long ago but are known from extensive preserved writings, while others disappeared with fewer traces.

What proves that all these tongues are related to each other and distinct from other language stocks? One obvious clue is shared vocabulary, as illustrated by the word table on page 250 and thousands of other examples. A second clue is similar word endings (so-called inflectional endings) used to form verb conjugations and noun declensions. These endings are illustrated by part of the conjugation of "to be" on page 254. It becomes easier to recognize such similarities when you realize that word roots and endings shared between related languages are generally not shared identically. Instead, a particular sound in one language is often replaced by another sound in the other language. Familiar examples are the frequent equivalence of English "th" and German "d" (English "thing" = German "ding," "thank" = "danke"), or of English "s" and Spanish "es" (English "school" = "escuela," "stupid" = "estupido").

Those resemblances among the Indo-European languages are detailed, but much grosser features of sounds and word formation set

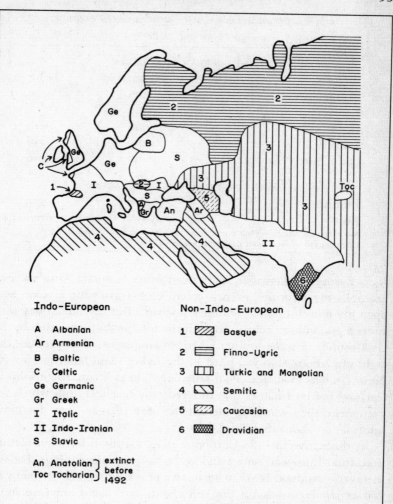

Figure 7. Language map of Europe and western Asia around 1492, just before the European discovery of the New World. There must have been other Indo-European language branches that had become extinct before then. However, the only ones for which we have long written texts are the Anatolian branch (including Hittite) and the Tocharian branch, whose homelands became occupied by speakers of Turkic and Mongolian languages before 1492.

Indo-European vs. non-Indo-European verb endings:
to be or not to be

INDO-EUROPEAN LANGUAGES		
English	(I) am	(he) is
Gothic	im	ist
Latin	sum	est
Greek	eimi	esti
Sanskrit	asmi	asti
Old Church Slavonic	jesmi	jesti
NON-INDO-EUROPEAN LANGUAGES		
Finnish	olen	on
Foré	miyuwe	miye

NOTE: Not just vocabulary but also verb and noun endings connect Indo-European languages and set them apart from other languages.

Indo-European languages apart from other language families. For example, my atrocious French accent embarrasses me as soon as I open my mouth to ask, "Où est le métro?" But my difficulties with French are nothing compared with my total inability to produce the click sounds of some southern African languages, or to produce the eight gradations of vowel pitch in the Lakes Plain languages of the New Guinea lowlands. Naturally, my Lakes Plain friends loved teaching me bird names that differed only in pitch from words for excrement, then watching me ask the next villager I met for more information about that "bird."

As distinctive to Indo-European as its sounds is its word formation. Indo-European nouns and verbs have various endings that we memorize assiduously when we learn a new language. (How many of you ex-scholars of Latin can still chant "amo amas amat amamus amatis amant"?) Each such ending conveys several types of information. For example, the "o" of "amo" specifies first person singular present active: the lover is I not my rival, one of me not two of me, I am giving not receiving love, and I am giving it now not yesterday. Heaven help the serenading lover who gets even a single one of those details wrong! But other languages, like Turkish, use a separate syllable or phoneme for each such type of information, while still other languages, like Vietnamese, virtually dispense with such variations of word form.

Given all these resemblances among Indo-European languages, how could the differences among them have arisen? A clue is that any language whose written documents span many centuries can be seen to change with time. For example, modern English-speakers find eighteenth-century English quaint but completely understandable; we can read Shakespeare (1564–1616), though we need notes to explain many of his words; but Old English texts, such as the poem *Beowulf* (around 700–750), are virtually a foreign language to us. (See example of the 23rd Psalm on page 275.) Hence, as speakers of one original language spread into different areas with limited contact, the independent changes of words and pronunciation in each area inevitably led to different dialects, such as those that have arisen in different parts of the U.S. in the few centuries since permanent English settlement began in 1607. With the passing of more centuries, dialects diverge to the point where their speakers can no longer understand each other and they now rank as distinct languages. One of the best-documented examples of this process is the development of the Romance languages from Latin. Surviving written texts from the eighth century onward show us how the languages of France, Italy, Spain, Portugal, and Rumania gradually diverged from Latin— and from each other.

The derivation of the modern Romance languages from Latin thus illustrates how groups of related languages develop from a shared ancestral tongue. Even if we had no surviving Latin texts, we could still reconstruct much of the Latin mother tongue by comparing traits in its daughter languages today. In the same way, one can reconstruct a family tree of all the Indo-European language branches, based partly on ancient texts and partly on inferences. Hence language evolution proceeds by descent and divergence, just as Darwin demonstrated for biological evolution. In their languages as well as their skeletons, modern Englishmen and Australians, who began to diverge with the colonization of Australia in 1788, are much more similar to each other than either are to the Chinese, from whom they diverged tens of thousands of years ago.

Given time, the languages within any part of the world will keep on diverging, held back only by contacts between adjacent peoples. An example of the end product is New Guinea, which had never been unified politically before European colonization, and where nearly one thousand mutually unintelligible languages—including dozens with no known relation to each other or to any other language in the world—are now spoken in an area the size of Texas. Thus, wherever you find the same language or related languages spoken over a wide area, you know that the clock of language evolution must have been restarted recently. That is, one language must have recently spread, eliminated other languages, and then started to differentiate all over again. Such a process accounts for the close similarities among southern Africa's Bantu languages, and among Austronesian languages of southeast Asia and the Pacific.

The Romance languages again provide our best-documented example. As of 500 B.C., Latin was confined to a small area around Rome and shared Italy with many other languages. The expansion of Latin-speaking Romans eradicated all those other languages of Italy, then eradicated entire branches of the Indo-European family elsewhere in Europe, like the continental Celtic languages. These sister branches were so thoroughly replaced by Latin that we know each of them only by scattered words, names, and inscriptions. With the subsequent overseas expansion of Spanish and Portuguese after 1492, the language spoken initially by a few hundred thousand Romans trampled hundreds of other languages out of existence, as it gave rise to the Romance languages spoken by half a billion people today.

If the Indo-European language family as a whole constituted a similar steamroller, we might expect to find trampled debris in the form of older non-Indo-European languages surviving here and there. The sole such vestige surviving in western Europe today is the Basque language of Spain, without known relations to any other language in the world. (The remaining non-Indo-European languages of modern Europe—Hungarian, Finnish, Estonian, and possibly Lapp—are relatively recent invaders of Europe from the east.) However, there were other languages spoken in Europe until Roman times, of which enough words or inscriptions have been preserved to identify them as non-Indo-European. The most extensively preserved of these vanished tongues is the mysterious Etruscan language of northwest Italy, for which we have a 281-line text written on a roll of linen that somehow ended up in Egypt as wrapping for a mummy. All such vanished non-Indo-European languages were part of the debris left from the Indo-European expansion.

Still more linguistic debris was swept up into the surviving Indo-

European languages themselves. To understand how linguists can recognize such debris, imagine that you, as a freshly arrived visitor from Outer Space, were given three books, one written in English by an Englishman, one by an American, and one by an Australian, about his or her country.

The language and most of the words in all three books would be the same. But if you compared the American book with the one about England, the American book would contain many place names that were obviously foreign to the basic language of the books—names like Massachusetts, Winnipesaukee, and Mississippi. The Australian book would contain more place names equally foreign to the language but unlike the American names—such as Woonarra, Goondi-windi, and Murrumbidgee. You might guess that English immigrants coming to America and Australia encountered natives who spoke different languages, and from whom the immigrants picked up names for local places and things. You'd even be able to infer something about the words and sounds of those unknown native languages. But we actually know the native American and Australian languages from which those borrowings took place, and we can confirm that your indirect inferences from the borrowed words alone would have been correct.

Linguists studying several Indo-European languages have similarly detected words borrowed from vanished, apparently non-Indo-European languages. For example, about one-sixth of Greek words whose derivations can be traced appear to be non-Indo-European. These words are just the sort that one might expect to have been borrowed by invading Greeks from the natives they encountered: place names like Corinth and Olympus, words for Greek crops like olive and vine, and names of gods or heroes like Athene and Odysseus. These words may be the linguistic legacy of Greece's pre-Indo-European population to the Greek speakers who overran them.

Thus, at least four types of evidence indicate that Indo-European languages are the products of an ancient steamroller. The evidence includes: the family-tree relationship of surviving Indo-European languages; the much greater linguistic diversity of areas like New Guinea that have not been recently overrun; the non-Indo-European languages that survived in Europe into Roman times or later; and the non-Indo-European legacy in several Indo-European languages.

*   *   *

GIVEN THIS EVIDENCE for an Indo-European mother tongue in the distant past, can one reconstruct something of this tongue? At first, the notion of learning how to write a vanished unwritten language seems absurd. In fact, linguists have been able to reconstruct much of the mother tongue by examining word roots shared among its daughter languages.

To take an example, if the word meaning "sheep" were totally different in each modern Indo-European language branch, we could conclude nothing about the word for "sheep" in the mother tongue. But if the word were similar in several branches, especially in ones as geographically distant as Indo-Iranian and Celtic, we might infer that the various branches had inherited the same root from the mother tongue. By knowing what sound shifts have taken place among the various daughter tongues, we could even reconstruct the form of the word root in the mother tongue.

As Figure 8 shows, the words for "sheep" in many Indo-European languages from India to Ireland really are very similar: "avis," "hawis," "ovis," "ois," "oi," etc. The modern English "sheep" is obviously from a different root, but English retains the original root in the word "ewe." Consideration of the sound shifts that the various Indo-European languages have undergone suggests that the original form was "owis."

Naturally, the same word root shared among several daughter languages doesn't automatically prove shared inheritance from the mother tongue. The word could also have spread later from one daughter language to another. Archaeologists skeptical of linguists' attempts to reconstruct mother tongues love to cite words like "coca-cola," shared among many modern European languages. The archaeologists claim that linguists would absurdly attribute "coca-cola" to the mother tongue of thousands of years ago. In fact, "coca-cola" illustrates how linguists distinguish recent borrowings from old inheritances: the word is obviously foreign ("coca" is actually from a Peruvian Indian word, "cola" West African); and it does not exhibit the same sound shifts among languages as do old Indo-European roots (in German it's still "Coca-Cola," not "Köcherköhler").

By such methods, linguists have been able to reconstruct much of the grammar and nearly two thousand word roots of the mother

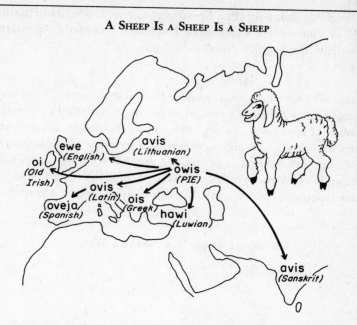

A SHEEP IS A SHEEP IS A SHEEP

Figure 8. In many modern Indo-European languages, as well as in some ancient ones that we know from preserved writings, the words meaning "sheep" are quite similar. These words must have been derived from an ancestral form that is inferred to have been "owis" and that was used in Proto-Indo-European (PIE), the unwritten mother tongue.

tongue, termed Proto-Indo-European but usually abbreviated as PIE. That's not to say that all words in modern Indo-European languages are descended from PIE: most are not, because there have been so many new inventions or borrowings (like the root "sheep" replacing the old PIE root "owis" in English). Our inherited PIE roots tend to be words for human universals that people surely were already naming thousands of years ago: words for the numbers and human relationships (as in the table on page 250); words for body parts and functions; and ubiquitous objects or concepts like "sky," "night," "summer," and "cold." Among the human universals thus reconstructed are such homely acts as "to break wind," with two distinct roots in PIE depending on whether one does it loudly or softly. The root for doing it loudly (PIE "perd") gave rise to a series of similar words in modern Indo-European languages ("perdet'," "pardate," etc.)—including English "fart" (see Figure 9).

So FAR, we've seen how linguists have been able to extract, from written languages, evidence of a preliterate mother tongue and steamroller. The obvious next questions are: when was PIE spoken, where was it spoken, and how was it able to overwhelm so many other languages? Let's begin with the matter of "when," another seemingly impossible question. It's bad enough that we have to infer the words of an unwritten language; how on earth do we determine when it was spoken?

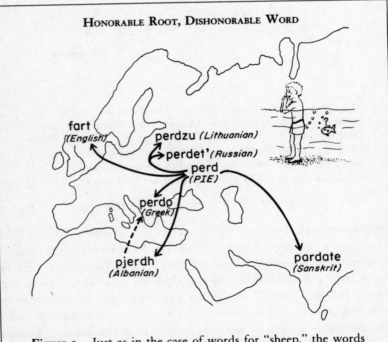

HONORABLE ROOT, DISHONORABLE WORD

fart *(English)*

perdzu *(Lithuanian)*

perdet' *(Russian)*

perd *(PIE)*

perdo *(Greek)*

pjerdh *(Albanian)*

pardate *(Sanskrit)*

Figure 9. Just as in the case of words for "sheep," the words that mean "to fart loudly" are similar among many written Indo-European languages. This suggests an ancestral form inferred to have been "perd" and used in Proto-Indo-European (PIE), the unwritten mother tongue.

We can at least start to narrow down the possibilities by examining the oldest written samples of Indo-European languages. For a long time, the oldest samples that scholars could identify were Iranian texts of around 1000–800 B.C., and Sanskrit texts probably composed around 1200–1000 B.C. but written down later. Texts of a Mesopotamian kingdom called Mitanni, written in a non-Indo-European language but containing some words obviously borrowed from a language related to Sanskrit, push the proven existence of Sanskrit-like languages back to nearly 1500 B.C.

The next breakthrough was the late-nineteenth-century discovery of a mass of ancient Egyptian diplomatic correspondence. Most of it was written in a Semitic language, but two letters in an unknown language remained a mystery until excavations in Turkey uncovered thousands of tablets in the same tongue. The tablets proved to be the archives of a kingdom that thrived between 1650 and 1200 B.C. and that we now refer to by the biblical name "Hittite."

In 1917 scholars were astonished by the announcement that the Hittite language proved on deciphering to belong to a previously unknown, very distinctive and archaic, now-vanished branch of the Indo-European family, termed Anatolian. Some obviously Hittite-like names mentioned in earlier letters of Assyrian merchants at a trading post near the Hittite capital's future site push the detective trail back to nearly 1900 B.C. This remains our first direct evidence for the existence of any Indo-European language.

Thus, as of 1917, two Indo-European branches—Anatolian and Indo-Iranian—had been shown to exist by around 1900 and 1500 B.C., respectively. A third early branch was established in 1952, when the young British cryptographer Michael Ventris showed that ancient Crete's and Greece's so-called Linear B writing, which had resisted deciphering since its rediscovery around 1900, was an early form of the Greek language. Those Linear B tablets date to around 1300 B.C. But Hittite, Sanskrit, and early Greek are very different from each other, certainly more so than modern French and Spanish, which diverged by one thousand years ago. That suggests that the Hittite, Sanskrit, and Greek branches must have split off from PIE by 2500 B.C. or earlier.

How much earlier do the differences between those branches imply? How can we obtain a calibration factor that converts "percentage difference between languages" into "time since the languages began to diverge"? Some linguists use the rate of word change in historically documented written languages, like the changes from Anglo-Saxon to Chaucer's English to Modern English. These calculations, which belong to a science called glottochronology (= chronology of languages), yield the rule of thumb that languages replace about 20 percent of their basic vocabulary every one thousand years.

Most scholars reject glottochronology calculations, on the grounds that word-replacement rates must vary with social circumstances and with the particular words themselves. However, the same scholars are generally still willing to make a seat-of-the-pants estimate. The usual conclusion from either glottochronology or pants' seats is that the PIE language community may have started to break up into several daughter language communities by 3000 B.C., surely by 2500 B.C. and not before 5000 B.C.

There's still another, completely independent approach to the dating problem: the science termed linguistic paleontology. Just as paleontologists try to discover what the past was like by looking for relics buried in the ground, linguistic paleontologists do it by looking for relics buried in languages.

To understand how this works, recall that linguists have reconstructed nearly two thousand words of PIE vocabulary. It's not surprising that these include words like "brother" and "sky," which must have existed and been named since the dawn of human language. But PIE shouldn't have had a word for "gun," since guns weren't invented till around A.D. 1300, long after PIE speakers had already scattered to speak distinct languages in Turkey and India. In fact, the word for "gun" uses different roots in different Indo-European languages: "gun" in English, "fusil" in French, "ruzhyo" in Russian, and so on. The reason is obvious: different languages couldn't possibly have inherited the same root for "gun" from PIE, and they each had to invent or borrow their own word when guns were invented.

The gun example suggests that we should take a series of inventions whose dates we know and see which of those do and which don't have reconstructed names in PIE. Anything—like the gun— that was invented after PIE began to break up shouldn't have a reconstructed name. Anything—like brother—that was known as a concept or invented before the break-up might have a name. (It doesn't *have* to have a name, because plenty of PIE words have surely become lost. We know the PIE words for "eye" and "eyebrow" but not "eyelid," although PIE speakers must have had eyelids.)

Perhaps the earliest major developments *without* PIE names are battle chariots, which became widespread between 2000 and 1500 B.C., and iron, whose use became important between 1200 and 1000 B.C. The lack of PIE terms for these relatively late inventions doesn't surprise us, since the distinctness of Hittite had already convinced us that PIE broke up long before 2000 B.C. Among earlier developments that do have PIE names, there are words for "sheep" and "goat," first domesticated by around 8000 B.C.; cattle (including separate words for cow, steer, and ox), domesticated by 6400 B.C.; horses, domesticated by around 4000 B.C.; and plows, invented around the time that horses were domesticated. The latest datable invention with a PIE name is the wheel, invented around 3300 B.C.

By such reasoning linguistic paleontology, even in the absence of any other evidence, would date the break-up of PIE as before 2000 B.C. but after 3300 B.C. This conclusion agrees well with the one reached by extrapolating the differences between Hittite, Greek, and Sanskrit backward in time. If we wish to find traces of the first Indo-Europeans, we shall be safe to concentrate on the archaeological record between 2500 and 5000 B.C., and perhaps slightly before 3000 B.C.

HAVING REACHED FAIR AGREEMENT about the "when" question, let's now ask: *where* was PIE spoken? Linguists have disagreed about the PIE homeland ever since they first began to appreciate its significance. Almost every possible answer has been proposed, from the North Pole to India, and from the Atlantic to the Pacific shores of Eurasia. As the archaeologist J. P. Mallory puts it, the question is not "Where do scholars locate the Indo-European homeland?" but "Where do they put it *now*?"

To understand why this problem has proved so hard, let's first try to solve it quickly by looking at a map (Figure 7, page 253). As of 1492, most surviving Indo-European branches were virtually confined to western Europe, and only Indo-Iranian extended east of the Caspian Sea. Hence western Europe would be the most parsimonious solution to the search for the PIE homeland: the solution that required the fewest people to move long distances.

Unfortunately for that solution, in 1900 a "new" but long-extinct Indo-European language was discovered in a triply unlikely location. First, the language (Tocharian, as it is now called) turned up in a secret chamber behind a wall in a Buddhist cave monastery. The chamber contained a library of ancient documents in the strange language, written around A.D. 600–800 by Buddhist missionaries and traders. Second, the monastery lay in Chinese Turkestan, east of all extant Indo-European speakers and about a thousand miles removed from the nearest ones. Finally, Tocharian was not related to Indo-Iranian, the geographically closest branch of Indo-European, but possibly instead to branches used in Europe itself, thousands of miles to the west. It's as if we suddenly discovered that the early medieval inhabitants of Scotland spoke a language related to Chinese.

Obviously, the Tocharians didn't reach Chinese Turkestan by helicopter. They surely walked or rode there, and we have to assume that Central Asia formerly had many other Indo-European languages that disappeared without the good fortune to be preserved by documents in secret chambers. A modern linguistic map of Eurasia (Figure 7) makes it obvious what must have happened to Tocharian and all those other lost Indo-European languages of central Asia. That whole area today is occupied by people speaking Turkic or Mongolian languages, descendants of hordes that overran the area from the time of at least the Huns to Genghis Khan. Scholars debate whether Genghis Khan's armies slaughtered 2,400,000 or only 1,600,000 people when they captured Harat, but scholars agree that such activities transformed the linguistic map of Asia. In contrast, most Indo-European languages known to have disappeared in Europe—like the Celtic languages Caesar found spoken in Gaul—were replaced by other Indo-European languages. The apparently European center of gravity of Indo-European languages as of 1492 was an artifact of recent linguistic holocausts in Asia. If the PIE homeland really was centrally located within what by A.D. 600 became the Indo-European realm, stretching from Ireland to Chinese Turkestan, then that homeland would have been in the Russian steppes north of the Caucasus, rather than in western Europe.

Just as the languages themselves gave us some clues to the time of PIE's break-up, so too they contain clues to the location of the PIE homeland. One clue is that the language family to which Indo-European has the clearest connections is Finno-Ugric, the family that includes Finnish and other languages native to the forest zone of north Russia (see Figure 7). Now it's true that the links between Finno-Ugric and Indo-European languages are enormously weaker than those between German and English, which stem from the fact that the English language was brought to England from northwest Germany only fifteen hundred years ago. The links are also much weaker than those between the Germanic and Slavic language branches of Indo-European, which probably diverged a few thousand years ago. Instead, the links suggest a much older propinquity between the speakers of PIE and of Proto-Finno-Ugric. But since Finno-Ugric comes from the North Russian forests, that suggests a PIE homeland in the Russian steppe south of the forests. In contrast, if PIE had arisen much further south (say, in Turkey), the closest affinities of Indo-European might have been with the ancient Semitic languages of the Near East.

A second clue to the PIE homeland is the non-Indo-European vocabulary swept up as debris into quite a few Indo-European languages. I mentioned that this debris is especially noticeable in Greek, and it's also conspicuous in Hittite, Irish, and Sanskrit. That suggests that those areas used to be occupied by non-Indo-Europeans and were later invaded by Indo-Europeans. If so, the PIE homeland wasn't Ireland or India (which almost no one suggests today anyway), but it also wasn't Greece or Turkey (which some scholars still do suggest).

Conversely, the modern Indo-European language still most similar to PIE is Lithuanian. Our first preserved Lithuanian texts, from around A.D. 1500, contain as high a fraction of PIE word roots as do Sanskrit texts of nearly three thousand years earlier. The conservatism of Lithuanian suggests that it has been subject to few disturbing influences from non-Indo-European languages and may have remained near the PIE homeland. Formerly, Lithuanian and other Baltic languages were more widely distributed in Russia, until Goths and Slavs pushed the Balts back to their current shrunken domain of Lithuania and Latvia. Thus, this reasoning too suggests a PIE homeland in Russia.

A third clue comes from the reconstructed PIE vocabulary. We already saw how its inclusion of words for things familiar in 4000 B.C., but not for things unknown until 2000 B.C., helps date the time when PIE was spoken. Might it also pinpoint the place where PIE was spoken? PIE includes a word for snow ("snoighwos"), suggesting a temperate rather than tropical location and providing the root of our English word "snow." Of the many wild animals and plants with PIE names (like "mus" = mouse), most are widespread in the temperate zone of Eurasia and help to pin down the homeland's latitude but not its longitude.

To me, the strongest clue from the PIE vocabulary is what it lacked rather than included: words for many crops. PIE speakers surely did some farming, since they had words for plow and sickle. But only one word for an unspecified grain has survived. In contrast, the reconstructed proto-Bantu language of Africa, and the proto-Austronesian language of southeast Asia, have many crop names. Proto-Austronesian was spoken even longer ago than PIE, so that modern Austronesian languages have had more time to lose those old names for crops than have the modern Indo-European languages. Despite that, the modern Austronesian languages still contain far more old names of crops. Hence PIE speakers probably actually had few crops, and their descendants borrowed or invented crop names as they moved to more agricultural areas.

But that conclusion presents us with a double puzzle. First, by 3500 B.C. farming had become the dominant way of life in almost all of Europe and much of Asia. That severely narrows down the possible choices for the PIE homeland: It must have been an unusual area where farming was not so dominant. And second, it begs the question why PIE speakers were able to expand. A major cause of the Bantu and Austronesian expansions was that the first speakers of those language families were farmers, spreading into areas occupied by hunter-gatherers whom they could outnumber or dominate. For PIE speakers to have been rudimentary farmers invading a farming Europe turns historical experience on its head. Thus, we cannot solve the "where" of Indo-European origins until we have come to grips with the hardest question: why?

IN EUROPE JUST BEFORE THE AGE OF WRITING, there were not one but two economic revolutions so far-reaching in impact that they could have caused a linguistic steamroller. The first was the arrival of farming and herding, which originated in the Near East around 8000 B.C., leapt from Turkey to Greece around 6500 B.C., and then spread north and west to reach Britain and Scandinavia. Farming and herding permitted a large increase in human population numbers over those previously sustainable by hunting and gathering alone. Colin Renfrew, Professor of Archaeology at Cambridge

University in England, recently published a thought-provoking book arguing that those farmers from Turkey were the PIE speakers who brought Indo-European languages to Europe.

My first reaction to reading Renfrew's book was "Of course, he must be right!" Farming *had* to produce a linguistic upheaval in Europe, just as it did in Africa and southeast Asia. This is especially likely since, as geneticists have shown, those first farmers made the biggest contribution to the genes of modern Europeans.

But—Renfrew's theory ignores or dismisses all the linguistic evidence. Farmers reached Europe thousands of years before the estimated arrival of PIE. The first farmers lacked, and PIE speakers possessed, innovations like plows, wheels, and domesticated horses. PIE is strikingly deficient in words for the crops that defined the first farmers. Hittite, the oldest known Indo-European language of Turkey, isn't the Indo-European language closest to pure PIE, as one might expect from Renfrew's Turkey-based theory, but is instead the most deviant one. Renfrew's theory rests on nothing more than a syllogism: farming probably caused a steamroller, the PIE steamroller requires a cause, so farming is assumed to have been that cause. Everything else suggests that farming instead brought to Europe the older languages that PIE overran, like Etruscan and Basque.

But around 5000–3000 B.C.—at the right time for PIE origins— there was a second economic revolution in Eurasia. This later revolution coincided with the beginnings of metallurgy and involved a greatly expanded use of domestic animals—not just for meat and hides, as humans had been using wild animals for a million years, but for new purposes that included producing milk and wool, pulling plows, pulling wheeled vehicles, and riding. The revolution is richly reflected in the PIE vocabulary, through words for "yoke" and "plow," "milk" and "butter," "wool" and "weave," and a host of words associated with wheeled vehicles ("wheel," "axle," "shaft," "harness," "hub," and "lynchpin").

The economic significance of this revolution was to increase human population and power far beyond the levels made possible by farming and herding alone. For instance, through milk and its products one cow gradually yielded many more calories than did its meat alone. Plowing let a farmer plant much more acreage than he could with a hoe or digging stick. Animal-drawn vehicles let people exploit far more land and still bring its produce to their village for processing.

For some of these advances it's hard to say where they arose, because they spread so quickly. For example, wheeled vehicles were unknown before 3300 B.C., but within a few centuries of that date they were widely recorded throughout Europe and the Middle East. But there is one crucial advance whose origin can be identified: the domestication of horses. Just before their domestication, wild horses were absent from the Mideast and southern Europe, rare in northern Europe, and abundant only in the steppes of Russia eastward. The first evidence of horse domestication is for the Sredny Stog culture around 4000 B.C., in the steppes just north of the Black Sea, where archaeologist David Anthony has identified wear marks on horses' teeth that indicate use of a bit for riding.

Throughout the world, wherever and whenever domestic horses have been introduced, they have yielded enormous benefits for human societies. For the first time in human evolution, people could travel overland faster than their own legs could propel them. Speed helped hunters run down their prey and helped herders manage their sheep and cattle over large areas. Most important, speed helped warriors to launch quick surprise raids on distant enemies and to withdraw again before the enemies had time to organize a counterattack. Hence throughout the world the horse revolutionized warfare and enabled horse-owning peoples to terrorize their neighbors. The stereotype that Americans hold of Great Plains Indians as fearsome mounted warriors was actually created only recently, within a few generations from 1660 to 1770. Since European horses reached the U.S. West in advance of Europeans themselves and other European goods, we can be sure that the horse alone was what transformed Plains Indian society.

Archaeological evidence makes clear that domestic horses had similarly transformed human society on the Russian steppe much earlier, around 4000 B.C. The steppe habitat of open grassland was hard for people to exploit until they could use horses to solve the problems of distance and transport. Human occupation of the Russian steppe accelerated with horse domestication and then exploded with the invention of ox-drawn wheeled vehicles around 3300 B.C. The steppe economy came to be based on the combination of sheep and cattle for meat, milk, and wool, plus horses and wheeled vehicles for transport, and supplemented by a little farming.

There is no evidence for intensive agriculture and food storage at those early steppe sites, in marked contrast to the abundant evidence at other European and Mideast sites around the same time. Steppe peoples lacked large permanent settlements and were evidently highly mobile—again in contrast to the villages with rows of hundreds of two-story houses in southeast Europe at the time. What the horsemen lacked in architecture they made up for in military zeal, as attested by their lavish tombs (for men only!), filled with enormous numbers of daggers and other weapons, and sometimes even with wagons and horse skeletons.

Thus, Russia's Dnieper River (see Figure 10, below) marked an abrupt cultural boundary: to the east, the well-armed horsemen;

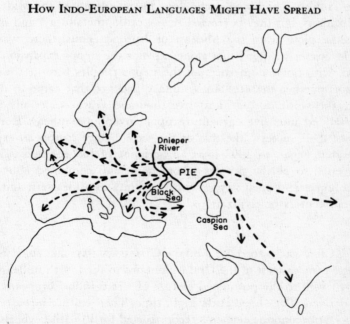

How Indo-European Languages Might Have Spread

Figure 10. This map shows how surviving Indo-European languages might have spread. The inferred homeland where Proto-Indo-European (PIE), the mother tongue, was spoken lay in the Russian steppes north of the Black Sea and east of the Dnieper River.

to the west, the rich farming villages with their granaries. That proximity of wolves and sheep spelled T-R-O-U-B-L-E. Once the invention of the wheel completed the horsemen's economic package, their artifacts indicate a very rapid spread for thousands of miles eastward through the steppes of Central Asia (see map). From that movement, the ancestors of the Tocharians may have arisen. The steppe peoples' spread westward is marked by the concentration of European farming villages nearest the steppes into huge defensive settlements, then the collapse of those societies, and the appearance of characteristic steppe graves in Europe as far west as Hungary.

Of the innovations that drove the steppe peoples' steamroller, the sole one for which they clearly get full credit is the domestication of the horse. They might also have developed wheeled vehicles, milking, and wool technology independently of the Mideast's civilizations, but they borrowed sheep, cattle, metallurgy, and probably the plow from the Mideast or Europe. Thus, there was no single "secret weapon" that alone explains the steppe expansion. Instead, with horse domestication the steppe peoples became the first to put together the economic-military package that came to dominate the world for the next five thousand years—especially after they added intensive agriculture upon invading southeast Europe. Hence their success, like that of the second-stage European expansion that began in 1492, was an accident of biogeography. They happened to be the peoples whose homeland combined abundant wild horses and open steppe with proximity to Mideastern and European centers of civilization.

As UCLA ARCHAEOLOGIST MARIJA GIMBUTAS HAS ARGUED, Russia's steppe peoples west of the Ural Mountains in the fourth millennium B.C. fit well to the postulated picture of Proto-Indo-Europeans that we derived. They lived at the right time. Their culture included the important economic elements reconstructed for PIE (like wheels and horses), and lacked the elements lacking from PIE (like battle chariots and many crop terms). They lived in the right place for PIE: the temperate zone, south of Finno-Ugric peoples, near the later homeland of Lithuanians and other Balts.

If the fit is so good, why does the steppe theory of Indo-European origins remain so controversial? There would have been no controversy if archaeologists had been able to demonstrate a rapid expansion of steppe culture from south Russia all the way to Ireland around 3000 B.C. But that didn't happen: direct evidence of the steppe invaders themselves extends no further west than Hungary. Instead, around and after 3000 B.C. one finds a bewildering array of other cultures developing in Europe and named for their artifacts (e.g., the Corded Ware and Battle-axe Culture). Those emerging western European cultures combine steppe elements like horses and militarism with old western European elements, especially settled agriculture. Such facts cause many archaeologists to discount the steppe hypothesis altogether, and to see the emerging western European cultures as local developments.

But there's an obvious reason why the steppe culture couldn't spread intact to Ireland. The steppe itself reaches its western limit in the plains of Hungary. That's where all subsequent steppe invaders of Europe, like the Mongols, stopped. To spread further, steppe society had to adapt to the forested landscape of western Europe—by adopting intensive agriculture or by taking over existing European societies and hybridizing with their peoples. Most of the genes of the resulting hybrid societies may have been the genes of Old Europe.

If steppe people imposed PIE, their mother tongue, on southeast Europe as far as Hungary, then it was the resulting daughter Indo-European culture, not the original steppe culture itself, that spread to derived granddaughter cultures elsewhere in Europe. Archaeological evidence of major cultural change suggests that such granddaughter cultures may have arisen throughout Europe and east to India between 3000 and 1500 B.C. Many non-Indo-European languages held out long enough to be preserved in writing (like Etruscan), and Basque still survives today. Thus, the Indo-European steamroller wasn't a single wave, but a long chain of events that has taken five thousand years.

As an analogy, consider how Indo-European languages came to dominate North and South America today. We have abundant written records to prove that they stem from invasions of Indo-European speakers from Europe. But those European immigrants didn't overrun the Americas in one step, and archaeologists don't find remains of unmodified European cultures throughout the sixteenth-century New World. That culture was useless on the U.S. frontier. Instead, the colonists' culture was a highly modified or hybrid one, that combined Indo-European languages and much of European technology (like guns and iron) with American Indian crops and (especially in Central and South America) Indian genes. Some areas of the New World have taken many centuries for Indo-European language and economy to master. The takeover didn't reach the Arctic until this century. It's reaching much of the Amazon only now, and the Andes of Peru and Bolivia promise to remain Indian for a long time yet.

Suppose that some future archaeologist should dig in Brazil, after written records have been destroyed and Indo-European languages have disappeared from Europe. The archaeologist will find European artifacts suddenly appearing on the coast of Brazil around 1530, but penetrating the Amazon only very slowly thereafter. The people whom the archaeologist finds living in the Brazilian Amazon will be a genetic mishmash of American Indians, African blacks, Europeans, and Japanese, speaking Portuguese. The archaeologist will be unlikely to realize that Portuguese was an intrusive language, contributed by invaders, to a hybrid local society.

EVEN AFTER THE PIE EXPANSION of the 4th millennium B.C., new interactions of horses, steppe peoples, and Indo-European languages continued to shape Eurasian history. PIE horse technology was primitive and probably involved little more than a rope bit and bareback rider. For thousands of years thereafter, the military value of horses continued to improve with inventions ranging from metal bits and horse-drawn battle chariots around 2000 B.C. to the horseshoes, stirrups, and saddles of later cavalry. While most of these advances didn't originate in the steppes, steppe peoples were still the ones who profited the most, because they always had more pasture and hence more horses.

As horse technology evolved, Europe was invaded by many more steppe peoples, among whom the Huns, Turks, and Mongols are best known. These peoples carved out a succession of huge, short-lived empires, stretching from the steppes to eastern Europe. But never again were steppe peoples able to impose their language on western Europe. They enjoyed their biggest advantage at the outset, when PIE bareback riders invaded a Europe entirely without domestic horses.

There was another difference between these later recorded invasions and the earlier unrecorded PIE invasion. The later invaders were no longer Indo-European speakers from the western steppes, but speakers of Turkic and Mongol languages from the eastern steppes. Ironically, horses were what enabled Turkish tribes from central Asia in the eleventh century A.D. to invade the land of the first written Indo-European language, Hittite. The most important innovation of the first Indo-Europeans was thus turned against their descendants. Turks are largely European in their genes, but non-Indo-European (Turkish) in their language. Similarly, an invasion from the east in A.D. 896 left modern Hungary largely European in its genes but Finno-Ugric in its language. By illustrating how a small invading force of steppe horsemen could impose their language on a European society, Turkey and Hungary provide models of how the rest of Europe came to speak Indo-European.

Eventually, steppe peoples in general, regardless of their language, ceased to win in the face of western Europe's advancing technology. When the end came, it was swift. In A.D. 1241 the Mongols achieved the largest steppe empire that ever existed, stretching from Hungary to China. But after about A.D. 1500 the Indo-European-speaking Russians began to encroach on the steppes from the west. It took only a few more centuries of tsarist imperialism to conquer the steppe horsemen who had terrorized Europe and China for over five thousand years. Today the steppes are divided between Russia and China, and only Mongolia remains as a vestige of steppe independence.

Much racist nonsense has been written about the supposed superiority of Indo-European peoples themselves. Nazi propaganda invoked a pure Aryan race. In fact, Indo-Europeans have never been unified since the PIE expansion of five thousand years ago, and even PIE speakers themselves may have been divided among related cultures. Some of the most bitter fighting and vilest deeds of recorded history pitted one Indo-European group against another. The Jews, Gypsies, and Slavs, whom the Nazis sought to exterminate, conversed in languages as Indo-European as that of their persecutors. Speakers of Proto-Indo-European merely happened to be in the right place at the right time to put together a useful package of technology. Through that stroke of luck, theirs was the mother tongue whose daughter languages came to be spoken by half the world today.

A PROTO-INDO-EUROPEAN FABLE

Owis Ekwoosque

Gwrreei owis, quesyo wlhnaa ne eest, ekwoons espeket, oinom ghe gwrrum woghom weghontm, oinomque megam bhorom, oinomque ghmmenm ooku bherontm.

Owis nu ekwomos ewewquet: "Keer aghnutoi moi ekwoons agontm nerm widntei."

Ekwoos tu ewewquont: "Kludhi, owei, keer ghe aghnutoi nsmei widntmos: neer, potis, owioom r wlhnaam sebhi gwhermom westrom qurnneuti. Neghi owioom wlhnaa esti."

Tod kekluwoos owis agrom ebhuget.

(The) Sheep and (the) Horses

On (a) hill, (a) sheep that had no wool saw horses, one (of them) pulling (a) heavy wagon, one carrying (a) big load, and one carrying (a) man quickly.

(The) sheep said to (the) horses: "My heart pains me, seeing (a) man driving horses."

(The) horses said: "Listen, sheep, our hearts pain us when we see (this): (a) man, the master, makes (the) wool of (the) sheep into (a) warm garment for himself. And (the) sheep has no wool."

Having heard this, (the) sheep fled into (the) plain.

To provide some sense of how Proto-Indo-European (PIE) might have sounded, I give a made-up fable in reconstructed PIE, together with an English translation. The fable was invented over a century ago by the linguist August Schleicher. The revised version that I give is based on one published by W. P. Lehmann and L. Zgusta in 1979 and takes account of added understanding of PIE gained since Schleicher's time. The version reproduced here has been slightly altered from that of Lehmann and Zgusta to make it more "user friendly" for nonlinguists, with the advice of Jaan Puhvel.

While PIE initially looks strange, many words will prove familiar on scrutiny because of similar English or Latin roots derived from PIE. For instance, "owis" means "sheep" (cf. "ewe," "ovine"); "wlhnaa" means "wool"; "ekwoos" means "horses" (cf. "equestrian,"

Latin "equus"); "ghmmenm" means "man" (cf. "human," Latin "hominem"); "que" means "and," as in Latin; "megam" means "big" (cf. "megabucks"); "keer" means "heart" (cf. "core," "cardiology"); "moi" means "to me"; and "widntei" and "widntmos" mean "see" (cf. "video"). The PIE text lacks definite and indefinite articles ("the" and "a") and places the verb at the end of the clause or sentence.

While this sample text will show what some linguists think PIE was like, don't take it as an exact sample. Remember: PIE was never written; scholars differ on details of how to reconstruct PIE; and the fable itself is invented.

How English has changed over the last 1000 years: the 23rd Psalm

Modern (1989)

The Lord is my shepherd, I lack nothing.

He lets me lie down in green pastures.

He leads me to still waters.

King James Bible (1611)

The Lord is my shepherd, I shall not want.

He maketh me to lie down in green pastures.

He leadeth me beside the still waters.

Middle English (1100–1500)

Our Lord gouerneth me, and nothyng shal defailen to me.

In the sted of pastur he sett me ther.

He norissed me upon water of fyllyng.

Old English (800–1066)

Drihten me raet, ne byth me nanes godes wan.

And he me geset on swythe good feohland.

And fedde me be waetera stathum.

CHAPTER 16

In Black and White

WHILE THE ANNIVERSARY OF ANY NATION'S FOUNDING IS TAKEN AS cause for its inhabitants to celebrate, Australians had special cause in 1988, their bicentennial year. Few groups of colonists faced such obstacles as those who landed with the First Fleet at the future site of Sydney in 1788. Australia was still *terra incognita:* the colonists had no idea what to expect or how to survive. They were separated from their mother country by a sea voyage of fifteen thousand miles, lasting eight months. Two and a half years of near-starvation would pass until a further supply fleet arrived from England. Many of the settlers were convicts who had already been traumatized by the most brutal aspects of brutal eighteenth-century life. Despite those beginnings, the settlers survived, prospered, filled a continent, built a democracy, and established a distinctive national character. It's no wonder that Australians felt pride as they celebrated their nation's founding.

Nevertheless, one set of protests marred the celebrations. The white settlers were not the first Australians. Instead, Australia had been settled around fifty thousand years ago by the ancestors of the people now usually referred to as "Australian Aborigines" and also known in Australia as "blacks." In the course of English settlement, most of those original inhabitants were killed by the settlers or died of other causes, leading some modern descendants of the survivors to stage bicentenary protests instead of celebrations. The celebrations focused implicitly on how Australia became white. I shall begin this chapter by focusing instead on how Australia ceased to be black, and how courageous English settlers came to commit genocide.

Lest white Australians take offense, I should make clear that I am not accusing their forefathers of having done something uniquely horrendous. Instead, my reason for discussing the extermination of the Aborigines is precisely because it isn't unique: it's a well-documented example of a phenomenon whose frequency few people appreciate. While our first association to the world "genocide" is likely to be the killings in Nazi concentration camps, those were not even the largest-scale genocide of this century. The Tasmanians and hundreds of other peoples were modern targets of successful smaller extermination campaigns. Numerous peoples scattered throughout the world are potential targets in the near future. Yet genocide is such a painful subject that either we'd rather not think about it at all, or else we'd like to believe that nice people don't commit genocide, only Nazis do. But our refusal to think about it has consequences: we've done little to halt the numerous episodes of genocide since World War II, and we're not alert to where it may happen next. Together with our destruction of our own environmental resources, our genocidal tendencies coupled to nuclear weapons now constitute the two most likely means by which the human species may reverse all its progress virtually overnight.

Despite increasing interest in genocide on the part of psychologists and biologists as well as some lay people, basic questions about it remain disputed. Do any animals routinely kill members of their own species, or is that a human invention without animal precedents? Throughout human history, has genocide been a rare aberration, or has it been common enough to rank as a human hallmark along with art and language? Is its frequency now increasing, because modern weapons permit push-button genocide and thereby reduce our instinctive inhibitions about killing fellow humans? Why have so many cases attracted so little attention? Are genocidal killers abnormal individuals, or are they normal people placed in unusual situations?

To understand genocide, we cannot proceed narrowly but must draw on biology, ethics, and psychology. Hence our exploration of genocide will begin by tracing its biological history, from our animal ancestors to the twentieth century. After asking how killers have reconciled genocide with their ethical codes, we can examine its psychological effects on the perpetrators, surviving victims, and onlookers. But before we search for answers to these questions, it is useful to start with the extermination of the Tasmanians, as a case study typical of a wide class of genocides.

TASMANIA IS A MOUNTAINOUS ISLAND similar in area to Ireland and lying two hundred miles off Australia's southeast coast. When discovered by Europeans in 1642, it supported about five thousand hunter-gatherers related to the Aborigines of the Australian mainland and with perhaps the simplest technology of any modern people. Tasmanians made only a few types of simple stone and wooden tools. Like the mainland Aborigines, they lacked metal tools, agriculture, livestock, pottery, and bows and arrows. Unlike the mainlanders, they also lacked boomerangs, dogs, nets, knowledge of sewing, and ability to start a fire.

Since the Tasmanians' sole boats were rafts capable of only short journeys, they had had no contact with any other humans since rising sea level cut off Tasmania from Australia ten thousand years ago. Confined to their private universe for hundreds of generations, they had survived the longest isolation in modern human history—an isolation otherwise depicted only in science fiction. When the white colonists of Australia finally ended that isolation, no two peoples on Earth were less equipped to understand each other than were Tasmanians and whites.

The tragic collision of these two peoples led to conflict almost as soon as British sealers and settlers arrived around 1800. Whites kidnapped Tasmanian children as laborers, kidnapped women as consorts, mutilated or killed men, trespassed on hunting grounds, and tried to clear Tasmanians off their land. Thus, the conflict quickly focused on lebensraum, which throughout human history has been among the commonest causes of genocide. As a result of the kidnappings, the native population of northeast Tasmania in November 1830 was reduced to seventy-two adult men, three adult women, and no children. One shepherd shot nineteen Tasmanians with a swivel gun loaded with nails. Four other shepherds ambushed a group of natives, killed thirty, and threw their bodies over a cliff remembered today as Victory Hill.

Naturally, Tasmanians retaliated, and whites counterretaliated in turn. To end the escalation, Governor Arthur in April 1828 ordered all Tasmanians to leave the part of their island already settled by Europeans. To enforce this order, government-sponsored groups called "roving parties," consisting of convicts led by police, hunted down and killed Tasmanians. With the declaration of martial law in November 1828, soldiers were authorized to kill on sight any Tasmanian in the settled areas. Next, a bounty was declared on the natives: five British pounds for each adult, two pounds for each child, caught alive. "Black catching," as it was called because of the Tasmanians' dark skins, became big business pursued by private as well as official roving parties. At the same time a commission headed by William Broughton, the Anglican archdeacon of Australia, was set up to recommend an overall policy toward the natives. After considering proposals to capture them for sale as slaves, poison or trap them, or hunt them with dogs, the commission settled on continued bounties and the use of mounted police.

In 1830 a remarkable missionary, George Augustus Robinson, was hired to round up the remaining Tasmanians and take them to Flinders Island, thirty miles away. Robinson was convinced that he was acting for the good of the Tasmanians. He was paid £300 in advance, £700 on completing the job. Undergoing real dangers and hardship, and aided by a courageous native woman named Truganini, he succeeded in bringing in the remaining natives—initially, by persuading them that a worse fate awaited them if they did not surrender, but later at gunpoint. Many of Robinson's captives died en route to Flinders, but about two hundred reached there, the last survivors of the former population of five thousand.

On Flinders Island Robinson was determined to civilize and christianize the survivors. His settlement was run like a jail, at a windy site with little fresh water. Children were separated from parents to facilitate the work of civilizing them. The regimented daily schedule included Bible reading, hymn singing, and inspection of beds and dishes for cleanness and neatness. However, the jail diet caused malnutrition, which combined with illness to make the natives die. Few

infants survived more than a few weeks. The government reduced
expenditures in the hope that the natives would die out. By 1869 only
Truganini, one other woman, and one man remained alive.

William Lanner, the last Tasmanian man.
Photograph by Wooley, from the collection of the
Tasmanian Museum and Art Gallery.

These last three Tasmanians attracted the interest of scientists,
who believed them to be a missing link between humans and apes.
Hence, when the last man, one William Lanner, died in 1869, competing teams of physicians, led by Dr. George Stokell from the Royal

Society of Tasmania and Dr. W. L. Crowther from the Royal College of Surgeons, alternately dug up and reburied Lanner's body, cutting off parts of it and stealing them back and forth from each other. Dr. Crowther cut off the head, Dr. Stokell the hands and feet, and someone else the ears and nose, as souvenirs. Dr. Stokell made a tobacco pouch out of Lanner's skin.

Before Truganini, the last woman, died in 1876, she was terrified of similar post-mortem mutilation and asked in vain to be buried at sea. As she had feared, the Royal Society dug up her skeleton and put it on public display in the Tasmanian Museum, where it remained until 1947. In that year the Museum finally yielded to complaints of poor taste and transferred Truganini's skeleton to a room where only scientists could view it. That too stimulated complaints of poor taste. Finally, in 1976—the centenary year of Truganini's death—her skeleton was cremated over the museum's objections, and her ashes were scattered at sea as she had requested.

While the Tasmanians were few in number, their extermination was disproportionately influential in Australian history, because Tasmania was the first Australian colony to solve its native problem and achieved the most nearly final solution. It had done so by apparently succeeding in getting rid of all its natives. (Actually, some children of Tasmanian women by white sealers survived, and their descendants today constitute an embarrassment to the Tasmanian government, which has not figured out what to do about them.) Many whites on the Australian mainland envied the thoroughness of the Tasmanian solution and wanted to imitate it, but they also learned a lesson from it. The extermination of the Tasmanians had been carried out in settled areas in full view of the urban press, and had attracted some negative comment. Hence the extermination of the much more numerous mainland Aborigines was instead effected at or beyond the frontier, far from urban centers.

The mainland governments' instrument of this policy, modeled on the Tasmanian government's roving parties, was a branch of mounted police termed "Native Police," who used search-and-destroy tactics to kill or drive out Aborigines. A typical strategy was to surround a camp at night and to shoot the inhabitants in an attack at dawn. White settlers also made widespread use of poisoned food to kill Aborigines. Another common practice was roundups in which captured Aborigines were kept chained together at the neck while being

Truganini, the last Tasmanian woman.
Photograph by Wooley, from the collection of the
Tasmanian Museum and Art Gallery.

marched to jail and held there. The British novelist Anthony Trol-
lope expressed the prevailing nineteenth-century British attitude toward Aborigines when he wrote, "Of the Australian black man we
may certainly say that he has to go. That he should perish without
unnecessary suffering should be the aim of all who are concerned in
the matter."

These tactics continued in Australia long into the twentieth century. In an incident at Alice Springs in 1928, police massacred thirty-one Aborigines. The Australian Parliament refused to accept a report on the massacre, and two Aboriginal survivors (rather than the police) were put on trial for murder. Neck chains were still in use and defended as humane in 1958, when the Commissioner of Police for the state of Western Australia explained to the Melbourne *Herald* that Aboriginal prisoners preferred being chained.

The mainland Aborigines were too numerous to exterminate completely in the manner of the Tasmanians. However, from the arrival of British colonists in 1788 until the 1921 census, the Aboriginal population declined from about 300,000 to 60,000.

Today, the attitudes of white Australians toward their murderous history vary widely. While government policy and many whites' private views have become increasingly sympathetic to the Aborigines, other whites deny responsibility for genocide. For instance, in 1982 one of Australia's leading news magazines, *The Bulletin,* published a letter by a lady named Patricia Cobern, who denied indignantly that white settlers had exterminated the Tasmanians. In fact, wrote Ms. Cobern, the settlers were peace-loving and of high moral character, while Tasmanians were treacherous, murderous, warlike, filthy, gluttonous, vermin-infested, and disfigured by syphilis. Moreover, they took poor care of their infants, never bathed, and had repulsive marriage customs. They died out because of all those poor health practices, plus a death wish and lack of religious beliefs. It was just a coincidence that, after thousands of years of existence, they happened to die out during a conflict with the settlers. The only massacres were of settlers by Tasmanians, not vice versa. Besides, the settlers only armed themselves in self-defense, were unfamiliar with guns, and never shot more than forty-one Tasmanians at one time.

To place these cases of the Tasmanians and the Australian Aborigines in perspective, consider three maps of the world (Figures 11, 12, and 13), depicting for three different time periods some mass killings that have been labeled as genocides. These maps beg a question for which there is no simple answer: how to define genocide. Etymologically, it means "group killing": the Greek root "genos," meaning race, and the Latin root "-cide," meaning killing (as in suicide, in-

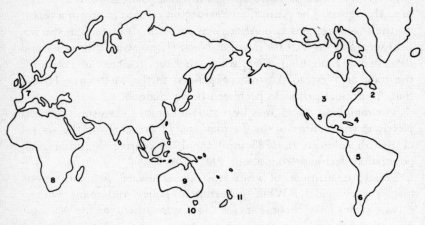

Figure 11

	DEATHS	VICTIMS	KILLERS	PLACE	DATE
1.	xx	Aleuts	Russians	Aleutian Islands	1745–1770
2.	x	Beothuk Indians	French, Micmacs	Newfoundland	1497–1829
3.	xxxx	Indians	Americans	U.S.A.	1620–1890
4.	xxxx	Caribbean Indians	Spaniards	West Indies	1492–1600
5.	xxxx	Indians	Spaniards	Central & South America	1498–1824
6.	xx	Araucanian Indians	Argentinians	Argentina	1870s
7.	xx	Protestants	Catholics	France	1572
8.	xx	Bushmen, Hottentots	Boers	South Africa	1652–1795
9.	xxx	Aborigines	Australians	Australia	1788–1928
10.	x	Tasmanians	Australians	Tasmania	1800–1876
11.	x	Moriois	Maoris	Chatham Islands	1835

x = less than 10,000; xx = 10,000 or more; xxx = 100,000 or more; xxxx = 1,000,000 or more

SOME GENOCIDES, 1900–1950

Figure 12

	DEATHS	VICTIMS	KILLERS	PLACE	DATE
1.	xxxxx	Jews, Gypsies, Poles, Russians	Nazis	Occupied Europe	1939–1945
2.	xxx	Serbs	Croats	Yugoslavia	1941–1945
3.	xx	Polish officers	Russians	Katyn	1940
4.	xx	Jews	Ukrainians	Ukraine	1917–1920
5.	xxxxx	Political opponents	Russians	Russia	1929–1939
6.	xxx	Ethnic minorities	Russians	Russia	1943–1946
7.	xxxx	Armenians	Turks	Armenia	1915
8.	xx	Hereros	Germans	Southwest Africa	1904
9.	xxx	Hindus, Moslems	Moslems, Hindus	India, Pakistan	1947

xx = 10,000 or more; xxx = 100,000 or more; xxxx = 1,000,000 or more; xxxxx = 10,000,000 or more

fanticide). The victims must be selected because they belong to a group, whether or not each victim as an individual has done something to provoke killing. As for the defining group characteristic, it may be racial (white Australians killing black Tasmanians), national (Russians killing fellow white Slavs, the Polish officers at Katyn in 1940), ethnic (the Hutu and Tutsi, two black African groups, killing each other in Rwanda and Burundi in the 1960s and 1970s), religious (Moslems and Christians killing each other in Lebanon in recent

SOME GENOCIDES, 1950–1990

Figure 13

	DEATHS	VICTIMS	KILLERS	PLACE	DATE
1.	xx	Indians	Brazilians	Brazil	1957–1968
2.	x	Aché Indians	Paraguayans	Paraguay	1970s
3.	xx	Argentine civilians	Argentine army	Argentina	1976–1983
4.	xx	Moslems, Christians	Christians, Moslems	Lebanon	1975–1990
5.	x	Ibos	North Nigerians	Nigeria	1966
6.	xx	Opponents	Dictator	Equatorial Guinea	1977–1979
7.	x	Opponents	Emperor Bokassa	Central African Republic	1978–1979
8.	xxx	South Sudanese	North Sudanese	Sudan	1955–1972
9.	xxx	Ugandans	Idi Amin	Uganda	1971–1979
10.	xx	Tutsi	Hutu	Rwanda	1962–1963
11.	xxx	Hutu	Tutsi	Burundi	1972–1973
12.	x	Arabs	Blacks	Zanzibar	1964
13.	x	Tamils, Sinhalese	Sinhalese, Tamils	Sri Lanka	1985
14.	xxxx	Bengalis	Pakistani army	Bangladesh	1971
15.	xxxx	Cambodians	Khmer Rouge	Cambodia	1975–1979
16.	xxx	Communists and Chinese	Indonesians	Indonesia	1965–1967
17.	xx	Timorese	Indonesians	East Timor	1975–1976

x = less than 10,000; xx = 10,000 or more; xxx = 100,000 or more; xxxx = 1,000,000 or more decades), or political (the Khmer Rouge killing their fellow Cambodians from 1975 to 1979).

While collective killing is thus the essence of genocide, one can argue over how narrow a definition to adopt. The word "genocide" is often used so broadly that it loses meaning and we become tired of hearing it. Even if it is to be restricted to large-scale cases of collective killing, ambiguities remain. Here is a sample of the ambiguities:

How many deaths are needed for a killing to count as genocide rather than mere murder? This is a totally arbitrary question. Australians killed all five thousand Tasmanians, and American settlers killed the last twenty Susquehanna Indians in 1763. Does the small number of available victims disqualify these killings as genocidal, despite the completeness of extermination?

Must genocide be carried out by governments, or do private acts also count? The sociologist Irving Horowitz distinguished private acts as "assassination," and defined genocide as "a structural and systematic destruction of innocent people by a state bureaucratic apparatus." However, there is a complete continuum from "purely" governmental killings (Stalin's purges of his opponents) to "purely" private killings (Brazilian land-development companies hiring professional Indian killers). American Indians were killed by private citizens and the U.S. army alike, while the Ibos in Northern Nigeria were killed both by street mobs and by soldiers. In 1835 the Te Ati Awa tribe of New Zealand Maoris succeeded in a bold plan to capture a ship, load it with supplies, invade the Chatham Islands, kill three hundred of the occupants (another Polynesian group called the Morioris), enslave the remainder, and thereby take over the islands. By Horowitz's definition, this and many other equally well-planned exterminations of one tribal group by another do not constitute genocide, because the tribes lacked a state bureaucratic apparatus.

If people die en masse as a result of callous actions not specifically designed to kill them, does that count as genocide? Well-planned genocides include that of Tasmanians by Australians, that of Armenians by Turks during World War I, and (most notably) those committed by the Nazis during World War II. At the other extreme, when the Choctaw, Cherokee, and Creek Indians of southeastern U.S. states were forced to resettle west of the Mississippi River in the 1830s, it was not President Andrew Jackson's specific intent that many Indians should die en route, but he also did not take the measures that would have been necessary to keep them alive. Their numerous deaths were instead merely an inevitable result of forced marches in winter with little or no food or clothing.

An unusually candid statement about the role of intent in genocide emerged when the Paraguayan government was charged with complicity in the disappearance of the Guayaki Indians, who had been enslaved, tortured, deprived of food and medicine, and massacred. Paraguay's defense minister replied quite simply that there had been no intent to destroy the Guayaki: "Although there are victims and victimizers, there is not the third element necessary to establish the crime of genocide—that is, 'intent.' Therefore, as there is no intent, one cannot speak of 'genocide.' " Brazil's Permanent Representative to the UN similarly rebutted charges of Brazilian genocide against Amazonian Indians: "... there was lacking the special malice or motivation necessary to characterize the occurrence of genocide. The crimes in question were committed for exclusively economic reasons, the perpetrators having acted solely to take possession of the lands of their victims."

Some mass killings, such as those of Jews and Gypsies by Nazis, were unprovoked: the slaughter was not in retaliation for previous murders committed by the slaughtered. In many other cases, however, a mass killing culminates a series of murders and countermurders. When a provocation is followed by massive retaliation all out of proportion to the provocation, how do we decide when "mere" retaliation becomes genocide? At the Algerian town of Sétif in May 1945, celebrations of the end of World War II developed into a race riot in which Algerians killed 103 French. The savage French response consisted of planes' destroying forty-four villages, a cruiser bombarding coastal towns, civilian commandos organizing reprisal massacres, and troops killing indiscriminately. The Algerian dead numbered fifteen hundred according to the French, fifty thousand according to the Algerians. The interpretations of this event differ as do the estimates of the dead: to the French, it was suppression of a revolt; to the Algerians, it was a genocidal massacre.

GENOCIDES PROVE AS HARD TO PIGEONHOLE in their motivation as in their definition. While several motives may operate simultaneously, it is convenient to divide them into four types. In the first two types there is a real conflict of interest over land or power, whether or not the conflict is also disguised in ideology. In the other two types such conflict is minimal, and the motivation is more purely ideological or psychological.

Perhaps the commonest motive for genocide arises when a militarily stronger people attempts to occupy the land of a weaker people, who resist. Among the innumerable straightforward cases of this sort are not only the killing of Tasmanians and Australian Aborigines by white Australians, but also the killings of American Indians by white Americans, of Araucanian Indians by Argentinians, and of Bushmen and Hottentots by the Boer settlers of South Africa.

Another common motive involves a lengthy power struggle within a pluralistic society, leading to one group's seeking a final solution by killing the other. Cases involving two different ethnic groups are the killing of Tutsi in Rwanda by Hutu in 1962–63, of Hutu in Burundi by Tutsi in 1972–73, of Serbs by Croats in Yugoslavia during World War II, of Croats by Serbs at the end of that war, and of Arabs in Zanzibar by blacks in 1964. However, the killer and killed may belong to the same ethnic group and may differ only in political views. Such was the case in history's largest known genocide, claiming an estimated twenty million victims in the decade 1929–39 and sixty-six million between 1917 and 1959: that committed by the U.S.S.R. government against political opponents among its own citizens. Political killings lagging far behind this record are the Khmer Rouge purge of several million fellow Cambodians during the 1970s, and Indonesia's killing of hundreds of thousands of Communists in 1965–67.

In these two just-described motives for genocide, the victims could be viewed as a significant obstacle to the killers' control of land or power. At the opposite extreme are scapegoat killings of a helpless minority blamed for frustrations of their killers. Jews were killed by fourteenth-century Christians as scapegoats for the bubonic plague, by early twentieth-century Russians as scapegoats for Russia's political problems, by Ukrainians after World War I as scapegoats for the Bolshevist threat, and by Nazis during World War II as scapegoats for Germany's defeat in World War I. When the U.S. Seventh Cavalry slaughtered several hundred Sioux Indians at Wounded Knee in 1890, the soldiers were taking belated revenge for the Sioux's annihilating counterattack on Custer's Seventh Cavalry force at the Battle of the Little Big Horn fourteen years previously. In 1943–44, at the height of Russia's suffering from the Nazi invasion, Stalin ordered the killing or deportation of six ethnic minorities who served as scapegoats: the Balkars, Chechens, Crimean Tatars, Ingush, Kalmyks, and Karachai.

Racial and religious persecutions have served as the remaining class of motives. While I do not claim to understand the Nazi mentality, the Nazis' extermination of Gypsies may have stemmed from relatively "pure" racial motivation, while scapegoating joined religious and racial motives in the extermination of Jews. The list of religious massacres is almost infinitely long. It includes the First Crusaders' massacre of all Moslems and Jews in Jerusalem when that city was finally captured in 1099, and the St. Bartholomew's Day massacre of French Protestants by Catholics in 1572. Of course, racial and religious motives have contributed heavily to genocides provoked by land struggles, power struggles, and scapegoating.

EVEN IF ONE ALLOWS for disagreements over definitions and motives, plenty of cases of genocide remain. Let us now see how far back in and before our history as a species the record of genocide extends.

Is it true, as often claimed, that man is unique among animals in killing members of his own species? For example, the distinguished biologist Konrad Lorenz, in his book *On Aggression,* argued that animals' aggressive instincts are held in check by instinctive inhibitions against murder. But in human history this equilibrium supposedly became upset by the invention of weapons: our inherited inhibitions were no longer strong enough to restrain our newly acquired powers of killing. This view of man as the unique killer and evolutionary misfit has been accepted by Arthur Koestler and many other popular writers.

Actually, studies in recent decades have documented murder in many, though certainly not all, animal species. Massacre of a neighboring individual or troop may be beneficial to an animal, if it can thereby take over the neighbor's territory, food, or females. But attacks also involve some risk to the attacker. Many animal species lack the means to kill their fellows, and of those species with the means, some refrain from using them. It may sound utterly repugnant to do a cost/benefit analysis of murder, but such analyses nevertheless help one understand why murder appears to characterize only some animal species.

In nonsocial species, murders are necessarily just of one individual by another. However, in social carnivorous species, like lions, wolves, hyenas, and ants, murder may take the form of coordinated attacks by members of one troop on members of a neighboring troop—i.e., mass killings or "wars." The form of war varies among species. Males may spare and mate with neighboring females, kill the infants, and drive off (langur monkeys) or even kill (lions) neighboring males; or both males and females may be killed (wolves). As one example, here is Hans Kruuk's account of a battle between two hyena clans in Tanzania's Ngorongoro Crater:

"About a dozen of the Scratching Rock hyenas . . . grabbed one of the Mungi males and bit him wherever they could—especially in the belly, the feet, and the ears. The victim was completely covered by his attackers, who proceeded to maul him for about ten minutes. . . . The Mungi male was literally pulled apart, and when I later studied the injuries more closely, it appeared that his ears were bitten off and so were his feet and testicles, he was paralyzed by a spinal injury, had large gashes in the hind legs and belly, and subcutaneous hemorrhages all over."

Of particular interest in understanding our genocidal origins is the behavior of two of our three closest relatives, gorillas and common chimpanzees. Two decades ago, any biologist would have assumed that our ability to wield tools and to lay concerted group plans made us far more murderous than apes—if indeed apes were murderous at all. Recent discoveries about apes suggest, however, that a gorilla or common chimp stands at least as good a chance of being murdered as does the average human. Among gorillas, for instance, males fight each other for ownership of harems of females, and the victor may kill the loser's infants as well as the loser himself. Such fighting is a major cause of death for infant and adult male gorillas. The typical gorilla mother loses at least one infant to infanticidal males in the course of her life. Conversely, 38 percent of infant gorilla deaths are due to infanticide.

Especially instructive, because it could be documented in detail, was the extermination of one of the common chimpanzee bands that Jane Goodall studied, carried out between 1974 and 1977 by another band. As of the end of 1973 the two bands were fairly evenly matched: the Kasakela band to the north, with eight mature males and occupying fifteen square kilometers; and the Kahama band to the south, with six mature males and occupying ten square kilometers. The first fatal incident occurred in January 1974, when six of the Kasakela adult males, one adolescent male, and one adult female left behind the young Kasakela chimps, traveled south, then moved silently and more quickly south when they heard chimp calls from that direction, until they surprised a Kahama male referred to as Godi. One Kasakela male pulled the fleeing Godi to the ground, sat on his head, and held down his legs while the others spent ten minutes hitting and biting him. Finally, one attacker threw a large rock at Godi, and the attackers then left. Although able to stand up, Godi was badly wounded, bleeding, and had puncture marks. He was never seen again and presumably died of his injuries.

The next month, three Kasakela males and one female again traveled south and attacked the Kahama male Dé, who was already weak from a previous attack or illness. The attackers pulled Dé out of a tree, stamped on him, bit and hit him, and tore off pieces of his skin. A Kahama estrus female with Dé was forced to return northward with the attackers. Two months later Dé was seen still alive but emaciated, with his spine and pelvis protruding, some fingernails and part of a toe torn off, and his scrotum shrunk to one-fifth of normal size. He was not seen thereafter.

In February 1975 five adult and one adolescent Kasakela males tracked down and attacked Goliath, an old Kahama male. For eighteen minutes they hit, bit, and kicked him, stamped on him, lifted and dropped him, dragged him over the ground, and twisted his leg. At the end of the attack Goliath was unable to sit up and was not seen again.

While the above attacks were aimed at Kahama males, in September 1975 the Kahama female Madam Bee was fatally injured after at least four nonfatal attacks over the course of the preceding year. The attack was carried out by four Kasakela adult males, while one adolescent male and four Kasakela females (including Madam Bee's kidnapped daughter) watched. The attackers hit, slapped, and dragged Madam Bee, stamped and pounded on her, threw her to the ground, picked her up and slammed her down, and rolled her downhill. She died five days later.

In May 1977 five Kasakela males killed the Kahama male Charlie, but details of the fight were not observed. In November 1977 six Kasakela males caught the Kahama male Sniff and hit, bit, and pulled him, dragged him by the legs, and broke his left leg. He was still alive the next day but was not seen again.

Of the remaining Kahama chimps, two adult males and two adult females disappeared from unknown causes, while two young females transferred to the Kasakela band, which proceeded to occupy the former Kahama territory. However, in 1979 the next band to the south, the larger Kalande band with at least nine adult males, began to encroach on Kasakela territory and may have accounted for several vanished or wounded Kasakela chimps. Similar intergroup assaults have been observed in the sole other long-term field study of common chimps, but not in long-term studies of pygmy chimps.

If one judges these murderous common chimps by the standards of human killers, it is hard not to be struck by their inefficiency. Even though groups of three to six attackers assaulted single victims, quickly rendered him or her defenseless, and continued the assault for ten to twenty minutes or more, the victim was always still alive at the end of that time. However, the attackers did succeed in immobilizing the victim and often causing eventual death. The pattern was that the victim initially crouched and may have tried to protect his head but then gave up any attempt at defense, and the attack continued beyond the point where the victim ceased moving. In this respect the interband attacks differ from the milder fights that often occur within a band. Chimps' inefficiency as killers reflects their lack of weapons, but it remains surprising that they have not learned to kill by strangling, although that would be within their capabilities.

Not only is each individual killing inefficient by our standards, but so is the whole course of chimp genocide. It took three years and ten months from the first killing of a Kahama chimp to the band's end, and all killings were of individuals, never of several Kahama chimps at once. In contrast, Australia's settlers often succeeded in eliminating a band of Aborigines in a single dawn attack. Partly, this inefficiency again reflects chimps' lack of weapons. Since all chimps are equally unarmed, killings can succeed only by several attackers' overpowering a single victim, whereas Australia's settlers had the advantage of guns over unarmed Aborigines and could shoot many at once. Partly, too, genocidal chimps are much inferior to humans in brainpower and hence in strategic planning. Chimps apparently cannot plan a night attack or a coordinated ambush by a split assault team.

However, genocidal chimps do seem to evince intent and unsophisticated planning. The Kahama killings resulted from Kasakela groups' proceeding directly, quickly, silently, and nervously toward or into Kahama territory, sitting in trees and listening for nearly an hour, and finally running to Kahama chimps that they detected. Chimps also share xenophobia with us: they clearly recognize members of other bands as different from members of their own band, and treat them very differently.

In short, of all our human hallmarks—art, spoken language, drugs, and the others—the one that has been derived most straightforwardly from animal precursors is genocide. Common chimps already carried out planned killings, extermination of neighboring bands, wars of territorial conquest, and abduction of young nubile females. If chimps were given spears and some instruction in their use, their killings would undoubtedly begin to approach ours in efficiency. Chimpanzee behavior suggests that a major reason for our human hallmark of group living was defense against other human groups, especially once we acquired weapons and a large enough brain to plan ambushes. If this reasoning is correct, then anthropologists' traditional emphasis on "man the hunter" as a driving force of human evolution might be valid after all—with the difference that we ourselves were our own prey as well as the predator that forced us into group living.

THUS, OF THE TWO PATTERNS of genocide commonest among humans, both have animal precedents: killing both men and women fits the common chimpanzee and wolf pattern, while killing men and sparing women fits the gorilla and lion pattern. Unprecedented even among animals, however, is a procedure adopted from 1976 to 1983 by the Argentine military, in the course of killing over 10,000 political opponents and their families, the *desaparecidos*. Victims included the usual men, nonpregnant women, and children down to the age of three or four years, who were often tortured before being killed. But Argentina's soldiers made a unique contribution to animal behavior when they arrested pregnant women: the women were kept alive until they delivered, and only then were they shot in the head, so that the newborn infant could be adopted by childless military parents.

Liliana Carmen Pereyra Azzarri (age twenty-one),
case 195 among the Argentine *desaparecidos* whom
human rights groups have sought to trace. In 1977
Liliana was abducted when she was five months
pregnant. She was kept alive in a torture center
(ESMA Military Academy) until she gave birth to
a boy in February 1978, whereupon she was killed
by a shotgun blast to the head from close quarters.
Her skeleton, found in a Mar de Plata cemetery where
other *desaparecidos* were buried, was identified in 1985.
Her son has not been found and may have been
taken by a military couple. Liliana's treatment exemplifies the concept of honor often invoked by the
former Argentine junta to justify its actions. I thank
the Abuelas de Plaza de Mayo for permission to reproduce Liliana's photograph.

If we are not unique among animals in our own propensity for murder, might our propensities nevertheless be a pathological fruit of modern civilization? Modern writers, disgusted by destruction of "primitive" societies by "advanced" societies, tend to idealize the former as noble savages who supposedly are peace-loving, or who commit only isolated murders rather than massacres. Erich Fromm believed the warfare of hunter-gatherer societies to be "characteristically unbloody." Certainly some preliterate peoples (Pygmies, Eskimos) seem less warlike than some others (New Guineans, Great Plains and Amazonian Indians). Even the warlike peoples—so it is claimed—practice war in a ritualized fashion and stop when only a few adversaries have been killed. But this idealization does not match my experience of the New Guinea highlanders, who are often cited as practicing limited or ritualized war. While most fighting in New Guinea consisted of skirmishes leaving no or few dead, groups sometimes did succeed in massacring their neighbors. Like other peoples, New Guineans tried to drive off or kill their neighbors on occasions when they found it advantageous, safe, or a matter of survival to do so.

When we consider early literate civilizations, written records testify to the frequency of genocide. The wars of the Greeks and Trojans, of Rome and Carthage, and of the Assyrians and Babylonians and Persians proceeded to a common end: the slaughter of the defeated irrespective of sex, or else the killing of the men and enslavement of the women. We all know the biblical account of how the walls of Jericho came tumbling down at the sound of Joshua's trumpets. Less often quoted is the sequel: Joshua obeyed the Lord's command to slaughter the inhabitants of Jericho as well as of Ai, Makkedah, Libnah, Hebron, Debir, and many other cities. This was considered so ordinary that the Book of Joshua devotes only a phrase to each slaughter, as if to say: of course he killed all the inhabitants, what else would you expect? The sole account requiring elaboration is of the slaughter at Jericho itself, where Joshua did something really unusual: he spared the lives of one family (because they had helped his messengers).

We find similar episodes in accounts of the wars of the Crusaders, Pacific islanders, and many other groups. Obviously, I'm not saying that slaughter of the defeated irrespective of sex has always followed crushing defeat in war. But either that outcome, or else milder versions like the killing of men and enslavement of women, happened often enough so that they must be considered more than a rare aberration in our view of human nature. Since 1950 there have been nearly twenty episodes of genocide, including two claiming over a million victims each (Bangladesh in 1971, Cambodia in the late 1970s) and four more with over a hundred thousand victims each (the Sudan and Indonesia in the 1960s, Burundi and Uganda in the 1970s) (see map on page 286).

Evidently, genocide has been part of our human and prehuman heritage for millions of years. In light of this long history, what about our impression that the genocides of the twentieth century are unique? There is little doubt that Stalin and Hitler set new records for number of victims, because they enjoyed three advantages over killers of earlier centuries: denser populations of victims, improved communications for rounding up victims, and improved technology for mass killing. As another example of how technology can expedite genocide, the Solomon Islanders of Roviana Lagoon in the Southwest Pacific were famous for their head-hunting raids, which depopulated neighboring islands. However, as my Roviana friends explained to me, these raids did not blossom until steel axes reached the Solomon Islands in the nineteenth century. Beheading a man with a stone axe is difficult, and the axe blade quickly loses its sharp edge and is tedious to resharpen.

A much more controversial question is whether technology also makes genocide psychologically easier today, as Konrad Lorenz has argued. His reasoning goes as follows. As humans evolved from apes, we depended increasingly for our food on killing animals. However, we also lived in societies of more and more individuals, between whom cooperation was essential. Such societies could not maintain themselves unless we developed strong inhibitions about killing fellow humans. Throughout most of our evolutionary history, our weapons operated only at close quarters, hence it was enough that we be inhibited from looking another person in the face and killing him/her. Modern push-button weapons bypassed these inhibitions by enabling us to kill without even seeing our victims' faces. Technology thus created the psychological prerequisites for the white-collar genocides of Auschwitz and Treblinka, of Hiroshima and Dresden.

I am uncertain whether this psychological argument really contributed significantly to the modern ease of genocide. The past frequency of genocide seems to have been at least as high as today's, though practical considerations limited the number of victims. To understand genocide further, we must leave dates and numbers and inquire about the ethics of killing.

THAT OUR URGE TO KILL is restrained by our ethics almost all the time is obvious. The puzzle is: what unleashes it?

Today, while we may divide the world's people into "us" and "them," we know that there are thousands of types of "them," all differing from each other as well as from us in language, appearance, and habits. To waste words on pointing this out seems silly: we all know it from books and television, and most of us also know it from firsthand experience of travel. It is hard to transfer ourselves back into the frame of mind prevailing throughout much of human history and already described in Chapter 13. Like chimpanzees, gorillas, and social carnivores, we lived in band territories. The known world was much smaller and simpler than it is today: there were only a few known types of "them," one's immediate neighbors.

For example, in New Guinea until recently, each tribe maintained a shifting pattern of warfare and alliance with each of its neighbors. A person might enter the next valley on a friendly visit (never quite without danger) or on a war raid, but the chances of being able to traverse a sequence of several valleys in friendship were negligible. The powerful rules about treatment of one's fellow "us" did not apply to "them," those dimly understood, neighboring enemies. As I walked between New Guinea valleys, people who themselves practiced cannibalism and were only a decade out of the Stone Age routinely warned me about the unspeakably primitive, vile, and cannibalistic habits of the people whom I would encounter in the next valley. Even Al Capone's gangs in twentieth-century Chicago made a policy of hiring out-of-town killers, so that the assassin could feel that he was killing one of "them" rather than of "us."

The writings of classical Greece reveal an extension of this tribal territorialism. The known world was larger and more diverse, but "us" Greeks were still distinguished from "them" barbarians. Our word "barbarian" is derived from the Greek *barbaroi*, which simply means non-Greek foreigners. Egyptians and Persians, whose level of civilization was like that of the Greeks, were nevertheless *barbaroi*.

The ideal of conduct was not to treat all men equally, but instead to reward one's friends and to punish one's enemies. When the Athenian author Xenophon wanted to express the highest praise for his admired leader Cyrus, Xenophon related how Cyrus always repaid his friends' good turns more generously, and how Cyrus retaliated on his enemies' misdeeds more severely (e.g., by gouging out their eyes or cutting off their hands).

Like the Mungi and Scratching Rocks hyena clans, humans practiced a dual standard of behavior: strong inhibitions about killing one of "us," but a green light to kill "them" when it was safe to do so. Genocide was acceptable under this dichotomy, whether one considers the dichotomy as an inherited animal instinct or as a uniquely human ethical code. We all still acquire in childhood our own arbitrary dichotomous criteria for respecting or scorning other humans. I recall a scene at Goroka airport in the New Guinea highlands, when my Tudawhe field assistants were standing awkwardly in torn shirts and bare feet, and an unshaven, unbathed white man with a strong Australian accent and hat crumpled over his eyes approached. Even before he had begun to sneer at the Tudawhes as "black bums, they won't be fit to run this country for a century," I had begun to think to myself, "Dumb Aussie redneck, why doesn't he go home to his goddamn sheep dip." There it was, a blueprint for genocide: I scorning the Australian, and he scorning the Tudawhes, based on collective characteristics taken in at a glance.

With time, this ancient dichotomizing has become increasingly unacceptable as a basis for an ethical code. Instead, there has been some tendency toward paying at least lip service to a universal code—i.e., one stipulating similar rules for treating different peoples. Genocide conflicts directly with a universal code.

Despite this ethical conflict, numerous modern genocidists have managed to take unabashed pride in their accomplishment. When Argentina's General Julio Argentino Roca opened the pampas for white settlement by ruthlessly exterminating the Araucanian Indians, a delighted and grateful Argentinian nation elected him president in 1880. How do today's genocidists wriggle out of the conflict between their actions and a universal code of ethics? They resort to one of three types of rationalizations, all of which are variations on a simple psychological theme: "Blame the victim!"

First, most believers in a universal code still consider self-defense justified. This is a usefully elastic rationalization, because "they" can invariably be provoked into some behavior adequate to justify self-defense. For instance, the Tasmanians delivered an excuse to genocidal white colonists by killing an estimated total of 183 colonists over 34 years, after being provoked by a far greater number of mutilations, kidnappings, rapes, and murders. Even Hitler claimed self-defense in starting World War II: he went to the trouble of faking a Polish attack on a German border post.

Possessing the "right" religion or race or political belief, or claiming to represent progress or a higher level of civilization, is a second traditional justification for doing anything, including genocide, to those possessing the wrong principle. When I was a student in Munich in 1962, unrepentant Nazis still explained to me matter-of-factly that the Germans had had to invade Russia, because the Russian people had adopted communism. My fifteen field assistants in New Guinea's Fakfak Mountains all looked pretty similar to me, but eventually they began explaining to me which of them were Moslems and who were Christians, and why the former (or the latter) were irredeemably lower humans. There is an almost universal hierarchy of scorn, according to which literate peoples with advanced metallurgy (e.g., white colonialists in Africa) look down on herders (e.g., Tutsi, Hottentots), who look down on farmers (e.g., Hutu), who look down on nomads or hunter-gatherers (e.g., Pygmies, Bushmen).

Finally, our ethical codes regard animals and humans differently. Thus, modern genocidists routinely compare their victims to animals in order to justify the killings. Nazis considered Jews to be subhuman lice; the French settlers of Algeria referred to local Moslems as *ratons* (rats); "civilized" Paraguayans described the Aché hunter-gatherers as rabid rats; Boers called Africans *bobbejaan* (baboons); and educated northern Nigerians viewed Ibos as subhuman vermin. The English language is rich with animal names used as pejoratives: you pig (ape, bitch, cur, dog, ox, rat, swine).

All three types of ethical rationalizations were employed by Australian colonists to justify exterminating Tasmanians. However, my fellow Americans and I can get better insight into the rationalization process by focusing on the case that we have been trained from childhood to rationalize: our not-quite-complete extermination of American Indians. A set of attitudes that we absorb goes roughly as follows:

To begin with, we don't discuss the Indian tragedy much—not nearly as much as the genocides of World War II in Europe, for instance. Our great national tragedy is instead viewed as the Civil War. Insofar as we stop to think about white-vs.-Indian conflict, we consider it as belonging to the distant past, and we describe it in military language: the Pequod War, Great Swamp Fight, Battle of Wounded Knee, Conquest of the West, etc. Indians, in our view, were warlike and violent even toward other Indian tribes, masters of ambush and treachery. They were famous for their barbarity, notably for the distinctively Indian practices of torturing captives and scalping enemies. They were few in number and lived as nomadic hunters, especially bison hunters. The Indian population of the United States as of 1492 is traditionally estimated at 1 million. This figure is so trivial, compared to the present U.S. population of 250 million, that the inevitability of whites' occupying this virtually empty continent becomes immediately apparent. Many Indians died from smallpox and other diseases. The aforementioned attitudes guided the Indian policy of the most admired U.S. presidents and leaders from George Washington onward (see quotations in "Indian Policies of Some Famous Americans," page 308).

These rationalizations rest on a transformation of historical facts. Military language implies declared warfare waged by adult male combatants. Actually, common white tactics were sneak attacks (often by civilians) on villages or encampments to kill Indians of any age and either sex. Within the first century of white settlement, governments were paying scalp bounties to semiprofessional killers of Indians. Contemporary European societies were at least as warlike and violent as Indian societies, when one considers the European frequency of rebellions, class wars, drunken violence, legalized violence against criminals, and total war, including destruction of food and property. Torture was exquisitely refined in Europe: think of drawing and quartering, burning at the stake, and the rack. While the precontact Indian population of North America is the subject of widely varying opinions, plausible recent estimates are about eighteen million, a population not reached by white settlers of the U.S. till around 1840. Although some Indians in the U.S. were seminomadic hunters without agriculture, most Indians in the U.S. were settled farmers living in villages. Disease may well have been the biggest killer of Indians, but some of the diseases were intentionally transmitted by whites, and the diseases still left plenty of Indians to kill by more direct means. It was only in 1916 that the last "wild" Indian in the United States (the Yahi Indian known as Ishi) died, and frank and unapologetic memoirs by the white killers of his tribe were still being published as recently as 1923.

In short, Americans romanticize the white versus Indian conflict as battles of grown men on horseback, fought by U.S. cavalry and cowboys against fierce nomadic bison hunters able to offer strong resistance. The conflict is more accurately described as one race of civilian peasant farmers exterminating another. We Americans remember with outrage our own losses at the Alamo (around 200 dead), on the battleship U.S.S. *Maine* (260 dead), and at Pearl Harbor (about 2,200 dead), the incidents that galvanized our support for the Mexican War, Spanish-American War, and World War II respectively. Yet these numbers of dead are dwarfed by the forgotten losses that we inflicted on the Indians. Introspection shows us how, in rewriting our great national tragedy, we like so many modern peoples reconciled genocide with a universal code of ethics. The solution was to plead self-defense and overriding principle, and to view the victims as savage animals.

OUR REWRITING OF AMERICAN HISTORY stems from the aspect of genocide that is of greatest practical importance in preventing it: its psychological effects on killers, victims, and third parties. The most puzzling question involves the effect, or rather the apparent noneffect, on third parties. On first thought, one might expect that no horror could grip public attention as much as the intentional, collective, and savage killing of many people. In reality, genocides rarely grip the public's attention in other countries, and even more rarely are interrupted by foreign intervention. Who among us paid much attention to the slaughter of Zanzibar's Arabs in 1964, or of Paraguay's Aché Indians in the 1970s?

Contrast our nonresponse to these and all the other genocides of recent decades with our strong reaction to the sole two cases of modern genocide that remain vivid in our imagination: that of the Nazis against the Jews, and (much less vivid for most people) that of the Turks against the Armenians. These cases differ in three crucial respects from the genocides we ignore: the victims were whites, with

Ishi, the last surviving Indian of the Yahi tribe of northern California. This photograph shows him starving and terrified on August 29, 1911, the day that he emerged from forty-one years of hiding in a remote canyon. Most of his tribe was massacred by white settlers between 1853 and 1870. In 1870 the sixteen survivors of the final massacre went into concealment in the Mount Lassen wilderness and continued to live as hunter-gatherers. In November 1908, when the survivors had dwindled to four, surveyors stumbled upon their camp and took all their tools, clothes, and winter food supply, with the result that three of the Yahis (Ishi's mother, his sister, and an old man) died. Ishi remained alone for three more years until he could stand it no longer and walked out to white civilization, expecting to be lynched there. In fact, he was employed by the University of California Museum at San Francisco and died of tuberculosis in 1916. The photo is from the archives of the Lowie Museum of Anthropology, University of California, Berkeley.

whom other whites identify; the perpetrators were our war enemies, whom we were encouraged to hate as evil (especially the Nazis); and there are articulate survivors in the United States who go to much effort to force us to remember. Thus, it takes a rather special constellation of circumstances to get third parties to focus on genocide.

The strange passivity of third parties is exemplified by that of governments, whose actions reflect collective human psychology. While the United Nations in 1948 adopted a Convention on Genocide that declared it a crime, the UN has never taken serious steps to prevent, halt, or punish it, despite complaints lodged before the UN against ongoing genocides in Bangladesh, Burundi, Cambodia, Paraguay, and Uganda. To a complaint lodged against Uganda at the height of Idi Amin's terror, the UN Secretary-General responded only by asking Amin himself to investigate. The United States is not even among the nations that ratified the UN Convention on Genocide.

Is our puzzling nonresponse because we did not know, or could not find out, about ongoing genocides? Certainly not: many genocides of the 1960s and 1970s received detailed publicity at the time, including those in Bangladesh, Brazil, Burundi, Cambodia, East Timor, Equatorial Guinea, Indonesia, Lebanon, Paraguay, Rwanda, Sudan, Uganda, and Zanzibar. (The casualties in Bangladesh and Cambodia each topped a million.) For example, in 1968 the Brazilian government filed criminal charges against 134 of the 700 employees of its Indian Protection Service for their acts in exterminating Amazonian Indian tribes. Among the acts detailed in the 5,115-page Figueiredo report by Brazil's attorney general, and announced at a press conference by Brazil's minister of the interior, were the following: killing of Indians by dynamite, machine guns, arsenic-laced sugar, and intentionally introduced smallpox, influenza, tuberculosis, and measles; kidnapping of Indian children as slaves; and the hiring of professional killers of Indians by land-development companies. Accounts of the Figueiredo report appeared in the American and British press, but failed to stimulate much reaction.

One might thus conclude that most people simply do not care about injustice done to other people, or regard it as none of their business. This is undoubtedly part of the explanation, but not all of it. Many people care passionately about some injustices, such as apartheid in South Africa; why not also about genocide? This question was addressed poignantly to the Organization of African States by Hutu victims of the Tutsi in Burundi, where somewhere between 80,000 and 200,000 Hutu were killed in 1972: "Tutsi apartheid is established more ferociously than the apartheid of Vorster, more inhumanly than Portuguese colonialism. Outside of Hitler's Nazi movement, there is nothing to compete with it in world history. And the peoples of Africa say nothing. African heads of state receive the executioner Micombero [president of Burundi, a Tutsi] and clasp his hand in fraternal greeting. Sirs, heads of state, if you wish to help the African peoples of Namibia, Zimbabwe, Angola, Mozambique, and Guinea-Bissau to liberate themselves from their white oppressors, you have no right to let Africans murder other Africans. . . . Are you waiting until the entire Hutu ethnic group of Burundi is exterminated before raising your voices?"

To understand this nonreaction of third parties, we need to appreciate the reaction of surviving victims. Psychiatrists who have studied witnesses of genocide, such as Auschwitz survivors, describe the effects on them as "psychological numbing." Most of us have experienced the intense and lasting pain that comes when a loved friend or relative dies a natural death, out of sight. It is virtually impossible for us to imagine the multiplied intensity of pain when one is forced to watch at close hand many loved friends and relatives being killed with extreme savagery. For the survivors, there is a shattering of the implicit belief system under which such savagery was forbidden; a sense of stigma that one must indeed be worthless to have been singled out for such cruelty; and a sense of guilt at surviving, when one's companions died. Just as intense physical pain numbs us, so does intense psychological pain: there is no other way to survive and remain sane. For me, these reactions were personified in a relative who survived two years in Auschwitz, and who remained practically unable to cry for decades afterward.

As for the reactions of the killers, those killers whose ethical code distinguishes between "us" and "them" may be able to feel pride, but those reared under a universal ethical code may share the numbing of their victims, exacerbated by the guilt of perpetration. Hundreds of thousands of Americans who fought in Vietnam suffered this numbing. Even the descendants of genocidists—descendants who have no individual responsibility—may feel a collective guilt, the mirror image of the collective labeling of victims that defines genocide. To reduce the pain of guilt, the descendants often rewrite history: witness the response of modern Americans, or that of Ms. Cobern and many other modern Australians.

We can now begin to understand better the nonreaction of third parties to genocide. Genocide inflicts crippling and lasting psychological damage on the victims and killers who experience it firsthand. But it also may leave deep scars on those who hear about it only secondhand, such as the children of Auschwitz survivors, or the psychotherapists who treat the survivors and Vietnam veterans. Therapists who have trained professionally to be able to listen to human misery often cannot bear to hear the sickening recollections of those involved in genocide. If paid professionals cannot stand it, who can blame the lay public for refusing to listen?

Consider the reactions of Robert Jay Lifton, an American psychiatrist who had already had much experience with survivors of extreme situations before he interviewed survivors of the Hiroshima A-bomb: ". . . now, instead of dealing with 'the atomic bomb problem,' I was confronted with the brutal details of actual experiences of human beings who sat before me. I found that the completion of each of these early interviews left me profoundly shocked and emotionally spent. But very soon—within a few days, in fact—I noticed that my reactions were changing. I was listening to descriptions of the same horrors, but their effect upon me had lessened. The experience was an unforgettable demonstration of the 'psychic closing off' we shall see to be characteristic of all aspects of atomic bomb exposure. . . ."

WHAT GENOCIDES can we expect from *Homo sapiens* in the future? There are plenty of obvious reasons for pessimism. The world abounds with trouble spots that seem ripe for genocide: South Africa, Northern Ireland, Yugoslavia, Sri Lanka, New Caledonia, and the Middle East, to name just a few. Totalitarian governments bent on genocide seem unstoppable. Modern weaponry permits one to kill ever larger numbers of victims, to be a killer while wearing a coat and tie, and even to effect a universal genocide of the human race.

At the same time, I see grounds for cautious optimism that the future need not be as murderous as the past. In many countries today, people of different races or religions or ethnic groups live together, with varying degrees of social justice but at least without open mass murder: for instance, Switzerland, Belgium, Papua New Guinea, Fiji, even the post-Ishi U.S.A. Some genocides have been successfully interrupted, reduced, or prevented by the efforts or anticipated reactions of third parties. Even the Nazi extermination of Jews, which we view as the most efficient and unstoppable of genocides, was thwarted in Denmark, Bulgaria, and every other occupied state where the head of the dominant church publicly denounced deportation of Jews before or as soon as it began. A further hopeful sign is that modern travel, TV, and photography enable us to see other people living ten thousand miles away as human, like us. Much as we damn twentieth-century technology, it is blurring the distinction between "us" and "them" that makes genocide possible. While genocide was considered socially acceptable or even admirable in the pre-first-contact world, the modern spread of international culture and knowledge of distant peoples has been making it increasingly hard to justify.

Still, the risk of genocide will be with us as long as we cannot bear to understand it, and as long as we delude ourselves with the belief that only rare perverts could commit it. Granted, it's hard not to go numb while reading about genocide. It's hard to imagine how we, and other nice ordinary people that we know, could bring ourselves to look helpless people in the face while killing them. I came closest to being able to imagine it when a friend whom I had long known told me of a genocidal massacre at which he had been a killer:

Kariniga is a gentle Tudawhe tribesman who worked with me in New Guinea. We shared life-threatening situations, fears, and triumphs, and I like and admire him. One evening after I had known Kariniga for five years, he told me of an episode from his youth. There had been a long history of conflict between the Tudawhes and a neighboring village of Daribi tribesmen. Tudawhes and Daribis seem quite similar to me, but Kariniga had come to view Daribis as inexpressibly vile. In a series of ambushes the Daribis finally succeeded in picking off many Tudawhes, including Kariniga's father, until the surviving Tudawhes became desperate. All the remaining Tudawhe men surrounded the Daribi village at night and set fire to the huts at dawn. As the sleepy Daribis stumbled down the steps of their burning huts, they were speared. Some Daribis succeeded in escaping to hide in the forest, where Tudawhes tracked down and killed most of them during the following weeks. But the establish-

ment of Australian government control ended the hunt before Kariniga could catch his father's killer.

Since that evening, I've often found myself shuddering as I recalled details of it—the glow in Kariniga's eyes as he told me of the
dawn massacre; those intensely satisfying moments when he finally
drove his spear into some of his people's murderers; and his tears of
rage and frustration at the escape of his father's killer, whom he still
hoped to kill someday with poison. That evening, I thought I understood how at least one nice person had brought himself to kill.
The potential for genocide that circumstances thrust on Kariniga lies
within all of us. As the growth of world population sharpens conflicts
between and within societies, humans will have more urge to kill
each other, and more effective weapons with which to do it. To listen
to first-person accounts of genocide is unbearably painful. But if we
continue to turn away and not understand it, when will it be our own
turn to become the killers, or the victims?

INDIAN POLICIES OF SOME FAMOUS AMERICANS

President George Washington. "The immediate objectives are the total
destruction and devastation of their settlements. It will be essential to
ruin their crops in the ground and prevent their planting more."

Benjamin Franklin. "If it be the Design of Providence to Extirpate
these Savages in order to make room for Cultivators of the Earth, it
seems not improbable that Rum may be the appointed means."

President Thomas Jefferson. "This unfortunate race, whom we had
been taking so much pains to save and to civilize, have by their
unexpected desertion and ferocious barbarities justified extermination and now await our decision on their fate."

President John Quincy Adams. "What is the right of the huntsman to
the forest of a thousand miles over which he has accidentally ranged
in quest of prey?"

President James Monroe. "The hunter or savage state requires a greater extent of territory to sustain it, than is compatible with the progress and just claims of civilized life ... and must yield to it."

President Andrew Jackson. "They have neither the intelligence, the industry, the moral habits, nor the desire of improvement which are essential to any favorable change in their condition. Established in the midst of another and a superior race, and without appreciating the causes of their inferiority or seeking to control them, they must necessarily yield to the force of circumstances and ere long disappear."

Chief Justice John Marshall. "The tribes of Indians inhabiting this country were savages, whose occupation was war, and whose subsistence was drawn from the forest. ... That law which regulates, and ought to regulate in general, the relations between the conqueror and conquered was incapable of application to a people under such circumstances. Discovery [of America by Europeans] gave an exclusive right to extinguish the Indian title of occupancy, either by purchase or by conquest."

President William Henry Harrison. "Is one of the fairest portions of the globe to remain in a state of nature, the haunt of a few wretched savages, when it seems destined by the Creator to give support to a large population and to be the seat of civilization?"

President Theodore Roosevelt. "The settler and pioneer have at bottom had justice on their side; this great continent could not have been kept as nothing but a game preserve for squalid savages."

General Philip Sheridan. "The only good Indians I ever saw were dead."

PART FIVE

REVERSING OUR PROGRESS OVERNIGHT

Our species is now at the pinnacle of its numbers, its geographic extent, its power, and the fraction of the Earth's productivity that it commands. That's the good news. The bad news is that we are also in the process of reversing all that progress much more rapidly than we created it. Our power threatens our own existence. We don't know whether we shall suddenly blow ourselves up before we would otherwise expire in a slow stew caused by global warming, pollution, habitat destruction, more mouths to feed, less food to feed those mouths, and extermination of other species that form our resource base. Are these dangers really new ones that have arisen since the Industrial Revolution, as widely assumed?

It's a common belief that species in a state of Nature live in balance with each other and with their environment. Predators don't exterminate their prey, nor do herbivores overgraze their plants. According to this view, humans are the unique misfit. If this were true, Nature would hold no lessons for us.

There is something to this view, insofar as species don't go extinct under natural conditions as rapidly as we are exterminating them now, except under rare circumstances. Such a rare event was the mass die-off sixty-five million years ago, possibly due to an asteroidal impact, that finished the dinosaurs. Since evolutionary multiplications of species are very slow, natural extinctions obviously must also be slow, otherwise we would have been left with no species long ago. Expressed alternatively, the vulnerable species get eliminated quickly, and what we see persisting in Nature are the robust species.

However, that broad conclusion still leaves us with many instructive examples of species' exterminating other species. Almost all known cases prove to combine two elements. First, the cases involve species reaching environments where they did not occur before, and where they encounter prey populations naïve to those invading predators. By the time that the ecological dust settles and a new equilibrium is reached, some of the newfound prey may have been exterminated. Second, the perpetrators of such exterminations prove to be so-called switching predators, which are not specialized to eat only a single prey species but can feed on many different ones. Although the predator exterminates some prey species, it survives by switching to others.

Such exterminations often occur when humans intentionally or accidentally transfer a species from one part of the globe to another. Rats, cats, goats, pigs, ants, and even snakes are among these transferred killers. For instance, during World War II a tree snake native to the Australian region was accidentally transported on ships or planes to the previously snake-free Pacific island of Guam. This predator has already exterminated or brought to the brink of extinction most of Guam's native forest-bird species, which had had no opportunity to evolve behavioral defenses against snakes. However, the snake is in no danger itself despite having virtually eliminated its bird prey, because it can switch to rats, mice, shrews, and lizards as victims. As another example, cats and foxes introduced into Australia by humans have been eating their way through Australia's small native marsupials and rats without endangering themselves, because there remain abundant rabbits and other prey species on which to feed.

We humans furnish the prime example of a switching predator. We eat everything from snails and seaweed to whales, mushrooms, and strawberries. We can overharvest some species to the point of extinction, and then just switch to other food. Hence a wave of extinctions has ensued every time that humans have reached a previously unoccupied part of the globe. The dodo, whose name has become synonymous with extinction, formerly lived on the island of Mauritius, half of whose land and freshwater bird species became extinct following the island's discovery in 1507. Dodos in particular were big, edible, flightless, and easily caught by hungry sailors. Hawaiian bird species similarly died out en masse following Hawaii's discovery by Polynesians fifteen hundred years ago, as did America's large mammal species after ancestral Indians arrived eleven thousand years ago. Extinction waves have also accompanied major improvements of hunting technology in lands long occupied by humans. For example, wild populations of the Arabian oryx, a beautiful antelope of the Near East, survived one million years of human hunting, only to succumb to high-powered rifles in 1972.

There are numerous animal precedents for our propensity to exterminate individual prey species but to sustain ourselves by switching to others. Is there any precedent for an animal population's destroying its entire resource base and eating its way into extinction? This outcome is uncommon, because animal numbers are regulated by many factors that tend automatically to lower birth rates or increase death rates when the animal is too numerous for its food supply, and vice versa when it is rare. For example, mortality due to external factors like predators, diseases, parasites, and starvation tends to increase at high population densities. High densities also trigger responses of the animal itself, such as infanticide, postponed breeding, and increased aggression. These responses and external factors generally reduce the animal's population and relieve its pressure on its resources before they can be exhausted.

Nevertheless, some animal populations actually have eaten themselves into extinction. One example involves the progeny of twenty-nine reindeer that were introduced in 1944 to St. Matthew Island in the Bering Sea. By 1963 they had multiplied to six thousand. But reindeer depend for food on slow-growing lichens, which on St. Matthew had no chance to recover from reindeer grazing, since the animals had nowhere to migrate. When a harsh winter struck in 1963–64, all the animals except forty-one females and one sterile male starved to death, leaving a doomed population on an island littered with thousands of skeletons. A similar example was the introduction of rabbits to Lisianski Island west of Hawaii in the first decade of this century. Within a decade the rabbits had eaten themselves into oblivion by consuming every plant on the island except two morning glories and a tobacco patch.

These and other similar examples of ecological suicide all involve populations that suddenly became free of the usual factors regulating their numbers. Rabbits and reindeer are normally subject to predators, and reindeer on continents use migration as a safety valve to leave an area and allow its vegetation to recover. But Lisianski and St. Matthew Islands lacked predators, and emigration was impossible, so that the animals bred and ate unchecked.

On reflection, it is clear that the entire human species has been equally successful in recently escaping from the former controls on our numbers. We eliminated predation on us long ago; twentieth-century medicine has greatly reduced our mortality from infectious disease; and some of our leading behavioral techniques of population control, such as infanticide, chronic war, and sexual abstinence, have become socially unacceptable. Our population is now doubling about every thirty-five years. Granted, that's not as fast as the St. Matthew reindeer. Island Earth is bigger than St. Matthew Island, and some of our resources are more elastic than lichens (though other resources, like oil, are less elastic). But the qualitative conclusion remains the same: no population can grow indefinitely.

Thus, our present ecological predicament has familiar animal precursors. Like many switching predators, we exterminate some prey species when we colonize a new environment or acquire new destructive power. Like some animal populations that suddenly escaped their former limits on growth, we risk destroying ourselves by destroying our resource base. What about the view that we were in a state of relative ecological equilibrium until the Industrial Revolution, and that only since then have we begun seriously to exterminate species and overexploit our environment? That Rousseauan fantasy will be taken up in the remaining three chapters of this book.

We'll first examine the widespread belief in a former Golden

Age, when we supposedly lived as noble savages practicing a conservation ethic and in harmony with Nature. In reality, mass extinctions have coincided with each major extension of human lebensraum during the last ten thousand years and possibly much longer. Our direct responsibility for the extinctions is clearest in the case of the most recent expansions, for which the evidence is still fresh: Europeans' expansion over the globe since 1492, and the somewhat earlier colonization of oceanic islands by Polynesians and the Malagasy. Older expansions such as the first human occupation of the Americas and Australia were also accompanied by mass extinctions, though the trail of evidence has had much more time to fade and hence conclusions about cause-and-effect are necessarily weaker.

It's not just that the Golden Age was blackened by mass extinctions. While no large human population has eaten itself out of existence, some populations on small islands have done so, and many large populations have damaged their resources to the point of economic collapse. The clearest examples come from isolated cultures, such as the Easter Island and Anasazi civilizations. But environmental factors also drove the major shifts in western civilization, including the successive collapses of Middle Eastern, then Greek, then Roman hegemony. Hence self-destructive abuse of our environment, far from being a modern invention, has long been a prime mover of human history.

We'll then look more closely at the biggest, most dramatic, and most controversial of these "Golden Age mass extinctions." Around eleven thousand years ago most of the large mammals of two entire continents, North America and South America, became extinct. Around the same time appears the first unequivocal evidence for human occupation of the Americas, by the ancestors of American Indians. It was the biggest expansion of human territory since *Homo erectus* spread out of Africa to colonize Europe and Asia a million years ago. The temporal coincidence between the first Americans and the last big American mammals, the lack of mass extinctions elsewhere in the world at that same time, and proofs that some of the now-extinct beasts were hunted have suggested what is termed the New World blitzkrieg hypothesis. According to this interpretation, as the first wave of human hunters multiplied and spread from Canada to Patagonia, they encountered big animals that had never seen humans before, and they exterminated as they marched. While this theory's critics are at least as numerous as its backers, we'll try to make sense of the debate.

Finally, we'll seek to put approximate numbers on the count of species that we have already driven into extinction. We shall start with the firmest numbers: the species whose extinctions occurred in modern times and were well documented, and for which the search for survivors has been so thorough as to leave no doubt that there are no survivors. Next come estimates of three less certain numbers: the modern species that haven't been seen alive for some time and that became extinct before anyone was aware of it; the modern species that haven't even been "discovered" and named; and the species that humans exterminated before the rise of modern science. That background will let us appraise the main mechanisms by which we exterminate, and the number of species that we are likely to exterminate within my sons' lifetime—if we proceed at our current rate.

CHAPTER 17

The Golden Age That
Never Was

*Every part of the earth is sacred to
my people. Every shining pine need-
le, every sandy shore, every mist in the
dark woods, every clearing and hum-
ming insect is holy in the memory and
experience of my people. . . . The white
man . . . is a stranger who comes in the
night and takes from the land whatever he needs. The earth is not his
brother but his enemy. . . . Continue
to contaminate your bed, and you will
one night suffocate in your own waste.*

—From a letter written in 1855 to President Franklin
Pierce by Chief Seattle of the Duwanish tribe
of American Indians.

E NVIRONMENTALISTS SICKENED BY THE DAMAGE THAT INDUSTRIAL
societies are wreaking on the world often look to the past as a Golden
Age. When Europeans began to settle America, the air and rivers
were pure, the landscape green, the Great Plains teeming with bison.
Today we breathe smog, worry about toxic chemicals in our drinking
water, pave over the landscape, and rarely see any large wild animal.
Worse is surely to come. By the time my infant sons reach retirement

age, half of the world's species will be extinct, the air radioactive, and the seas polluted with oil.

Undoubtedly, two simple reasons go a long way toward explaining our worsening mess: modern technology has far more power to cause havoc than did the stone axes of the past, and far more people are alive now than ever before. But a third factor may also have contributed: a change in attitudes. Unlike modern city dwellers, at least some preindustrial peoples—like the Duwanish, whose chief I quoted—depend on and revere their local environment. Stories abound of how such peoples are in effect practicing conservationists. As a New Guinea tribesman once explained to me, "It's our custom that if a hunter one day kills a pigeon in one direction from the village, he waits a week before hunting for pigeons again, and then goes in the opposite direction." We're only beginning to realize how sophisticated the conservationist policies of so-called primitive peoples actually are. For instance, well-intentioned foreign experts have made deserts out of large areas of Africa. In those same areas, local herders had thrived for uncounted millennia by making annual nomadic migrations, which ensured that land never became overgrazed.

The nostalgic outlook shared till recently by most of my environmentalist colleagues and me is part of a human tendency to view the past as a Golden Age in many other respects. A famous exponent of this outlook was the eighteenth-century French philosopher Jean-Jacques Rousseau, whose *Discourse on the Origin of Inequality* traced our degeneration from the Golden Age to the human misery that Rousseau saw around him. When eighteenth-century European explorers encountered preindustrial peoples like Polynesians and American Indians, those peoples became idealized in European salons as "noble savages" living in a continued Golden Age, untouched by such curses of civilization as religious intolerance, political tyranny, and social inequality.

Even now, the days of classical Greece and Rome are widely considered to be the Golden Age of western civilization. Ironically, the Greeks and Romans also saw themselves as degenerates from a past Golden Age. I can still recite half-conscious those lines of the Roman poet Ovid that I memorized in tenth-grade Latin, "Aurea prima sata est aetas, quae vindice nullo. . . .": "First came the Golden Age, when men were honest and righteous of their own free will." Ovid went on to contrast those virtues with the rampant treachery and warfare of his own times. I have no doubt that any humans still alive in the radioactive soup of the twenty-second century will write equally nostalgically about our own era, which will then seem untroubled by comparison.

Given this widespread belief in a Golden Age, some recent discoveries by archaeologists and paleontologists have come as a shock. It's now clear that preindustrial societies have been exterminating species, destroying habitats and undermining their own existence for thousands of years. Some of the best-documented examples involve Polynesians and American Indians, the very peoples most often cited as exemplars of environmentalism. Needless to say, this revisionist view is hotly contested, not only in the halls of academia but also among lay people in Hawaii, New Zealand, and other areas with large Polynesian or Indian minorities. Are the new "discoveries" just one more piece of racist pseudo-science by which white settlers seek to justify dispossessing indigenous peoples? How can the discoveries be reconciled with all the evidence for conservationist practices by modern preindustrial peoples? If the discoveries are correct, can we use them as case histories to help us predict the fate that our own environmental policies may bring upon us? Can the recent findings explain some otherwise mysterious collapses of ancient civilizations, like those of Easter Island or the Maya Indians?

Before we can answer these controversial questions, we need to understand the new evidence belying the assumed past Golden Age of environmentalism. Let's first consider evidence for past waves of exterminations, then evidence for past destruction of habitats.

WHEN BRITISH COLONISTS began to settle New Zealand in the 1800s, they found no native land mammals except bats. That wasn't surprising: New Zealand is a remote island lying much too far from the continents for flightless mammals to reach. However, the colonists' plows uncovered the bones and eggshells of large birds that were then already extinct but that the Maori (the earlier Polynesian settlers of New Zealand) remembered by the name "moa." From complete skeletons, some of them evidently recent and still retaining skin and feathers, we have a good idea how moas must have looked alive: ostrichlike birds comprising a dozen species, and ranging from little ones "only" three feet high and forty pounds in weight up to giants of five hundred pounds and ten feet tall. Their food habits can be inferred from preserved gizzards containing twigs and leaves of dozens of plant species, showing them to have been herbivores. They used to be New Zealand's equivalents of big mammalian herbivores like deer and antelope.

While the moas are New Zealand's most famous extinct birds, many others have been described from fossil bones, totaling at least twenty-eight species that disappeared before Europeans arrived. Quite a few besides the moas were big and flightless, including a big duck, a giant coot, and an enormous goose. These flightless birds were descended from normal birds that had flown to New Zealand and that had then evolved to lose their expensive wing muscles in a land free of mammalian predators. Others of the vanished birds, such as a pelican, a swan, a giant raven, and a colossal eagle, were perfectly capable of flight.

Weighing up to thirty pounds, the eagle was by far the biggest and most powerful bird of prey in the world when it was alive. It dwarfed even the largest hawk now in existence, tropical America's harpy eagle. The New Zealand eagle would have been the sole predator capable of attacking adult moas. Although some moas were nearly twenty times heavier than the eagle, it still could have killed them by taking advantage of the moas' erect two-legged posture, crippling them with an attack on the long legs, then killing them with an attack on the head and long neck, and finally remaining for many days to consume the carcasses, just as lions take their time at consuming a giraffe. The eagle's habits may explain the many headless moa skeletons that have been found.

Up to this point I've discussed New Zealand's big extinct animals. But fossil hunters have also discovered the bones of small animals the size of mice and rats. Scampering or crawling on the ground were at least three species of flightless or weak-flying songbirds, several frogs, giant snails, many giant cricketlike insects up to double the weight of a mouse, and strange mouselike bats that rolled up their wings and ran. Some of these little animals were completely extinct by the time that Europeans arrived. Others still survived on small offshore islands near New Zealand, but their fossil bones show that they were formerly abundant on the New Zealand mainland. Collectively, all these now-extinct species that had evolved in isolation would have provided New Zealand with the ecological equivalents of the continents' flightless mammals that had never arrived: moas instead of deer, flightless geese and coots instead of rabbits, big crickets and little songbirds and bats instead of mice, and colossal eagles instead of leopards.

Fossils and biochemical evidence indicate that the moas' ancestors had reached New Zealand millions of years ago. When and why, after surviving for so long, did the moas finally become extinct? What disaster could have struck so many species as different as crickets, eagles, ducks, and moas? Specifically, were all these strange creatures still alive when the ancestors of the Maoris arrived around A.D. 1000?

At the time that I first visited New Zealand in 1966, the received wisdom was that moas had died out because of a change in climate, and that any moa species surviving to greet the Maoris were on their figurative last legs. New Zealanders took it as dogma that Maoris were conservationists and didn't exterminate the moas. There is still no doubt that Maoris, like other Polynesians, used stone tools, lived mainly by farming and fishing, and lacked the destructive power of modern industrial societies. At most, it was assumed, Maoris might have given the coup de grace to populations already on the verge of extinction. But three sets of discoveries have demolished this conviction.

First, much of New Zealand was covered with glaciers or cold tundra during the last Ice Age, ending about ten thousand years ago. Since then, the New Zealand climate has become much more favorable, with warmer temperatures and the spread of magnificent forests. The last moas died with their gizzards full of food, and enjoying the best climate that they had seen for tens of thousands of years.

Second, radiocarbon-dated bird bones from dated Maori archaeological sites prove that all known moa species were still present in abundance when the first Maoris stepped ashore. So were the extinct geese, ducks, swan, eagle, and other birds now known only from fossil bones. Within a few centuries, the moas and most of those other birds were extinct. It would have been an incredible coincidence if individuals of dozens of species that had occupied New Zealand for millions of years selected the precise geological moment of human arrival as the occasion to expire in synchrony.

Finally, more than a hundred large archaeological sites are

known—some of them covering dozens of acres—where Maoris cut up prodigious numbers of moas, cooked them in earth ovens, and discarded the remains. They ate the meat, used the skins for clothing, fashioned bones into fishhooks and jewelry, and blew out the eggs for use as water containers. During the nineteenth century moa bones were carted away from these sites by the wagonload. The number of moa skeletons in known Maori moa-hunter sites is estimated to be between 100,000 and 500,000, about ten times the number of moas likely to have been alive in New Zealand at any instant. Maoris must have been slaughtering moas for many generations.

Hence it is now clear that Maoris exterminated moas, at least partly by killing them, partly by robbing their nests of eggs, and probably partly as well by clearing some of the forests in which moas lived. Anyone who has hiked in New Zealand's rugged mountains will initially be incredulous at this thought. Just picture those travel posters of New Zealand's Fiordland, with its steep-walled gorges ten thousand feet deep, its four hundred inches of annual rainfall, and its cold winters. Even today, full-time professional hunters armed with telescopic rifles and operating from helicopters can't control the numbers of deer in those mountains. How could the few thousand Maoris living on New Zealand's South Island and Stewart Island, armed only with stone axes and clubs and operating on foot, have hunted down the last moas?

But there would have been a crucial difference between deer and moas. Deer have been selected for tens of thousands of generations to flee from human hunters, while moas had never seen humans until Maoris arrived. Like the naïve animals of the Galápagos Islands today, moas were probably tame enough for a hunter to walk up to one and club it. Unlike deer, moas may have had sufficiently low reproductive rates for a few hunters visiting a valley only once every couple of years to kill moas faster than they could breed. That's precisely what's happening today to New Guinea's largest surviving native mammal, a tree kangaroo in the remote Bewani Mountains. In areas settled by people, tree kangaroos are nocturnal, incredibly shy, live in trees, and are thus far harder to hunt than moas would have been. Despite all that, and despite the very small human population of the Bewanis, the cumulative effects of occasional hunting parties— literally one visit per valley per several years—have sufficed to bring this kangaroo to the verge of extinction. Having seen it happen to tree kangaroos, I now have no difficulty understanding how it happened to moas.

Not just moas but also all of New Zealand's other extinct bird species were still alive when Maoris landed. Most were gone a few centuries later. The larger ones—the swan and pelican, the flightless goose and coot—were surely hunted for food. The giant eagle, however, may have been killed by Maoris in self-defense. What do *you* think happened when that eagle, specialized at crippling and killing two-legged prey between three and ten feet tall, did when it saw its first six-foot-tall Maoris? Even today, Manchurian eagles trained for hunting occasionally kill their human handlers, but the Manchurian birds were mere dwarfs beside New Zealand's giant, which was preadapted to become a man killer.

Surely, though, neither self-defense nor hunting for food explains the rapid disappearance of New Zealand's peculiar crickets, snails, wrens, and bats. Why were so many of those species exterminated, either throughout their range or else everywhere except on some offshore islands? Deforestation may be part of the answer, but the major reason was the other hunters that Maoris intentionally or accidentally brought with them: rats! Just as moas that evolved in the absence of humans were defenseless against humans, so, too, small insular animals that evolved in the absence of rats were defenseless against rats. We know that the rat species spread by Europeans played a major role in modern exterminations of many bird species on Hawaii and other previously rat-free oceanic islands. For example, when rats finally reached Big South Cape Island off New Zealand in 1962, within three years they exterminated or decimated the populations of eight bird species and a bat. That's why so many New Zealand species are restricted today to rat-free islands, the sole places where they could survive when the rat tide accompanying the Maoris swept over the New Zealand mainland.

Thus, when the Maoris landed, they found an intact New Zealand biota of creatures so strange that we would dismiss them as science-fiction fantasies if we did not have their fossilized bones to convince us of their former existence. The scene was as close as we shall ever get to what we might see if we could reach another fertile planet on which life had evolved. Within a short time, much of that community had collapsed in a biological holocaust, and some of the remaining community collapsed in a second holocaust following the arrival of

Europeans. The end result is that New Zealand today has about half of the bird species that greeted the Maoris, and many of the survivors are either now at risk of extinction or else confined to islands with few introduced mammalian pests. A few centuries of hunting had sufficed to end millions of years of moa history.

NOT ONLY ON NEW ZEALAND but on all other remote Pacific islands where archaeologists have looked recently, bones of many now-extinct bird species have been found at sites of the first settlers, proving there too that the bird extinctions and human colonizations were somehow related. From all the main islands of Hawaii, paleontologists Storrs Olson and Helen James of the Smithsonian Institution have identified fossil bird species that disappeared during the Polynesian settlement that began around A.D. 500. The fossils include not only small honey creepers related to species still present but also bizarre flightless geese and ibises with no living close relatives at all. While Hawaii is notorious for its bird extinctions following European settlement, this earlier extinction wave had been unknown until Olson and James began publishing their discoveries in 1982. The known extinctions of Hawaiian birds before Captain Cook's arrival now total the incredible number of at least fifty species, nearly one-tenth the number of bird species breeding on mainland North America.

That's not to say that all these Hawaiian birds were hunted out of existence. Although geese probably were indeed exterminated by overhunting, like the moas, small songbirds are more likely to have been eliminated by rats that arrived with the first Hawaiians, or else by destruction of forests that Hawaiians cleared for agriculture. Similar discoveries of extinct birds at archaeological sites of early Polynesians have also been made on Tahiti, Fiji, Tonga, New Caledonia, the Marquesas Islands, Chatham Islands, Cook Islands, Solomon Islands, and Bismarck Archipelago.

An especially intriguing collision of birds and Polynesians took place on Henderson Island, an extremely remote speck of land lying in the tropical Pacific Ocean 125 miles east of Pitcairn Island, which is famous for its own isolation. (Recall that Pitcairn is so remote that the mutineers who wrested the H.M.S. *Bounty* from Captain Bligh lived undetected on Pitcairn for eighteen years until the island was rediscovered.) Henderson consists of jungle-covered coral riddled with crevices and totally unsuited for agriculture. Naturally, the island is now uninhabited and has been ever since Europeans first saw it in 1606. Henderson has often been cited as one of the world's most pristine habitats, totally unaffected by humans.

It was therefore a big surprise when Olson and fellow paleontologist David Steadman recently identified bones of two large species of pigeons, one smaller pigeon, and three seabirds that had gone extinct on Henderson some time between five hundred and eight hundred years ago. The same six species or close relatives had already been found in archaeological sites on several inhabited Polynesian islands, where it was clear how they could have been exterminated by people. The apparent contradiction of birds' also being exterminated by humans on uninhabited, seemingly uninhabitable Henderson was solved by the discovery there of former Polynesian sites with hundreds of cultural artifacts, proving that the island had actually been occupied by Polynesians for several centuries. At those same sites, along with the bones of the six bird species that were exterminated on Henderson, were the bones of other bird species that survived, plus many fish.

Thus, those early Polynesian colonists of Henderson evidently subsisted mainly on pigeons, seabirds, and fish until they had decimated the bird populations, at which point they had destroyed their food supply and either starved or else abandoned the island. The Pacific contains at least eleven other "mystery" islands besides Henderson, islands that were uninhabited on European discovery but that showed archaeological evidence of former occupation by Polynesians. Some of those islands had been settled for hundreds of years before their human population finally died out or left. All were small or in other respects marginally suitable for agriculture, leaving human settlers heavily dependent on birds and other animals for food. Given the widespread evidence for overexploitation of wild animals by early Polynesians, not only Henderson but the other mystery islands as well may represent the graveyards of human populations that ruined their own resource base.

LEST I LEAVE the impression that Polynesians were in any way unique as preindustrial exterminators, let's now jump nearly halfway around the globe to the world's fourth-largest island, Madagascar, lying in the Indian Ocean off the coast of Africa. When Portuguese explorers arrived around 1500, they found Madagascar already occupied by people now called the Malagasy. On geographic grounds, you might have expected their language to be related to African languages spoken a mere two hundred miles to the west, on the coast of Mozambique. Astonishingly, though, it actually proved to belong to a group of languages spoken on the Indonesian island of Borneo, on the opposite side of the Indian Ocean thousands of miles to the northeast. Physically, the Malagasy range in appearance from typical Indonesians to typical blacks of East Africa. These paradoxes are due to the Malagasy's having arrived between one thousand and two thousand years ago, as a result of Indonesian traders' voyaging around the Indian Ocean coastline to India and eventually to East Africa. In Madagascar they proceeded to build a society based on herding cattle, goats, and pigs, farming, and fishing, and linked to the East African coast by Muslim traders.

As interesting as Madagascar's people are the wild animals that it has—and those that it lacks. Living in enormous abundance on the nearby African mainland are many species of large and conspicuous beasts that run on the ground and are active by day—the antelopes, ostriches, zebras, baboons, and lions that draw modern tourists to East Africa. None of these animals, and no animals remotely equivalent to them, have occurred on Madagascar in modern times. They were kept out by the two hundred miles of sea separating Madagascar from Africa, just as the sea also kept Australia's marsupials from reaching New Zealand. Instead, Madagascar supports two dozen species of small, monkeylike primates called lemurs, weighing only up to twenty pounds and mostly active at night and living in trees. Various species of rodents, bats, insectivores, and relatives of mongooses also occur, yet the largest still weighs only about twenty-five pounds.

However, littering Madagascar's beaches are proofs of vanished giant birds, in the form of countless eggshells the size of a soccer ball. Eventually, bones turned up not only of the birds that laid those eggs but also of a remarkable suite of vanished large mammals and reptiles. The egg makers were half-a-dozen species of flightless birds up to ten feet tall and weighing up to one thousand pounds, like moas and ostriches but more massively built and hence now termed elephant birds. The reptiles were two species of giant land tortoises with shells about a yard long, and formerly very common, as indicated by the abundance of their bones. More diverse than either these large birds or reptiles were a dozen species of lemurs up to the size of a gorilla, and all larger than or at least as large as the largest surviving lemur species. To judge from the small size of the eye orbits in their skulls, all or most of the extinct lemurs were probably diurnal rather than nocturnal. Some of them evidently lived on the ground like baboons, while others climbed in trees like orangutans and koala bears.

As if all this were not enough, Madagascar also yielded the bones of an extinct "pygmy" hippopotamus ("only" the size of a cow), an aardvark, and a big mongoose-related carnivore built like a short-legged puma. Taken together, these extinct large animals formerly gave Madagascar the functional equivalents of the surviving large beasts for which tourists still flock to African game parks—just as did New Zealand's moas and other strange birds. The tortoises, elephant birds, and pygmy hippos would have been the herbivores in place of antelope and zebras; the lemurs would have replaced the baboons and great apes; and the mongoose-related carnivore made do for a leopard or scaled-down lion.

What happened to all these big extinct mammals, reptiles, and birds? We can be confident that at least some of them were alive to delight the eyes of the first arriving Malagasy, who used elephant-bird eggshells as water containers and discarded butchered bones of the pygmy hippo and some of the other species in their garbage heaps. In addition, the bones of all the other extinct species are known from fossil sites only a few thousand years old. Since they must have evolved and survived for millions of years until then, it is unlikely that all those animals had the foresight to give up the ghost just in those last few moments before hungry humans showed up. In fact, a few may still have been holding out in remote parts of Madagascar when Europeans arrived, since the seventeenth-century French governor Flacourt was given descriptions of an animal suggestive of the gorilla-sized lemur. The elephant birds may have survived long enough to become known to Arab traders in the Indian Ocean, and to give rise to the account of the rok (a giant bird) in the tale of Sindbad the Sailor.

Certainly some and probably all of Madagascar's vanished giants were somehow exterminated by the activities of the early Malagasy. It's not hard to understand why the elephant birds went extinct, since their eggshells made such convenient two-gallon jerrycans. While the Malagasy were herders and fishermen rather than big-game hunters, the other big animals would have been as easy prey as New Zealand's moas, since they had never seen humans before. That's presumably why the easy-to-see, easy-to-catch lemurs big enough to be worth the effort of butchering them—the large, diurnal, terrestrial species—all went extinct, while the small, nocturnal, tree-living ones all survived.

However, unintended by-products of Malagasy activities probably killed more big animals than did hunting. Fires lit to clear forest for pasture and to stimulate growth of new grass each year would have destroyed habitats on which the beasts depended. Grazing cattle and goats also transformed habitats, as well as competing directly with grazing tortoises and elephant birds for food. Introduced dogs and pigs would have preyed on ground-dwelling animals, their young, and their eggs. By the time the Portuguese arrived, Madagascar's once-abundant elephant birds had all been reduced to eggshells covering the beaches, skeletons in the ground, and vague memories of roks.

MADAGASCAR AND POLYNESIA merely provide well-documented examples of the extinction waves that probably unfolded on all large oceanic islands colonized by people before the European expansion of the last five hundred years. All such islands where life had evolved in the absence of humans used to have unique species of big animals that modern zoologists never saw alive. Mediterranean islands like Crete and Cyprus had pygmy hippos and giant tortoises (just as did Madagascar), as well as dwarf elephants and dwarf deer. The West Indies lost monkeys, ground sloths, a bear-sized rodent, and owls of several sizes: normal, giant, colossal, and titanic. It seems likely that these big birds, mammals, and tortoises too somehow succumbed to the first Mediterranean peoples or American Indians to reach their islands. Nor were birds, mammals, and tortoises the only victims: lizards, frogs, snails, and even large insects disappeared as well, comprising thousands of species when one adds up all oceanic islands. Olson describes these insular extinctions as "one of the swiftest and most profound biological catastrophes in the history of the world." However, we won't be sure that humans were responsible until the bones of the last animals and the remains of the first people have been dated more exactly for other islands, as has already been done for Polynesia and Madagascar.

In addition to these preindustrial extermination waves on islands, other species may have fallen victim to extermination waves on continents in the more distant past. About eleven thousand years ago, around the probable time that the first ancestors of American Indians reached the New World, most large species of mammals became extinct throughout all of North and South America. A long-standing debate has raged over whether these big mammals were done in by Indian hunters, or whether they just happened to succumb to climate changes around the same time. I'll explain in the next chapter why I personally think that hunters did it. However, it's much harder to pinpoint dates and causes of events that happened around eleven thousand years ago than it is for recent events like the Maori versus moa collision within the past one thousand years. Similarly, within the past fifty thousand years Australia was colonized by the ancestors of today's Aboriginal Australians *and* lost most of its species of big animals. Those animals included giant kangaroos, the "marsupial lion," and "marsupial rhinos" (known as diprotodonts), plus giant lizards, snakes, crocodiles, and birds. However, we still don't know whether the arrival of Australia's humans somehow caused the disappearance of Australia's big animals. Although it's now reasonably certain that the first preindustrial peoples to reach islands wrought havoc among island species, the jury is still out on the question whether this also happened on continents.

FROM ALL THIS EVIDENCE that the Golden Age was tarnished by exterminations of species, let's now turn to evidence for destruction of habitats. Three dramatic examples involve famous archaeological puzzles: the giant stone statues of Easter Island, the abandoned pueblos of the U.S. Southwest, and the ruins of Petra.

An aura of mystery has clung to Easter Island ever since it and its Polynesian inhabitants were "discovered" by the Dutch explorer Jakob Roggeveen in 1722. Lying in the Pacific Ocean 2,300 miles west of Chile, Easter surpasses even Henderson as one of the world's most isolated scraps of land. Hundreds of statues, weighing up to eighty-five tons and up to thirty-seven feet tall, were carved from volcanic quarries, somehow transported several miles, and raised to an upright position on platforms by people without metal or wheels and with no power source other than human muscle. Even more statues remain unfinished in the quarries, or lie finished but abandoned between the quarries and platforms. The scene today is as if the carvers and movers had suddenly walked off the job, leaving an eerily silent landscape.

When Roggeveen arrived, many statues were still standing, though new ones were no longer being carved. By 1840 all the erected statues had been deliberately toppled by the Easter Islanders themselves. How were such huge statues transported and erected, why were they eventually toppled, and why had carving ceased?

The first of those questions was answered when living Easter Islanders showed Thor Heyerdahl how their ancestors had used logs as rollers to transport the statues and then as levers to erect them. The other questions were solved by subsequent archaeological and paleontological studies that revealed Easter's gruesome history. When Polynesians settled Easter around A.D. 400, the island was covered by forest that they gradually proceeded to clear, in order to plant gardens and to obtain logs for canoes and for erecting statues. By around 1500 the human population had built up to about 7,000 (over 150 per square mile), about 1,000 statues had been carved, and at least 324 of those statues had been erected. But—the forest had been destroyed so thoroughly that not a single tree survived.

An immediate result of this self-inflicted ecological disaster was that the islanders no longer had the logs needed to transport and erect statues, so carving ceased. But deforestation also had two indirect consequences that brought starvation: soil erosion, hence lower crop yields, plus lack of timber to build canoes, hence less protein available from fishing. As a result, the population was now greater than Easter could support, and island society collapsed in a holocaust of internecine warfare and cannibalism. A warrior class took over; spear points manufactured in huge quantities came to litter the landscape; the defeated were eaten or enslaved; rival clans pulled down each other's statues; and people took to living in caves for self-protection. What had once been a lush island supporting one of the world's most remarkable civilizations deteriorated into the Easter Island of today:

a barren grassland littered with fallen statues, and supporting less than one-third of its former population.

OUR SECOND CASE STUDY of preindustrial habitat destruction involves the collapse of one of the most advanced Indian civilizations of North America. When Spanish explorers reached the U.S. Southwest, they found gigantic multistory dwellings (pueblos) standing uninhabited in the middle of treeless desert. For example, the 650-room dwelling at Chaco Canyon National Monument in New Mexico was 5 stories high, 670 feet long, and 315 feet wide, making it the largest building ever erected in North America until topped by steel skyscrapers in the late nineteenth century. Navajo Indians in the region knew of the vanished builders only as "Anasazi," meaning "the Ancient Ones."

Archaeologists subsequently established that construction of the Chaco pueblos began shortly after A.D. 900, and that occupation ceased in the twelfth century. Why did the Anasazi erect a city in a barren wasteland, of all unpromising places? Where did they obtain their firewood, or the sixteen-foot-long wooden beams (200,000 of them!), that supported the roofs? Why did they then abandon the city that they had built at such enormous effort?

The conventional view, analogous to the claim that Madagascar's elephant birds and New Zealand's moas died out from natural changes in climate, attributes the abandonment of Chaco Canyon to a drought. However, a different interpretation emerges from the work of paleobotanists Julio Betancourt, Thomas Van Devender, and their colleagues, who used an ingenious technique to decipher changes in Chaco vegetation through time. Their method relied on the little rodents called packrats, which gather plants and other materials into shelters (termed middens) that they eventually abandon after fifty to one hundred years but that remain well preserved under desert conditions. The plants can be identified centuries later, and the midden can be dated by radiocarbon techniques. Thus, each midden is virtually a time capsule of the local vegetation.

By this method, Betancourt and Van Devender were able to reconstruct the following course of events. At the time that the Chaco pueblos were erected, they were not surrounded by barren desert but by pinyon-juniper woodland, with ponderosa pine forest nearby. This discovery at once solves the mystery of where the firewood and timber came from, and disposes of the apparent paradox of an advanced civilization rising from barren desert. As occupation continued at Chaco, however, the woodland and forest were cleared until the environment became the treeless wasteland that it remains today. The Indians were then having to go over ten miles to get firewood, and over twenty-five miles to get pine logs. When the pine forest had been felled, they built an elaborate road system to haul spruce and fir logs from mountain slopes over fifty miles away, relying on nothing more than their own muscle power. In addition, the Anasazi had solved the problems of agriculture in a dry environment by building irrigation systems to concentrate available water into valley bottoms. As deforestation caused progressively increasing erosion and water runoff, and as irrigation channels gradually dug gullies into the ground, the water table may finally have dropped below the level of the Anasazi fields, making irrigation without pumps impossible. Thus, while drought may have made some contribution to the Anasazi abandonment of Chaco Canyon, a self-inflicted ecological disaster was also a major factor.

OUR REMAINING EXAMPLE of preindustrial habitat destruction illuminates the gradual geographic shift in the power center of ancient western civilizations. Recall that the first center of power and innovation was the Mideast, where so many crucial developments arose—agriculture, animal domestication, writing, imperial states, battle chariots, and others. Ascendancy shifted between Assyria, Babylon, Persia, and occasionally Egypt or Turkey, but remained in or near the Mideast. With the overthrow of the Persian Empire by Alexander the Great, ascendancy moved finally westward, at first to Greece, then to Rome, and later to western and northern Europe. Why did the Mideast, Greece, and Rome in turn lose their primacy? (The transient current importance of the Mideast, resting as it does on the single resource of oil, merely emphasizes by contrast the region's modern weakness in other respects.) Why do modern superpowers include the United States and the USSR, Germany and England, Japan and China, but no longer Greece and Persia?

This geographic shift in power is too big and lasting a pattern to have arisen by accident. A plausible hypothesis attributes it to each ancient center of civilization in turn ruining its resource base. The Mideast and Mediterranean were not always the degraded landscape that they appear today. In ancient times much of this area was a lush mosaic of wooded hills and fertile valleys. Thousands of years of deforestation, overgrazing, erosion, and valley siltation converted this heartland of western civilization into the relatively dry, barren, infertile landscape that predominates today. Archaeological surveys of ancient Greece have revealed several cycles of population growth alternating with population crashes and local abandonment of human settlement. In the growth phases, terracing and dams initially protected the landscape until felling of forests, clearing of steep slopes for agriculture, overgrazing by too many livestock, and planting of crops at too short intervals overwhelmed the system. The result each time was massive erosion of the hills, flooding of the valleys, and the collapse of local human society. One such event coincided with (and may have caused) the otherwise mysterious collapse of Greece's glorious Mycenean civilization, after which Greece fell back for several centuries into a dark age of illiteracy.

The support for this view of ancient environmental destruction comes from sources such as contemporary accounts and archaeological evidence. Yet a few sequences of photographs would constitute more decisive tests than all that anecdotal evidence combined. If we had snapshots of the same Greek hillside taken at thousand-year intervals, we could identify the plants, measure the ground cover, and calculate the shift from forest to goat-proof shrubs. We could thereby put numbers on the extent of environmental degradation.

Enter middens to the rescue again. While the Mideast doesn't have packrats, it does have rabbit-sized, marmotlike animals called hyraxes that build packrat-like middens. (Surprisingly, the closest living relatives of hyraxes may be elephants.) Three Arizona scientists—Patricia Fall, Cynthia Lindquist, and Steven Falconer—studied hyrax middens at Jordan's famous lost city of Petra, which typifies the paradox of ancient western civilization. Petra is now especially familiar to moviegoing aficionados of Steven Spielberg and George Lucas, whose film *Indiana Jones and the Last Crusade* shows Sean Connery and Harrison Ford searching for the Holy Grail in Petra's magnificent rock tombs and temples amidst the desert sand. Anyone who sees those scenes of Petra must wonder how such a wealthy city

could have arisen and supported itself in such a bleak landscape. In
fact, there was already a Neolithic village near the site of Petra before
7000 B.C., and farming and herding appeared there soon after. Under
the Nabataean kingdom, of which it was the capital, Petra thrived as
a commercial center controlling trade between Europe, Arabia, and
the Orient. The city grew ever larger and richer under Roman, then
Byzantine, control. But it was subsequently abandoned and so completely forgotten that its ruins were not rediscovered until 1812. What
caused Petra's collapse?

Each hyrax midden at Petra yielded remains of up to one hundred
plant species, and the habitat prevailing when each midden's owner
was alive could be calibrated by comparing pollen proportions in the
midden with those in modern habitats. From the middens, the following trajectory was reconstructed for the degradation of Petra's
environment:

Petra lies in an area of dry Mediterranean climate not unlike that
of the wooded mountains behind my home in Los Angeles. The
original vegetation would have been a woodland dominated by oak
and pistachio trees. By Roman and Byzantine times, most of the trees
had been felled, and the surroundings had been degraded to an open
steppe, as expressed in the fact that only 18 percent of midden pollen
came from trees, the rest from low plants. (For comparison, trees
contribute 40–85 percent of the pollen in modern Mediterranean
forests, 18 percent in forest-steppe.) By A.D. 900, a few centuries after
Byzantine control of the Petra area ended, two-thirds of the remaining trees had disappeared. Even shrubs, herbs, and grasses had declined, converting the environment into the desert that we see now.
Surviving trees today have their lower branches pruned off by goats
and are scattered on goat-proof cliffs or in groves protected from
goats.

Juxtaposing these data from hyrax middens with archaeological
and literary data yields the following interpretation. Deforestation
from Neolithic to imperial times was driven by the clearing of land
for agriculture, browsing by sheep and goats, gathering of firewood,
and wood needs for house construction. Even Neolithic houses not
only were supported by massive timbers but also consumed up to
thirteen tons of firewood per house to make the plaster for the walls
and floor. The imperial population explosion quickened the pace of
forest destruction and overgrazing. Elaborate systems of channels,

pipes, and cisterns were needed to collect and store water for the orchards and city.

After Byzantine authority collapsed, orchards were abandoned and the population crashed, but land degradation continued as the remaining inhabitants became dependent on intensive grazing. The insatiable goats began to eat their way through the shrubs, herbs, and grasses. The Ottoman government decimated surviving woodlands before the First World War, to obtain the wood needed for the Hejaz Railway. I and many other moviegoers thrilled at the sight of Arab guerrillas led by Lawrence of Arabia (a.k.a. Peter O'Toole) blowing up that railway in wide-screen Technicolor, without realizing that we were watching the last act in the destruction of Petra's forests.

Petra's ravaged landscape today is a metaphor for what happened to the rest of the cradle of western civilization. The modern surrounds of Petra could no more feed a city that commanded the world's main trade routes than the modern surrounds of Persepolis could feed the capital of a superpower such as the Persian Empire once was. The ruins of those cities, and of Athens and Rome, are monuments to states that destroyed their means of survival. Nor are Mediterranean civilizations the only literate societies that committed ecological suicide. The collapse of Classic Maya civilization in Central America, and of Harappan civilization in India's Indus Valley, are other obvious candidates for ecodisasters due to an expanding human population overwhelming its environment. While courses in the history of civilization often dwell on kings and barbarian invasions, deforestation and erosion may in the long run have been more important shapers of human history.

THESE ARE SOME of the recent discoveries making the supposed past Golden Age of environmentalism look increasingly mythical. Let's now go back to the larger issues I raised at the outset. First, how can these discoveries of past environmental damage be reconciled with accounts of conservationist practices by so many modern preindustrial peoples? Obviously, not all species have been exterminated, and not all habitats have been destroyed, so the Golden Age couldn't have been all black.

I suggest the following answer to this paradox. It's still true that small, long-established, egalitarian societies tend to evolve conservationist practices, because they've had plenty of time to get to know their local environment and to perceive their own self-interest. Instead, damage is likely to occur when people suddenly colonize an unfamiliar environment (like the first Maoris and Easter Islanders); or when people advance along a new frontier (like the first Indians to reach America), so that they can just move beyond the frontier when they've damaged the region behind; or when people acquire a new technology whose destructive power they haven't had time to appreciate (like modern New Guineans, now devastating pigeon populations with shotguns). Damage is also likely in centralized states that concentrate wealth in the hands of rulers, who are out of touch with their environment. And some species and habitats are more susceptible to damage than others—such as flightless birds that never had seen humans (like moas and elephant birds), or the dry, fragile, unforgiving environments in which both Mediterranean civilization and Anasazi civilization arose.

Second, are there any practical lessons that we can learn from these recent archaeological discoveries? Archaeology is often viewed as a socially irrelevant academic discipline that becomes a prime target for budget cuts whenever money gets tight. In fact, archaeological research is one of the best bargains available to government planners. All over the world, we're launching developments that have great potential for doing irreversible damage, and that are really just more powerful versions of ideas put into operation by past societies. We can't afford the experiment of developing five counties in five different ways and seeing which four counties get ruined. Instead, it will cost us much less in the long run if we hire archaeologists to find out what happened the last time than if we go making the same mistakes again.

Here's just one example. The American Southwest has over 100,000 square miles of pinyon-juniper woodland that we are exploiting more and more for firewood. Unfortunately, the U.S. Forest Service has little data available to help it calculate sustainable yields and recovery rates in that woodland. Yet the Anasazi already tried the experiment and miscalculated, with the result that the woodland still hasn't recovered in Chaco Canyon after over eight hundred years. Paying some archaeologists to reconstruct Anasazi firewood consumption would be cheaper than committing the same mistake and ruining 100,000 square miles of the United States, as we may now be doing.

Finally, let's face the touchiest question. Today, environmentalists view people who exterminate species and destroy habitats as morally bad. Industrial societies have jumped at any excuse to denigrate preindustrial peoples, in order to justify killing them and appropriating their land. Are the purported new finds about moas and Chaco Canyon vegetation just pseudo-scientific racism that in effect is saying, Maoris and Indians don't deserve fair treatment because they were bad?

What has to be remembered is that it's always been hard for humans to know the rate at which they can safely harvest biological resources indefinitely, without depleting them. A significant decline in resources may not be easy to distinguish from a normal year-to-year fluctuation. It's even harder to assess the rate at which new resources are being produced. By the time that the signs of decline are clear enough to convince everybody, it may be too late to save the species or habitat. Thus, preindustrial peoples who couldn't sustain their resources were guilty not of moral sins, but of failures to solve a really difficult ecological problem. Those failures were tragic, because they caused a collapse in life-style for the people themselves.

Tragic failures become moral sins only if one should have known better from the outset. In that regard there are two big differences between us and eleventh-century Anasazi Indians: scientific understanding, and literacy. We know, and they didn't know, how to draw graphs that plot sustainable resource population size as a function of resource harvesting rate. We can read about all the ecological disasters of the past; the Anasazi couldn't. Yet our generation continues to hunt whales and clear tropical rain forest as if no one had ever hunted moas or cleared pinyon-juniper woodland. The past was still a Golden Age, of ignorance, while the present is an Iron Age of willful blindness.

From this point of view it's beyond understanding to see modern societies repeating the past's suicidal ecological mismanagement, with much more powerful tools of destruction in the hands of far more people. It's as if we hadn't already run that particular film many times before in human history, and as if we didn't know the inevitable outcome. Shelley's sonnet "Ozymandias" evokes Persepolis, Tikal, and Easter Island equally well; perhaps it will someday evoke to others the ruins of our own civilization:

I met a traveller from an antique land
Who said: "Two vast and trunkless legs of stone
Stand in the desert. Near them, on the sand,
Half sunk, a shattered visage lies, whose frown,
And wrinkled lip, and sneer of cold command,
Tell that its sculptor well those passions read
Which yet survive, stamped on these lifeless things,
The hand that mocked them and the heart that fed;
And on the pedestal these words appear—
'My name is Ozymandias, king of kings:
Look on my works, ye Mighty, and despair!'
Nothing beside remains. Round the decay
Of that colossal wreck, boundless and bare
The lone and level sands stretch far away."

CHAPTER 18

Blitzkrieg and Thanksgiving in the New World

THE UNITED STATES DEVOTES TWO NATIONAL HOLIDAYS, COLUMBUS Day and Thanksgiving Day, to celebrating dramatic moments in the European "discovery" of the New World. No holidays commemorate the much earlier discovery by Indians. Yet archaeological excavations suggest that, in drama, that earlier discovery dwarfs the adventures of Christopher Columbus and of the Plymouth Pilgrims. Within perhaps as little as a thousand years of finding a way through an Arctic ice sheet to cross the present Canada-U.S. border, Indians had swept down to the tip of Patagonia and populated two productive and unexplored continents. The Indians' march southward was the greatest range expansion in the history of *Homo sapiens*. Nothing remotely like it can ever happen again on our planet.

The sweep southward was marked by another drama. When Indian hunters arrived, they found the Americas teeming with big mammals that are now extinct: elephantlike mammoths and mastodonts, ground sloths weighing up to three tons, armadillolike glyptodonts weighing up to one ton, bear-sized beavers, and sabertooth cats, plus American lions, cheetahs, camels, horses, and many others.

Had those beasts survived, today's tourists in Yellowstone National Park would be watching mammoths and lions along with the bears and bison. The question of what happened at that moment of hunters-meet-beasts is still highly controversial among archaeologists and paleontologists. According to the interpretation that seems most plausible to me, the outcome was a "blitzkrieg" in which the beasts were quickly exterminated—possibly within a mere ten years at any given site. If that view is correct, it would have been the most concentrated extinction of big animals since an asteroid collision knocked off the dinosaurs sixty-five million years ago. It would also have been the first of the series of blitzkriegs that marred our supposed Golden Age of environmental innocence, and that have remained a human hallmark ever since.

THAT DRAMATIC CONFRONTATION came as the finale to a long epic in which humans, spreading out of their center of origins in Africa, occupied all the other habitable continents. Our African ancestors expanded to Asia and Europe around a million years ago, and from Asia to Australia around fifty thousand years ago, leaving North and South America as the last habitable continents still without *Homo sapiens*.

From Canada to Tierra del Fuego, American Indians today are physically more homogeneous than the inhabitants of any other continent, implying that they arrived too recently to have become very diverse genetically. Even before archaeology uncovered evidence of the first Indians, it was clear that they must have originated from Asia, because modern Indians look similar to Asiatic Mongoloids. Much recent evidence from genetics and anthropology has made that conclusion certain. A glance at a map shows that by far the easiest route from Asia to America is across the Bering Strait separating Siberia from Alaska. The last land bridge across the strait existed (with a few brief interruptions) from about twenty-five thousand to ten thousand years ago.

However, colonization of the New World required more than a land bridge: there had to be people living at the Siberian end of the bridge. Because of its harsh climate the Siberian Arctic, too, was not colonized until late in human history. Those colonists must have come from the cold temperate zones of Asia or eastern Europe, as exemplified by Stone Age hunters who lived in what is now the Ukraine and who built their houses out of neatly stacked bones of mammoths. By at least twenty thousand years ago there were mammoth hunters in the Siberian Arctic as well, and by around twelve thousand years ago stone tools similar to those of the Siberian hunters appear in Alaska's archaeological record.

After traversing Siberia and the Bering Strait, the Ice Age hunters were still separated by one more barrier from their future hunting grounds in the U.S.: a broad ice cap like that covering Greenland today, but stretching coast-to-coast across Canada. At intervals during the Ice Ages a narrow, ice-free, north-south corridor opened through this ice cap, just east of the Rocky Mountains. One such corridor closed around twenty thousand years ago, but there had apparently as yet been no human in Alaska waiting to cross it. However, when the corridor next opened around twelve thousand years ago, the hunters must have been ready, for their telltale stone tools soon thereafter appear not only at the south end of the corridor near Edmonton (Alberta) but also elsewhere south of the ice cap. At that point, hunters met America's elephants and other great beasts, and the drama began.

Archaeologists term these pioneering ancestral Indians the Clovis people, since their stone tools were first recognized at an excavation near the town of Clovis, ten miles inside New Mexico from the Texas border. However, Clovis tools or ones similar to them have been found in all forty-eight contiguous states of the U.S., and from Edmonton in the north to Mexico. Vance Haynes, a University of Arizona archaeologist, has emphasized that the tools are much like those of the earlier eastern European and Siberian mammoth hunters, with one conspicuous exception: the flattish, two-faced stone spear points were "fluted" on each face as a result of a longitudinal groove having been chipped out to make it easier to bind the stone point to the shaft. It isn't clear whether the fluted points were mounted on spears to throw by hand, on darts to hurl by a throwing stick, or on lances to thrust. Somehow, though, the points were propelled into big mammals with enough force for the points sometimes to snap in half, or else to penetrate bone. Archaeologists have dug up skeletons of mammoths and bison with Clovis points inside the rib cages, including a mammoth from southern Arizona containing a total of eight points. At excavated Clovis sites, mammoths are by far the commonest prey (to judge from their bones), but other victims include bison, mastodont, tapir, camel, horse, and bear.

Among the startling discoveries about Clovis people is the speed of their spread. All Clovis sites in the U.S. dated by the most advanced radiocarbon techniques were occupied for only a few centuries, in the period just before 11,000 years ago. A human site even at the southern tip of Patagonia is dated at about 10,500 years. Thus, within about a millennium of emerging from the ice-free corridor at Edmonton, humans had spread from coast to coast and over the entire length of the New World.

Equally startling is the rapid transformation of Clovis culture. Around 11,000 years ago Clovis points are abruptly replaced by a smaller, more finely made model now known as Folsom points (after a site near Folsom, New Mexico, where they were first identified). The Folsom points are often found associated with bones of an extinct wide-horned bison, never with the mammoths preferred by Clovis hunters.

There may be a simple reason why Folsom hunters switched from mammoths to bison: there weren't any mammoths left. There also weren't any more mastodonts, camels, horses, giant ground sloths, and several dozen other types of big mammals. In all, North America lost an astonishing 73 percent, South America 80 percent, of their genera of big mammals around this time. Many paleontologists don't blame this American extinction spasm on Clovis hunters, since there is no surviving evidence of mass slaughter—only the fossilized bones of a few butchered carcasses here and there. Instead, those paleontologists attribute the extinctions to changes of climate and habitats at the end of the Ice Ages, just around the time that Clovis hunters arrived. That reasoning puzzles me for several reasons: ice-free habitats for mammals expanded rather than contracted as glaciers yielded to grass and forest; big American mammals had already survived the ends of at least twenty-two previous Ice Ages without such an extinction spasm; and there were far fewer extinctions in Europe and Asia when the glaciers of those continents melted around the same time.

If changing climate had been the cause, one might have expected opposite effects on species preferring hot and cold climates. Instead, radiocarbon-dated fossils from the Grand Canyon show that the Shasta ground sloth and Harrington's mountain goat, derived from areas of hot and cold climates respectively, both died out within a century or two of each other, around 11,100 years ago. The sloths were common until just before their sudden extinction. In their softball-sized dung balls, still well-preserved in some southwestern U.S. caves, botanists identified remains of plants on which the last sloths chomped: Mormon tea and globe mallow, which still occur around those caves today. It is highly suspicious that both those well-fed sloths and the goats of the Grand Canyon disappeared just after Clovis hunters reached Arizona. Juries have convicted murderers on the grounds of less compelling circumstantial evidence. If climate really was what did in the sloths, we would have to credit those supposedly stupid beasts with unsuspected intelligence, since they all chose to drop dead simultaneously at just the right instant to deceive some twentieth-century scientists into blaming Clovis hunters.

A more plausible explanation of this "coincidence" is that it really was a case of cause and effect. It was Paul Martin, a geoscientist at the University of Arizona, who described the dramatic outcome of hunter-meets-elephant as a "blitzkrieg." According to his view, the first hunters to emerge from the ice-free corridor at Edmonton thrived and multiplied, because they found an abundance of tame, easy-to-hunt big mammals. As the mammals were killed off in one area, the hunters and their offspring kept fanning out into new areas that still had abundant mammals, and kept exterminating the mammal populations at the front of their advance. By the time the hunters' front finally reached the south tip of South America, most of the big mammal species of the New World had been exterminated.

MARTIN'S THEORY has attracted lots of vigorous criticism, most of it centering on four doubts: Could a band of a hundred hunters arriving at Edmonton breed fast enough to populate a hemisphere in a thousand years? Could they spread fast enough to cover the nearly eight thousand miles from Edmonton to Patagonia in that time? Were Clovis hunters really the first people in the New World? And could Stone Age hunters really have pursued hundreds of millions of big mammals so efficiently that not a single individual survived, while nevertheless leaving little fossil evidence of their hunts?

Take first the question of breeding rates. Populations of modern hunter-gatherers on even their best hunting grounds number only about one per square mile. Hence, once the whole western hemisphere had been settled, its population of hunter-gatherers would have been at most ten million, since the New World's area outside of Canada and other areas covered by glaciers in Clovis times is about ten million square miles. In modern instances when colonists have arrived at an uninhabited land (e.g., when the H.M.S. *Bounty* mutineers reached Pitcairn Island), their population growth has been as rapid as 3.4 percent per year. That growth rate, which corresponds to each couple's having four surviving children and a mean generation time of 20 years, would multiply 100 hunters into 10 million in only 340 years. Thus, Clovis hunters should easily have been able to multiply to 10 million within a millennium.

Could the descendants of the Edmonton pioneers have reached the southern tip of South America in 1,000 years? The overland straight-line distance is slightly under 8,000 miles, so that they would have had to average 8 miles a year. That's a trivial task: any fit hunter or huntress could have fulfilled the year's quota in a day and not moved for another 364 days. The quarry from which a Clovis tool was made can often be identified by its local type of stone, and we know in that way that individual tools traveled up to 200 miles. Some of the nineteenth-century Zulu migrations in southern Africa are known to have covered nearly 3,000 miles in a mere 50 years.

Were Clovis hunters the first humans to spread south of the Canadian ice sheet? That's a harder question, and it's extremely controversial among archaeologists. Primacy claims for Clovis are inevitably based on negative evidence: there are no unequivocal human remains or artifacts with universally accepted pre-Clovis dates anywhere in the New World south of the former Canadian ice sheet. Mind you, there are dozens of *claims* of sites with pre-Clovis human evidence, but all or almost all of them are marred by serious questions about whether the material used for radiocarbon dating was contaminated by older carbon, or whether the dated material was really associated with the human remains, or whether the tools supposedly made by humans were just naturally shaped rocks. The two most nearly convincing of those claimed pre-Clovis sites are Meadowcroft Rock Shelter in Pennsylvania, dated to about sixteen thousand years ago, and the Monte Verde site in Chile, dated to at least thirteen thousand years ago. Monte Verde is described as having amazingly good preservation of many types of human artifacts, but those results haven't yet been published in detail, so they can't yet be properly evaluated. At Meadowcroft there has been an unresolved debate about whether the radiocarbon dates are in error, especially because the plant and animal species from the site are ones expected to have been living there only much more recently than sixteen thousand years ago.

In contrast, the evidence for Clovis people is undeniable, is to be found in all forty-eight contiguous states, and is accepted by all archaeologists. Evidence for the still earlier settlement of the other habitable continents by more primitive humans is also unequivocal and universally accepted. At one Clovis site after another, you can see a level with Clovis artifacts and bones of numerous large extinct mammal species; immediately above (i.e., younger than) the Clovis level, a level with Folsom artifacts but with the bones of not a single large extinct mammal except for bison; and immediately below the Clovis level, levels spanning thousands of years before Clovis times, reflecting benign environmental conditions, and full of the bones of large extinct mammals, but with not a single human artifact. How could people possibly have settled the New World in pre-Clovis times and *not* left behind the usual trail of abundant evidence that convinces archaeologists, like stone tools, hearths, occupied caves, and occasionally skeletons, with unequivocal radiocarbon dates? How could there have been pre-Clovis people who left no trace of their presence at Clovis sites, despite such favorable living conditions? How could people have gotten from Alaska to Pennsylvania or Chile, as if by helicopter, without leaving good evidence of their presence in all the intervening territory? For these reasons, I find it more plausible that the dates given for Meadowcroft and Monte Verde are somehow wrong than that they are correct. The Clovis-first interpretation makes good sense; the pre-Clovis interpretation just doesn't make sense to me.

THE OTHER HOTLY CONTESTED ARGUMENT over Martin's blitzkrieg theory concerns the supposed overhunting and extermination of big mammals. It seems hard to imagine how Stone Age hunters could kill a mammoth at all, much less hunt all mammoths to extinction.

Even if the hunters *could* slaughter mammoths, why would they *want* to? And where are all the skeletons now?

Certainly, when we stand under a mammoth skeleton in a museum, the thought of using a stone-tipped spear to attack such a gigantic tusked beast feels utterly suicidal. Yet modern Africans and Asians with equally simple weapons do succeed in killing elephants, often hunting as a group relying on ambush or fire, but sometimes stalking an elephant as a single hunter armed with a spear or poisoned arrow. These modern elephant hunters still rate as amateur dabblers, compared to the mammoth hunters of Clovis times, heirs to hundreds of thousands of years of hunting experience with stone tools. Museum artists like to depict late Stone Age hunters as naked brutes risking their lives to hurl boulders at an enraged charging mammoth, with one or two hunters already lying trampled to death on the ground. That's absurd. If any hunters had died in a typical mammoth hunt, mammoths would have exterminated hunters, rather than vice versa. Instead, a more realistic picture is of warmly clad professionals safely spearing a terrified mammoth ambushed in a narrow streambed.

Recall also that the big mammals of the New World had probably never seen humans before Clovis hunters, if the hunters indeed were the first people to reach the New World. We know from Antarctica and the Galápagos how tame and unafraid are animals that evolved in the absence of humans. When I visited New Guinea's isolated Foja Mountains, which lack any human population, I found the large tree kangaroos so tame that I could approach within a few yards of them. Probably the New World's large mammals were equally naïve and were killed off before they could have time to evolve a fear of man.

Could Clovis hunters have killed mammoths fast enough to exterminate them? Assume again that an average square mile supports one hunter-gatherer and (by comparison with elephants in Africa today) one mammoth, and that one-quarter of the Clovis population consisted of adult male hunters who each killed a mammoth every two months. That means six mammoths killed per four square miles per year, so the mammoths would have had to reproduce their numbers in less than a year to keep up with the killing. Yet modern elephants are slow breeders that take about twenty years to reproduce their numbers, and few other large mammal species breed fast enough to reproduce their numbers in less than three years. Thus, it could plausibly have taken Clovis hunters only a few years to exterminate the large mammals locally and to move on to the next area. Archaeologists trying to document the slaughter today are searching for needles in a fossil haystack: a few years' worth of butchered mammoth bones among the bones of all the mammoths that died naturally over hundreds of thousands of years. It's no wonder that so few mammoth carcasses with Clovis points among the ribs have been found.

Why would a Clovis hunter even want to kill a mammoth every two months, when a five-thousand-pound mammoth yielding twenty-five hundred pounds of meat would provide ten pounds of meat per day per person for two months for the hunter, his wife, and two children? Ten pounds may sound like gross gluttony, but it actually approaches the daily meat ration per person on the U.S. frontier in the last century. That's assuming that Clovis hunters really ate all twenty-five hundred pounds of mammoth meat. But to keep the meat for two months would require drying it: would you go to the work of drying a ton of meat, when you could instead just go kill a fresh mammoth? As Vance Haynes noted, Clovis mammoth kills prove to have been only partly butchered, suggesting very wasteful and selective utilization of meat by people living amidst an abundance of game. Some hunting probably wasn't for meat at all but for ivory, hides, or just machismo. Seals and whales have similarly been hunted in modern times for oil or fur, leaving the meat to rot. In New Guinea fishing villages I often see the discarded carcasses of large sharks, killed only for their fins to make the delicacy shark's fin soup.

We are all too familiar with the blitzkriegs by which modern European hunters nearly exterminated bison, whales, seals, and many other large animals. Recent archaeological discoveries on many oceanic islands have shown that such blitzkriegs were an outcome whenever earlier hunters reached a land with animals naïve to humans. Since the collision between humans and large naïve animals has always ended in an extermination spasm, how could it have been otherwise when Clovis hunters entered a naïve New World?

*   *   *

THIS END, though, would hardly have been foreseen by the first hunters to arrive at Edmonton. It must have been a dramatic moment when, after entering the ice-free corridor from an overpopulated, overhunted Alaska, they emerged to see herds of tame mammoths, camels, and other beasts. In front of them stretched the Great Plains to the horizon. As they began to explore, they must soon have realized (unlike Christopher Columbus and the Plymouth Pilgrims) that there were no people at all in front of them, and that they had truly arrived first in a fertile land. Those Edmonton Pilgrims, too, had cause to celebrate a Thanksgiving Day.

CHAPTER 19

The Second Cloud

UNTIL OUR OWN GENERATION, NO ONE HAD GROUNDS TO WORRY whether the next human generation would survive or enjoy a planet worth living on. Ours is the first generation to be confronted with these questions about its children's future. We devote much of our lives to training our children to support themselves and to get along with other people. Increasingly, we're asking ourselves whether all those efforts of ours might be wasted.

These concerns arise because of two clouds hanging over us—clouds that would have similar consequences but that we view very differently. One, the risk of a nuclear holocaust, first revealed itself in the cloud over Hiroshima. Everyone agrees that the risk is real, since there are huge stockpiles of nuclear weapons and since politicians throughout history have occasionally made dumb miscalculations. Everyone agrees that, if a nuclear holocaust does happen, it will be bad for us and might even kill us all. This risk shapes much of current world diplomacy. The only thing about which we disagree is how best to handle it—e.g., whether we should aim for complete or partial nuclear disarmament, nuclear balance, or nuclear superiority.

The other cloud is the risk of an environmental holocaust, of which one often-discussed potential cause is the gradual extinction of most of the world's species. In contrast to the case with nuclear holocaust, there is almost complete disagreement about whether the risk of a mass extinction is real and about whether it would really do us much harm if it happened. For instance, one of the most frequently cited estimates is that humans have caused about 1 percent of the world's bird species to go extinct within the last few centuries. At one extreme, many thoughtful people—especially economists and industrial leaders, but also some biologists and many lay people—think that that loss of 1 percent would be inconsequential, even if it really happened. In fact, such people reason that 1 percent is a gross *over*estimate, that most species are superfluous to us, and that it would do us no harm to lose ten times more species. At the opposite extreme, many other thoughtful people—especially conservation biologists and a growing number of lay people belonging to environmentalist movements—think that the 1 percent figure is a gross *under*estimate, and that mass extinction would undermine the quality or possibility of human life. Obviously, it will make a big difference to our children which of these two extreme views is closer to the truth.

The risks of a nuclear holocaust and of an environmental holocaust constitute the two really pressing questions facing the human race today. Compared to these two clouds, our usual obsessions with cancer, AIDS, and diet pale into insignificance, because those problems don't threaten the survival of the human species. If the nuclear and environmental risks should not materialize, we'll have plenty of leisure time to solve bagatelles like cancer. If we fail to avert those two risks, solving cancer won't have helped us anyway.

How many species have humans really driven into extinction already? How many more are likely to go extinct within our kids' lifetimes? If more do go extinct, so what? How much do wrens contribute to our gross national product? Aren't all species destined to go extinct sooner or later? Is the claimed mass-extinction crisis a hysterical fantasy, a real risk for the future, or a proven event that's already well underway?

We need to go through three steps if we are to arrive at realistic estimates of the numbers involved in the mass-extinction debate. First, let's see how many species have gone extinct in modern times (i.e., since 1600). Second, let's estimate how many more extinctions had already occurred before 1600. As the third step, let's try to predict how many further ones are likely to happen within the life spans of ourselves, our children, and our grandchildren. Finally, let's ask what difference it all makes to us anyway.

THE FIRST STEP, that of calculating the number of species that have gone extinct in modern times, seems easy when one initially thinks about it. Just take some group of plants or animals, count up in a catalog the total number of species, mark off the ones known to have become extinct since 1600, and add them up. As a group on which to try this exercise, birds have the advantage that they're easy to see and identify, and hordes of bird-watchers watch them. As a result, more is known about them than about any other group of animals.

Approximately 9,000 species of birds exist today. Only one or two previously unknown species are still being discovered each year, so virtually all living birds have already been named. The leading agency concerned with the status of the world's birds—the International Council for Bird Preservation (ICBP)—lists 108 species of birds, plus many additional subspecies, as having become extinct since 1600. Virtually all of these extinctions were caused in one way or another by humans—more of that later. One hundred eight is about 1 percent of that total number of bird species, 9,000. That's where the 1 percent figure I mentioned earlier comes from.

Before we take that as the final word on the number of modern bird extinctions, let's understand how the number of 108 was arrived at. The ICBP decides to list a species as extinct only after that bird has been specifically looked for in areas where it was previously known to occur or might have turned up, and after it has not been found for many years. In many cases, birders have watched a population dwindle down to a few individuals and have followed the fates of those last individuals. For example, the most recent subspecies of bird to have gone extinct in the United States was the dusky seaside sparrow that lived in marshes near Titusville, Florida. As its population shrank owing to destruction of the marshes where it lived, wildlife agencies put identification bands on the few remaining sparrows so that they could be individually recognized. When only six remained, they were brought into captivity in order to protect and breed them.

Unfortunately, one after another died. The last individual, and with it the subspecies itself, died on June 16, 1987.

Thus, there is no doubt that the dusky seaside sparrow is extinct. Equally little doubt attaches to the many other subspecies and the 108 full species of birds listed as extinct. The full species listed as having vanished in North America since European settlement, and the years in which the last individual of each died, are: the great auk (1844), spectacled cormorant (1852), Labrador duck (1875), passenger pigeon (1914), and Carolina parakeet (1918). The great auk also formerly occurred in Europe, but no other European bird species is listed as having gone extinct since 1600, though some species have disappeared within Europe while surviving on other continents.

What about all those remaining bird species that did not fulfill the ICBP's rigorous criteria for extinction? Can we be certain that they still exist? For most North American and European birds the answer is "yes." Hundreds of thousands of fanatical bird-watchers monitor all bird species on these continents every year. The rarer the species, the more fanatical is the annual search for it. No North American or European bird species could possibly drift into extinction unnoticed. There is only one North American bird species whose current existence is uncertain: Bachman's warbler, last definitely recorded in 1977; but the ICBP hasn't given up hope for it because of more recent unconfirmed records. (The ivory-billed woodpecker may also be extinct, but the North American population is "only" a subspecies; a few individuals of the other subspecies of this woodpecker survive in Cuba.) Thus, the number of extinctions of North American bird species since 1600 is surely not fewer than five nor more than six. Every species but Bachman's warbler can be assigned to one of two categories: "definitely extinct" or "definitely in existence." Similarly, the number of European bird species extinct since 1600 is surely one—not two, not zero, but one.

Hence we have an exact, unequivocal answer to the question how many North American and European bird species have gone extinct since 1600. If we could be equally definite for other groups of species, our first step in assessing the mass-extinction debate would be complete. Unfortunately, this cut-and-dried situation does not apply to plants and other groups of animals, nor does it apply elsewhere in the world—least of all in the tropics, where the overwhelming majority of species lives. Most tropical countries have few or no bird-watchers, and no annual monitoring of birds. Many tropical areas have never again been monitored since they were first explored biologically many years ago. The status of many tropical species is unknown, because no one has seen them again or specifically looked for them since they were discovered. For instance, among the New Guinea birds that I study, Brass's friarbird is known only from eighteen specimens shot at one lagoon on the Idenburg River between March 22 and April 29, 1939. No scientist has revisited that lagoon, so we know nothing about the current status of Brass's friarbird.

At least we know where to look for that friarbird. Many other species have been described from specimens collected by nineteenth-century expeditions that provided only vague indications of the collecting site—e.g., "South America." Try resolving the status of some rare species when you have only that broad hint of where to look! The songs, behavior, and habitat preferences of such species are unknown. Hence we don't know where to seek them, or how to identify them if we glimpse or hear them.

Thus, the status of many tropical species cannot be classified either as "definitely extinct" or "definitely in existence," but just as "unknown." Instead, it becomes a matter of chance which species happens to attract the attention of some ornithologist, becomes the object of a specific search, and hence may be recognized as possibly extinct.

Here's an example. The Solomon Islands are another of my favorite birding areas in the tropical Pacific Ocean, and will be recalled by older Americans and Japanese as the site of some of the fiercest fighting in World War II. (Remember Guadalcanal, Henderson Field, President Kennedy's PT boat, the Tokyo Express?) The ICBP lists one Solomon bird species, Meek's crowned pigeon, as extinct. But when I tabulated all recent observations of all 164 known Solomon bird species, I noticed that 12 of those 164 species hadn't been encountered since 1953. Some of those 12 species are surely extinct, because they were formerly abundant and conspicuous, or because Solomon Islanders told me that those birds had been exterminated by cats.

Twelve species possibly extinct out of 164 still may not sound like much to worry about. However, the Solomons are in much better shape environmentally than most of the remaining tropical world, because they have relatively few people, few bird species, little economic development, and much natural forest. More typical of the tropics is Malaysia, which is rich in species and has had most of its lowland forest cut down. Biological explorers had identified 266 fish species dependent on fresh water in Malaysia's forest rivers. A recent search that lasted four years was able to find only 122 of those 266 species—fewer than half. The other 144 Malaysian freshwater fish species must either be extinct, rare, or very local. They reached that status before anyone noticed it.

Malaysia is typical of the tropics in the pressure it faces from humans. Fish are typical of all species other than birds, in that they attract only patchy scientific attention. The estimate that Malaysia has already lost (or nearly lost) half of its freshwater fish is therefore a reasonable ballpark figure for the status of plants, invertebrates, and vertebrates other than birds in much of the rest of the tropics.

That's one complication in trying to pinpoint the number of extinctions since 1600: the status of many or most named species is unknown. But there's a further complication. So far, we have been trying to assess extinctions only of those species that had already been discovered and described (named). Could any species have gone extinct before they were even described?

Of course there could, since sampling procedures suggest that the actual number of the world's species is near thirty million, but fewer than two million species have been described. Two examples illustrate the certainty of additional extinctions before description. Botanist Alwyn Gentry surveyed the plants of an isolated ridge in Ecuador called Centinela, where he found thirty-eight new species confined to that ridge. Shortly afterward, the ridge was logged and those plants were exterminated. On Grand Cayman Island in the Caribbean, zoologist Fred Thompson discovered two new species of land snails confined to forest on a limestone ridge that was completely cleared a few years later for a housing development.

The accident that Gentry and Thompson visited those ridges before rather than after they were cleared means that we have names for those extinct species. But most tropical areas that are being developed aren't first surveyed by biologists. There must have been land snails on Centinela, and plants and snails on innumerable other tropical ridges, that we exterminated before we discovered them.

In short, the problem of determining the number of modern extinctions seems at first to be simple and to lead to modest estimates—e.g., just five or six extinct bird species in all of North America plus

Europe. On reflection, though, we appreciate two reasons why published lists of species known to be extinct must be gross underestimates of the actual number of extinctions. First, by definition the published lists consider only named species, whereas the great majority of species (except in well-studied groups like birds) haven't even been named. Second, outside of North America and Europe and except for birds, the published lists consist only of those few named species that some biologist happened to get interested in for one or another reason and found to be extinct. Among all those remaining species of unknown status, many are likely to be extinct or nearly so: e.g., about half in the case of Malaysian freshwater fishes.

Now LET'S MOVE ON to the second step in evaluating the mass-extinction debate. Our estimates up to this point have concerned only those species exterminated since A.D. 1600, when scientific classification of species was beginning. These exterminations have taken place because the world's human population has grown in numbers, reached previously uninhabited areas, and invented increasingly destructive technologies. Did these factors spring up suddenly in 1600, after several million years of human history? Were there no exterminations before 1600?

Of course not. Until fifty thousand years ago, humans were confined to Africa plus the warmer areas of Europe and Asia. Between then and A.D. 1600 our species underwent a massive geographic expansion that took us to Australia and New Guinea around fifty thousand years ago, then to Siberia and most of North and South America, and finally to most of the world's remote oceanic islands only around 2000 B.C. We also underwent a massive expansion in numbers, from perhaps a few million people fifty thousand years ago to about half a billion in 1600. And our destructiveness also increased with the development of improved hunting skills in the last fifty thousand years, polished stone tools and agriculture in the last ten thousand years, and metal tools in the last six thousand years.

In every area of the world that paleontologists have studied and that humans first reached within the last fifty thousand years, human arrival approximately coincided with massive prehistoric extinctions. For Madagascar, New Zealand, Polynesia, Australia, the West Indies, the Americas, and Mediterranean islands, I've described those extinctions in the preceding two chapters. Ever since scientists became aware of these prehistoric extinction waves associated with human arrival, they've argued over whether people were the cause or just happened to arrive while animals were succumbing to climate changes. In the case of the extinction waves on Polynesian islands, there is now no reasonable doubt that Polynesian arrival in one way or another caused the extinctions. Bird extinctions and Polynesian arrival coincided within a few centuries at a time when no big climate change was happening, and bones of thousands of roasted moas have been found in Polynesian ovens. The coincidence of timing is equally convincing for Madagascar. But the causes of the earlier extinctions, especially those in Australia and the Americas, are still being debated.

As I explained for America's extinctions in the previous chapter, the evidence seems to me overwhelming that humans also played a role in those prehistoric extinctions outside Polynesia and Madagascar. In each part of the world an extinction wave occurred after the first arrival of humans, but didn't occur simultaneously in other areas undergoing similar climate swings, and didn't occur in the same area whenever similar climate swings had occurred previously.

So I doubt that climate did it. Instead, all of you who have visited Antarctica or the Galápagos know how tame the animals there are, being unaccustomed to humans until recently. Photographers can still walk up to those naïve animals as easily as hunters used to. I assume that the first arriving hunters similarly walked up to naïve mammoths and moas elsewhere in the world, while rats that came with the first hunters walked up to naïve little birds of Hawaii and other islands.

It's not just in those areas of the world previously unoccupied by humans that prehistoric humans probably exterminated species. Within the last twenty thousand years there were also extinctions in the areas long occupied by humans—in Eurasia, where woolly rhinos, mammoths, and giant deer ("Irish elk") died out, and in Africa, which lost its giant buffalo, giant hartebeest, and giant horse. These big beasts may also have been among the victims of prehistoric humans who had already been hunting them for a long time but who now were able to hunt them with better weapons than ever before. Eurasia's and Africa's big mammals weren't naïve toward humans, but they disappeared for the same two simple reasons that California's grizzly bear and Britain's bears, wolves, and beavers succumbed only in recent times, after thousands of years of human persecution. Those reasons were more people and better weaponry.

Can we at least estimate how many species were involved in these prehistoric extinctions? No one has ever tried to guess the number of plants, invertebrates, and lizards exterminated by prehistoric habitat destruction. But virtually all oceanic islands explored by paleontologists have yielded remains of recently extinct bird species. Extrapolation to those islands not yet paleontologically explored suggests that about two thousand bird species—one-fifth of all the birds that existed a few thousand years ago—were island species already exterminated prehistorically. That doesn't count birds that may have been exterminated prehistorically on the continents. Among genera of large mammals, about 73, 80, and 86 percent respectively became extinct in North America, South America, and Australia at the time of or after human arrival.

THE REMAINING STEP in evaluating the mass-extinction debate is to predict the future. Is the peak of the extinction wave that we've caused already past, or is most still to come? There are a couple of ways to assess this question.

A simple way is to reason that tomorrow's extinct species will be drawn from today's endangered species. How many species that still exist have populations already reduced to dangerously low levels? The ICBP estimates that at least 1,666 bird species are either endangered or at imminent risk of extinction—almost 20 percent of the world's surviving birds. I said "at least 1,666," because this number is an underestimate for the same reason I mentioned that the ICBP's estimate of extinct species was an underestimate. Both numbers are based just on those species whose status caught a scientist's attention, rather than on a reappraisal of the status of all bird species.

The alternative way of predicting what's to come is to understand the mechanisms by which we exterminate species. Human-caused extinctions may continue accelerating until human population and technology reach a plateau, but neither shows any signs of plateauing. Our population, which grew tenfold from half a billion in 1600 to over five billion now, is still growing at close to 2 percent per year. Every day brings new technological advances for changing the earth and its denizens. There are four main mechanisms by which our growing population exterminates species: by overhunting, species introduction, habitat destruction, and ripple effects. Let's see if these four mechanisms have plateaued.

Overhunting—killing animals faster than they can breed—is the main mechanism by which we've exterminated big animals, from mammoths to California grizzly bears. (The latter appears on the flag of California, the state in which I live, but many of my fellow Californians don't recall that we exterminated our state's symbol long ago). Have we already killed off all big animals that we might kill off? Obviously not. While the low numbers of whales led to an international ban on whaling for commercial purposes, Japan thereupon announced its decision to triple the rate at which it killed whales "for scientific reasons." We've all seen photos of the accelerating slaughter of Africa's elephants and rhinos for their ivory and horns respectively. At current rates of change, not just elephants and rhinos but most populations of most other large mammals of Africa and Southeast Asia will be extinct outside game parks and zoos in a decade or two.

The second mechanism by which we exterminate is through intentionally or accidentally introducing certain species to parts of the world where they didn't previously occur. Familiar examples of introduced species now firmly established in the U.S. are Norway rats, European starlings, boll weevils, and the fungi causing Dutch elm disease and chestnut blight. Europe too has acquired introduced species, of which the misnamed Norway rat is an example (it originated in Asia, not Norway). When species are introduced from one region to another, they often proceed to exterminate some of the native species they encounter, by eating them or causing diseases. The victims evolved in the absence of the introduced pests and never developed defenses against them. American chestnut trees have been virtually exterminated in this way by chestnut blight, an Asian fungus to which Asian chestnut trees are resistant. Similarly, goats and rats have exterminated many plants and birds on oceanic islands.

Have we already spread all possible pests all around the world? Obviously not: there are many islands still free of goats and Norway rats, and many insects and diseases to try to keep out of many countries by quarantines. The U.S. Department of Agriculture has been trying at great expense, but apparently without success, to forestall the arrival of killer bees and Mediterranean fruit flies. In fact, what will probably prove to be the biggest extinction wave caused by an introduced predator in modern times has just started in Africa's Lake Victoria, home to hundreds of species of remarkable fishes found nowhere else in the world. A large predatory fish called the Nile perch, intentionally introduced in a misguided effort to establish a new fishery, is now eating its way through the lake's unique fishes.

Habitat destruction is the third means by which we exterminate. Most species occur in just a certain type of habitat: marsh warblers live in marshes, while pine warblers live in pine forests. If one drains marshes or cuts forests, one eliminates the species dependent on those habitats just as certainly as if one were to shoot every individual of the species. For example, when all the forest on Cebu Island in the Philippines was logged, nine of the ten birds unique to Cebu became extinct.

In the case of habitat destruction, the worst is still to come because we are just starting in earnest to destroy tropical rain forests, the world's most species-rich habitats. The rain forests' biological richness is legendary: e.g., over fifteen hundred beetle species living in a single rain-forest tree species in Panama. Rain forests cover only 6 percent of the Earth's surface but harbor about half of its species. Each area of rain forest has large numbers of species unique to that area. To mention only some exceptionally rich rain forests now being destroyed, the felling of Brazil's Atlantic forest and Malaysia's lowland forest is already almost complete, and those of Borneo and the Philippines will be mostly logged within the next two decades. By the middle of the next century, the only large tracts of tropical rain forest likely to be still surviving will be in parts of Zaire and the Amazon Basin.

Every species depends on other species for food and for providing its habitat. Thus, species are connected to each other like branching rows of dominoes. Just as toppling one domino in a row will topple some others, so too the extermination of one species may lead to the loss of others, which may in turn push still others over the brink. This fourth mechanism of extinction may be described as a ripple effect. Nature consists of so many species, connected to each other in such complex ways, that it's virtually impossible to foresee where the ripple effects from the extinction of any particular species may lead.

For example, fifty years ago no one anticipated that the extinction of big predators (jaguars, pumas, and harpy eagles) on Panama's Barro Colorado Island would lead to the extinction there of little antbirds, and to massive changes in the tree-species composition of the island's forest. But it did, because the big predators used to eat medium-sized predators like peccaries, monkeys, and coatimundis, and medium-sized seedeaters like agoutis and pacas. With the disappearance of the big predators, there was a population explosion of the medium-sized predators, which proceeded to eat up the antbirds and their eggs. The medium-sized seedeaters also exploded in abundance and ate large seeds that had fallen on the ground, thereby suppressing the propagation of tree species producing large seeds and favoring instead the spread of competing tree species with small seeds. That shift in forest tree composition is expected in turn to cause an explosion of mice and rats feeding on small seeds, and hence to an explosion in hawks, owls, and ocelots preying on those small rodents. Thus, the extinction of three uncommon species of big predators will have triggered a rippling series of changes in the whole plant and animal community, including the extinction of many other species.

Through these four mechanisms—overhunting, species introductions, habitat destruction, and ripple effects—probably over half of existing species will be extinct or endangered by the middle of the next century, when this year's crop of human babies reaches age sixty. Like many fathers today, I often wonder how I'll describe to my twin sons the world that I grew up in and that they will never see. By the time they are old enough to come with me to New Guinea, one of the world's biological treasure houses, where I've worked for the past twenty-five years, most of New Guinea's eastern highlands will be deforested.

When one adds these extinctions we've already caused to those that we're about to cause, it's clear that the current extinction wave is surpassing the asteroid collision that wiped out the dinosaurs. Mammals, plants, and many other types of species survived that collision nearly unscathed, while the current wave is impacting everything from leeches and lilies to lions. Thus, the claimed extinction crisis is neither a hysterical fantasy nor just a serious risk for the future. Instead, it's an event that has already been accelerating for fifty thousand years and will start to approach completion in our children's lifetimes.

Let's finally consider two arguments that accept the reality of the extinction crisis but dismiss its significance. First, isn't extinction a natural process anyway? If so, why make a big deal about the extinctions happening now?

The answer to this first argument is that the current human-caused extinction rate is far higher than the natural rate. If the estimate that half the world's total of 30 million species will go extinct in the next century is correct, then species are now going extinct at a rate of about 150,000 per year, or 17 per hour. The world's 9,000 bird species are going extinct at a rate of at least 2 per year. But bird extinctions under natural conditions were at a rate of less than 1 per century, so the present rate is at least 200 times normal. Dismissing the extinction crisis on the grounds that extinction is natural would be just like dismissing genocide on the grounds that death is the natural fate of all humans.

The second argument is a simple one: So what? We care about our children, not about beetles and snail darters; who cares if ten million beetle species go extinct? The answer to this argument is equally simple. Like all species, we depend on other species for our existence, in many ways. Some of the most obvious ways are that other species produce the oxygen we breathe, absorb the carbon dioxide we exhale, decompose our sewage, provide our food, maintain the fertility of our soil, and provide our wood and paper.

Then couldn't we just preserve those particular species that we need, and let other species go extinct? Of course not, because the species we need also depend on other species. Just as Panama's ant-birds couldn't have anticipated their need for jaguars, the ecological row of dominoes is much too complex for us to have figured out which dominoes we can dispense with. For instance, could anyone please answer these three questions: Which ten tree species produce most of the world's paper pulp? For each of those ten tree species, which are the ten bird species that eat most of its insect pests, the ten insect species that pollinate most of its flowers, and the ten animal species that spread most of its seeds? Which other species do these ten birds, insects, and animals depend on? You'd have to be able to answer those three impossible questions if you were the president of a timber company trying to figure out which species you could afford to let go extinct.

If you're trying to evaluate some proposed development project that would bring in a million dollars but might exterminate a few species, it's still tempting to prefer the certain profit over the uncertain risk. Then consider the following analogy. Suppose someone offers you a million dollars in return for the privilege of painlessly cutting out two ounces of your valuable flesh. You figure that two ounces is only one thousandth of your body weight, so you'll still have nine-hundred-ninety-nine thousandths of your body left, which is plenty. That's fine if the two ounces come from your spare body fat and if they'll be removed by a skilled surgeon. But what if the surgeon just hacks two ounces from any conveniently accessible part of your body, or doesn't know which parts are essential? You might then find that the two ounces came from your urethra. And if you plan to sell off most of your body, as we now plan to sell off most of our planet's natural habitats, you're certain eventually to lose your urethra.

To CONCLUDE, let's place matters in perspective by comparing the two clouds I mentioned at the outset as hanging over our future. A nuclear holocaust is certain to prove disastrous, but it isn't happening now, and it may or may not happen in the future. An environmental holocaust is equally certain to prove disastrous, but it differs in that it's already well underway. It started tens of thousands of years ago, is now causing more damage than ever before, is in fact accelerating, and will climax within about a century if unchecked. The only uncertainties are whether the resulting disaster would strike our children or our grandchildren, and whether we choose to adopt now the many obvious countermeasures.

Epilogue:
Nothing Learned,
and Everything Forgotten?

Let's now draw together the themes of this book by tracing our rise over the last three million years, as well as our incipient reversal of all our progress more recently.

The first indications that our ancestors were in any respect unusual among animals were our extremely crude stone tools that began to appear in Africa by around two and a half million years ago. The quantities of tools suggest that they were beginning to play a regular, significant role in our livelihood. Among our closest relatives, in contrast, the pygmy chimpanzee and gorilla don't use tools, while the common chimpanzee occasionally makes some rudimentary ones but hardly depends on them for its existence.

Nevertheless, those crude tools of ours did not trigger any quantum jump in our success as a species. For another million and a half years, we remained confined to Africa. Around a million years ago we did manage to spread to warm areas of Europe and Asia, thereby becoming the most widespread of the three chimpanzee species but still much less widespread than lions. Our tools progressed only at an infinitely slow rate, from extremely crude to very crude. By a hundred thousand years ago, at least the human populations of Europe and western Asia, the Neanderthals, were regularly using fire. Yet in other respects we continued to rate as just another species of big mammal. We had developed not a trace of art, agriculture, or high technology. It's unknown whether we had developed language, drug addictions, or our strange modern sexual habits and life cycle, but Neanderthals rarely lived beyond age forty and hence may not yet have evolved female menopause.

Clear evidence of a Great Leap Forward in our behavior appears suddenly in Europe around forty thousand years ago, coincident with the arrival of anatomically modern *Homo sapiens* from Africa via the Near East. At that point, we began displaying art, technology based on specialized tools, cultural differences from place to place, and cultural innovation with time. This leap in behavior had undoubtedly been developing outside Europe, but the development must have been rapid, since the anatomically modern *Homo sapiens* populations living in southern Africa 100,000 years ago were still just glorified chimpanzees as judged by the debris in their cave sites. Whatever caused the leap, it must have involved only a tiny fraction of our genes, because we still differ from chimps in only 1.6 percent of our genes, and most of that difference had already developed long before our leap in behavior. The best guess I can make is that the leap was triggered by the perfection of our modern capacity for language.

Although we usually think of the Cro-Magnons as the first bearers of our noblest traits, they also bore the two traits that lie at the root of our current problems: our propensities to murder each other en masse and to destroy our environment. Even before Cro-Magnon times, fossil human skulls punctured by sharp objects and cracked to extract the brains bear witness to murder and cannibalism. The suddenness with which Neanderthals disappeared after Cro-Magnons arrived hints that genocide had now become efficient. Our efficiency at destroying our own resource base is suggested by extinctions of almost all large Australian animals following our colonization of Australia fifty thousand years ago, and of some large Eurasian and African mammals as our hunting technology improved. If the seeds of self-destruction have been so closely linked with the rise of advanced civilizations in other solar systems as well, it becomes easy to understand why we have not been visited by any flying saucers.

At the end of the last Ice Age around ten thousand years ago, the pace of our rise quickened. We occupied the Americas, coincident with a mass extinction of big mammals that we may have caused. Agriculture emerged soon thereafter. Some thousands of years later, the first written texts start to document the pace of our technical inventiveness. They also show that we were already addicted to drugs, and that genocide had become routine and admired. Habitat destruction began undermining many societies, and the first Polynesian and Malagasy settlers caused mass exterminations of species. From A.D. 1492 onward, the worldwide expansion of literate Europeans lets us trace our rise and fall in detail.

Within the last few decades we have developed the means to send radio signals to other stars, and also to blow ourselves up overnight. Even if we don't blunder into that quick end, our harnessing of much of the Earth's productivity, our exterminations of species, and our damage to our environment are accelerating at a rate that cannot be sustained for even another century. One might object that, if we look around us, we see no obvious sign that the climax of our history will come soon. In fact, the signs become obvious if one looks and then extrapolates. Starvation, pollution, and destructive technology are increasing; usable farmland, food stocks in the sea, other natural products, and environmental capacity to absorb wastes are decreasing. As more people with more power scramble for fewer resources, something has to give way.

So what is likely to happen?

There are many grounds for pessimism. Even if every human now alive were to die tomorrow, the damage that we have already inflicted on our environment would ensure that its degradation will continue for decades. Innumerable species already belong to the "living dead," with populations fallen to levels from which they cannot recover, even though not all individuals have died yet. Despite all our past self-destructive behavior from which we could have learned, many people who should know better dispute the need for limiting our population and continue to assault our environment. Others join that assault for selfish profit or out of ignorance. Even more people are too caught up in the desperate struggle for survival to be able to enjoy the luxury of weighing the consequences of their actions. All these facts suggest that the juggernaut of destruction has already reached unstoppable momentum, that we too are among the living dead, and that our future is as bleak as that of the other two chimpanzees.

This pessimistic view is captured by a cynical sentence that Arthur Wichmann, a Dutch explorer and professor, penned in another context in 1912. Wichmann had devoted a decade of his life to writing a monumental three-volume treatise on the history of New Guinea exploration. In 1,198 pages he evaluated every source of information about New Guinea that he could find, from the earliest reports filtering through Indonesia to the great expeditions of the nineteenth and early twentieth centuries. He grew disillusioned as he realized that successive explorers committed the same stupidities again and again: unwarranted pride in overstated accomplishments, refusal to acknowledge disastrous oversights, ignoring the experience of previous explorers, consequent repetition of previous errors, hence a long history of unnecessary sufferings and deaths. Looking back on this history, Wichmann predicted that future explorers would continue to repeat the same errors. The bitter last sentence that concluded Wichmann's last volume was: "Nothing learned, and everything forgotten!"

Despite all the grounds I've mentioned for being equally cynical about humanity's future, my view is that our situation isn't hopeless. We are the only ones creating our problems, so it's completely within our power to solve them. While our language and art and agriculture aren't quite unique, we really are unique among animals in our capacity to learn from the experience of others of our species living in distant places or in the distant past. Among the hopeful signs, there are many realistic, often-discussed policies by which we could avoid disaster, such as by limiting human population growth, preserving natural habitats, and adopting other environmental safeguards. Many governments are already doing some of these obvious things in some cases.

For example, awareness of environmental problems is spreading, and environmental movements are gaining political clout. Developers don't win all the battles, nor do shortsighted economic arguments always prevail. Many countries have lowered their rate of population growth in recent decades. While genocide hasn't vanished, the spread of communications technology has at least the potential for reducing our traditional xenophobia, and for making it harder to regard distant peoples as subhumans unlike ourselves. I was seven years old when the A-bombs were dropped on Hiroshima and Nagasaki, so I remember well the sense of an imminent risk of nuclear holocaust that prevailed for several decades thereafter. But nearly half a century has now passed without any further military use of nuclear weapons. The risk of a nuclear holocaust now seems more remote than at any other time since August 9, 1945.

My own outlook is conditioned by my experiences since 1979 as consultant to the Indonesian government, on setting up a nature reserve system in Indonesian New Guinea (called Irian Jaya province). On the face of it, Indonesia doesn't seem a promising place to hope for much success in preserving our shrinking natural habitats. Instead, Indonesia exemplifies the problems of tropical third-world countries in acute form. With over 180 million inhabitants, it is the world's fifth most populous country, as well as one of the poorer ones. The population is growing rapidly; nearly half of all Indonesians are under fifteen years old. Some provinces with inordinately high population density are exporting their population surpluses to the less populated provinces (such as Irian Jaya). There are no armies of bird-watchers, no broad-based indigenous environmental movements. The government is not a democracy in the western sense, and corruption is viewed as pervasive. Indonesia depends on logging of its virgin rain forests, second only to exploitation of oil and natural gas as a source of its foreign exchange.

For all these reasons, one might not expect preservation of species and habitats to be a national priority pursued seriously in Indonesia. When I first went to Irian Jaya, I was frankly doubtful that an effective conservation program would result. Fortunately, my Wichmann-like cynicism proved wrong. Thanks to the leadership of a core of Indonesians convinced of conservation's value, Irian Jaya now has the beginnings of a nature reserve system comprising 20 percent of the province's area. Nor do those reserves exist just on paper. As my work proceeded, I was pleasantly surprised to come across sawmills abandoned because they conflicted with nature reserves, park guards out on patrol, and management plans being drawn up. All these measures were adopted not out of idealism but out of cold-blooded, correct perception of Indonesia's natural self-interest. If Indonesia can do it, so can other countries with similar obstacles to environmentalism, as well as much richer countries with broad-based environmental movements.

We don't need novel, still-to-be invented technologies to solve our problems. We just need more governments to do many more of the same obvious things that some governments are already doing in some cases. Nor is it true that the average citizen is powerless. There are many causes of extinction that citizen groups have helped scale back in recent years—for instance, commercial whaling, hunting big cats for fur coats, and importing wild-caught chimpanzees, to mention just a few examples. In fact, this is one area where it's particularly easy for a modest donation by the average citizen to have a big impact, because all conservation organizations now have such modest budgets. For instance, the annual combined budget for *all* primate conservation projects that World Wildlife Fund supports throughout the world is only a few hundred thousand dollars. An extra thousand dollars means an extra project on some endangered monkey, ape, or lemur that might otherwise have been ignored. Pages 390–91 suggest some specific starting points for interested readers.

While I do see us facing serious problems with an uncertain prognosis, I'm cautiously optimistic. Even the cynical last sentence of Wichmann's book proved false: New Guinea explorers since Wichmann really have learned from the past and avoided the disastrous stupidities of their predecessors. A motto more appropriate for our future than Wichmann's motto comes from the memoirs of the statesman Otto von Bismarck. As he reflected on the world around him toward the end of his long life, he too had reasons to be cynical. Possessing a keen intellect and working at the center of European politics for decades, Bismarck had witnessed a history of unnecessarily repeated errors as gross as those pervading the early history of New Guinea exploration. But Bismarck still considered it worthwhile to write his memoirs, to draw lessons from history, and to dedicate his memoirs "to [my] children and grandchildren, towards an understanding of the past, and for guidance for the future."

This is also the spirit in which I dedicated this book to my young sons and their generation. If we will learn from our past that I have traced, our own future may yet prove brighter than that of the other two chimpanzees.

Acknowledgments

It's a pleasure for me to acknowledge the contributions of many people to this book. From my parents and from my teachers at Roxbury Latin School, I learned to pursue interests along many lines simultaneously. My debt to my many New Guinea friends will be obvious from the frequency with which I cite their experiences. I owe an equal debt to my many scientist friends and professional colleagues, who patiently explained the subtleties of their subjects and read my drafts. Earlier versions of most of the chapters appeared as articles in *Discover* magazine and in *Natural History* magazine. I've been fortunate to have John Brockman as my agent, and as my editors Leon Jaroff, Fred Golden, Gil Rogin, Paul Hoffman, and Marc Zabludoff at *Discover,* Alan Ternes and Ellen Goldensohn at *Natural History,* Thomas Miller at HarperCollins, Neil Belton at Hutchinson Radius Publishers, and my wife, Marie Cohen.

Further Readings

THESE SUGGESTIONS are for readers interested in reading further. In addition to key books and papers, I have also tended to favor recent references that provide comprehensive listings of the earlier literature. A journal title is followed by the volume number, followed after a colon by the first and last page number, and then the year of publication in parentheses.

CHAPTER 1. *A Tale of Three Chimps*

The literature on deducing relationships among humans and other primates by means of the DNA clock consists of technical articles in scientific journals. C. G. Sibley and J. E. Ahlquist present their studies in three papers: "The phylogeny of the hominoid primates, as indicated by DNA-DNA hybridization," *Journal of Molecular Evolution* 20: 2–15 (1984); "DNA hybridization evidence of hominoid phylogeny: results from an expanded data set," *Journal of Molecular Evolution* 26: 99–121 (1987); and C. G. Sibley, J. A. Comstock, and J. E. Ahlquist, "DNA hybridization evidence of hominoid phylogeny: a reanalysis of the data," *Journal of Molecular Evolution* 30: 202–236 (1990). Sibley's and Ahlquist's many studies of bird relationships by means of the same DNA methods are summarized in two books: C. G. Sibley and J. E. Ahlquist, *Phylogeny and Classification of Birds* (New Haven:

Yale University Press, 1990); and C. G. Sibley and B. L. Monroe, Jr., *Distribution and Taxonomy of the Birds of the World* (New Haven: Yale University Press, 1990).

Similar conclusions about human and primate relationships were obtained by DNA comparisons using a different method (termed the tetraethyl-ammonium chloride method, rather than the hydroxyapatite method used by Sibley and Ahlquist). The results were described by A. Caccone and J. R. Powell, "DNA divergence among hominoids," *Evolution* 43: 925–942 (1989). A paper by the same authors explains how percentage similarity among DNAs can be calculated from DNA mixed melting points: A. Caccone, R. DeSalle, and J. R. Powell, "Calibration of the changing thermal stability of DNA duplexes and degree of base pair mismatch," *Journal of Molecular Evolution* 27: 212–216 (1988).

The above papers compare the entire genetic material (DNA) of two species by means of mixed melting points in order to obtain a single measure of overall similarity. Alternatively, a much more laborious method yielding much more detailed information about a tiny fraction of each species' DNA consists of determining the actual sequence of molecular units comprising that portion of DNA. Five studies stemming from a single laboratory and applying that method to human and primate relationships are: M. M. Miyamoto et al., "Phylogenetic relations of humans and African apes from DNA sequence in the Ψ-globin region," *Science* 238: 369–373 (1987); M. M. Miyamoto et al., "Molecular systematics of higher primates: genealogical relations and classification," *Proceedings of the National Academy of Sciences* 85: 7627–7631 (1988); M. Goodman et al., "Molecular phylogeny of the family of apes and humans," *Genome* 31: 316–335 (1989); M. M. Miyamoto and M. Goodman, "DNA systematics and evolution of primates," *Annual Reviews of Ecology and Systematics* 21: 197–220; and M. Goodman et al., "Primate evolution at the DNA level and a classification of hominoids," *Journal of Molecular Evolution* 30: 260–266 (1990). The same principle is applied to relationships among Lake Victoria's cichlid fishes by A. Meyer et al., "Monophyletic origin of Lake Victoria cichlid fishes suggested by mitochondrial DNA sequences," *Nature* 347: 550–553 (1990).

Two papers that vigorously criticize the DNA clock in general, and Sibley's and Ahlquist's application of it to human-primate relationships in particular, are: J. Marks, C. W. Schmidt, and V. M. Sarich, "DNA hybridization as a guide to phylogeny: relationships of the Hominoidea," *Journal of Human Evolution* 17: 769–786 (1988); and V. M. Sarich, C. W. Schmidt, and J. Marks, "DNA hybridization as a guide to phylogeny: a critical analysis," *Cladistics* 5: 3–32 (1989). In my view, the criticisms by Marks, Schmidt, and Sarich have been adequately answered. The good agreement between conclusions about human/primate relationships based on the DNA clock as measured by Sibley and Ahlquist, the DNA clock as measured by Caccone and Powell, and DNA sequencing further supports the correctness of these conclusions.

Other papers on the DNA clock are in two issues of the *Journal of Molecular Evolution* that also include some of the above-cited papers: volume 30, numbers 3 and 5 (1990).

CHAPTER 2. *The Great Leap Forward*

Among the many books providing detailed accounts of human evolution, the recent one that I found the most useful is by Richard Klein, *The Human Career* (Chicago: University of Chicago Press, 1989). Illustrated and less technical accounts are by Roger Lewin, *In the Age of Mankind* (Washington, D.C.: Smithsonian Books, 1988), and Brian Fagan, *The Journey from Eden* (New York: Thames and Hudson, 1990).

Two books presenting multiauthored technical accounts of recent human evolution are edited by Fred H. Smith and Frank Spencer, *The Origins of Modern Humans* (New York: Liss, 1984) and by Paul Mellars and Chris Stringer, *The Human Revolution: Behavioural and Biological Perspectives on the Origins of Modern Humans* (Edinburgh: Edinburgh University Press, 1989). Some recent articles on the dating and geography of human evolution are by C. B. Stringer and P. Andrews, "Genetic and fossil evidence for the origin of modern humans," *Science* 239: 1263–1268 (1988); H. Valladas et al., "Thermoluminescence dating of Mousterian 'proto-Cro-Magnon' remains from Israel and the origin of modern man" *Nature* 331: 614–616 (1988); C. B. Stringer et al., "ESR dates for the hominid burial site of Es Skhul in Israel," *Nature* 338: 756–758 (1989); J. L. Bischoff et al., "Abrupt Mousterian-Aurignacian boundaries at c. 40 ka bp: accelerator ^{14}C dates from l'Arbreda Cave (Catalunya, Spain)," *Journal of Archaeological Science* 16: 563–576 (1989); V. Cabrera-Valdes and J. Bischoff, "Accelerator ^{14}C dates for Early Upper Paleolithic (Basal Aurignacian) at El Castillo Cave (Spain)," *Journal of Archaeological Science* 16: 577–584 (1989); E. L. Simons, "Human origins," *Science* 245: 1343–1350 (1989); and R. Grün et al., "ESR dating evidence for early modern humans at Border Cave in South Africa," *Nature* 344: 537–539 (1990).

Three books with many beautiful illustrations of Ice Age art are by Randall White, *Dark Caves, Bright Visions* (New York: American Museum of Natural History, 1986); Mario Ruspoli, *Lascaux: The Final Photographs* (New York: Abrams, 1987); and Paul G. Bahn and Jean Vertut, *Images of the Ice Age* (New York: Facts on File, 1988).

Matthew H. Nitecki and Doris V. Nitecki, *The Evolution of Human Hunting* (New York: Plenum Press, 1986), provide a series of chapters by various authors on that subject.

The question whether Neanderthals really did bury their dead is debated in an article by R. H. Gargett, "Grave shortcomings: the evidence for Neanderthal burial," and in accompanying responses published in *Current Anthropology* 30: 157–190 (1989).

Three sources that will provide an entrance into the literature on the linked questions of human vocal-tract anatomy and whether Neanderthals could speak are: a book by Philip Lieberman, *The Biology and Evolution of Language* (Cambridge, Mass.: Harvard University Press, 1984); another book by E. S. Crelin, *The Human Vocal Tract* (New York: Vantage Press, 1987); and an article by B. Arensburg et al., "A Middle Palaeolithic human hyoid bone," *Nature* 338: 758–760 (1989).

CHAPTER 3. *The Evolution of Human Sexuality*

CHAPTER 4. *The Science of Adultery*

For anyone interested in an evolutionary approach to behavior in general (including reproductive behavior), two books are a must: E. O. Wilson, *Sociobiology* (Cambridge, Mass.: Harvard University Press, 1975), and John Alcock, *Animal Behavior,* 4th ed. (Sunderland: Sinauer, 1989).

Outstanding books that discuss the evolution of sexual behavior include Donald Symons, *The Evolution of Human Sexuality* (Oxford: Oxford University Press, 1979); R. D. Alexander, *Darwinism and Human Affairs* (Seattle: University of Washington Press, 1979); Napoleon A. Chagnon and William Irons, *Evolutionary Biology and Human Social Behavior* (North Scituate, Mass.: Duxbury, Press, 1979); Tim Halliday, *Sexual Strategies* (Chicago: University of Chicago Press, 1980); Glenn Hausfater and Sarah Hrdy, *Infanticide* (Hawthorne, N.Y.: Aldine, 1980); Sarah Hrdy, *The Woman That Never Evolved* (Cambridge, Mass.: Harvard University Press, 1981); Nancy Tanner, *On Becoming Human* (New York: Cambridge University Press, 1981); Frances Dahlberg, *Woman the Gatherer* (New Haven: Yale University Press, 1981); Martin Daly and Margo Wilson, *Sex, Evolution, and Behavior* (Boston: Willard Grant Press, 1983); Bettyann Kevles, *Females of the Species* (Cambridge, Mass.: Harvard University Press, 1986); and Hanny Lightfoot-Klein, *Prisoners of Ritual: An Odyssey into Female Genital Circumcision in Africa* (Binghamton: Harrington Park Press, 1989).

Books dealing specifically with primate reproductive biology include C. E. Graham, *Reproductive Biology of the Great Apes* (New York: Academic Press, 1981); B. B. Smuts et al., *Primate Societies* (Chicago: University of Chicago Press, 1986); Jane Goodall, *The Chimpanzees of Gombe* (Cambridge, Mass.: Harvard University Press, 1986); Toshisada Nishida, *The Chimpanzees of the Mahale Mountains, Sexual and Life History Strategies* (Tokyo: University of Tokyo Press, 1990); and Takayoshi Kano, *The Last Ape: Pygmy Chimpanzee Behavior and Ecology* (Stanford: Stanford University Press, 1991).

Articles on the evolution of sexual physiology and behavior include the following: R. V. Short, "The evolution of human reproduction," *Proceed-*

ings of the Royal Society (*London*), series B 195: 3–24 (1976); R. V. Short, "Sexual selection and its component parts, somatic and genetical selection, as illustrated by man and the great apes," *Advances in the Study of Behavior* 9: 131–158 (1979); N. Burley, "The evolution of concealed ovulation," *American Naturalist* 114: 835–858 (1979); A. H. Harcourt et al., "Testis weight, body weight, and breeding system in primates," *Nature* 293: 55–57 (1981); R. D. Martin and R. M. May, "Outward signs of breeding," *Nature* 293: 7–9 (1981); M. Daly and M. I. Wilson, "Whom are newborn babies said to resemble?," *Ethology and Sociobiology* 3: 69–78 (1982); M. Daly, M. Wilson, and S. J. Weghorst, "Male sexual jealousy," *Ethology and Sociobiology* 3: 11–27 (1982); A. F. Dixson, "Observations on the evolution and behavioral significance of 'sexual skin' in female primates," *Advances in the Study of Behavior* 13: 63–106 (1983); S. J. Andelman, "Evolution of concealed ovulation in vervet monkeys (*Cercopithecus aethiops*)," *American Naturalist* 129: 785–799 (1987); and P. H. Harvey and R. M. May, "Out for the sperm count," *Nature* 337: 508–509 (1989).

Chapter 4 discussed several examples illustrating how birds combine extramarital sex with apparent monogamy. Detailed examples of such studies are presented in papers by D. W. Mock, "Display repertoire shifts and extra-marital courtship in herons," *Behaviour* 69: 57–71 (1979); P. Mineau and F. Cooke, "Rape in the lesser snow goose," *Behaviour* 70: 280–291 (1979); D. F. Werschel, "Nesting ecology of the Little Blue Heron: promiscuous behavior," *Condor* 84: 381–384 (1982); M. A. Fitch and G. W. Shuart, "Requirements for a mixed reproductive strategy in avian species," *American Naturalist* 124: 116–126 (1984); and R. Alatalo et al., "Extra-pair copulations and mate guarding in the polyterritorial pied flycatcher, *Ficedula hypoleuca*," *Behavior* 101: 139–155 (1987).

CHAPTER 5. *How We Pick Our Mates and Sex Partners*

Not surprisingly, this topic has called forth much scientific study. Some papers exemplifying the literature on mate choice by humans are the following: E. Walster et al., "Importance of physical attractiveness in dating behavior," *Journal of Personality and Social Psychology* 4: 508–516 (1966); J. N. Spuhler, "Assortative mating with respect to physical characteristics," *Eugenics Quarterly* 15: 128–140 (1968); E. Berscheid and K. Dion, "Physical attractiveness and dating choice: a test of the matching hypothesis," *Journal of Experimental Social Psychology* 7: 173–189 (1971); S. G. Vandenberg, "Assortative mating, or who marries whom?," *Behavior Genetics* 2: 127–157 (1972); G. E. DeYoung and B. Fleischer, "Motivational and personality trait relationships in mate selection," *Behavior Genetics* 6: 1–6 (1976); E. Crognier, "Assortative mating for physical features in an African population from Chad," *Journal of Human Evolution* 6: 105–114 (1977); P. N. Bentler and M. D. Newcomb, "Longitudinal study of marital success and failure,"

Journal of Consulting and Clinical Psychology 46: 1053–1070 (1978); R. C. Johnson et al., "Secular change in degree of assortatative mating for ability?," *Behavior Genetics* 10: 1–8 (1980); W. E. Nance et al., "A model for the analysis of mate selection in the marriages of twins," *Acta Geneticae Medicae Gemellologiae* 29: 91–101 (1980); D. Thiessen and B. Gregg, "Human assortative mating and genetic equilibrium: an evolutionary perspective," *Ethology and Sociobiology* 1: 111–140 (1980); D. M. Buss, "Human mate selection," *American Scientist* 73: 47–51 (1985); A. C. Heath and L. J. Eaves, "Resolving the effects of phenotype and social background on mate selection," *Behavior Genetics* 15: 75–90 (1985); and A. C. Heath et al., "No decline in assortative mating for educational level," *Behavior Genetics* 15: 349–369 (1985). Also relevant is a book by B. I. Murstein, *Who Will Marry Whom? Theories and Research in Marital Choice* (New York: Springer, 1976).

The literature on mate choice by animals is at least as extensive as that for humans. A good starting point is a book edited by Patrick Bateson, *Mate Choice* (Cambridge, Mass.: Cambridge University Press, 1983). Bateson's own studies on Japanese quail are summarized in chapter 11 of that book, and also in his papers "Sexual imprinting and optimal outbreeding," *Nature* 273: 659–660 (1978) and "Preferences for cousins in Japanese quail," *Nature* 295: 236–237 (1982). Studies of mice and rats that grow up to prefer the perfumes of their mothers or fathers are described by T. J. Fillion and E. M. Blass, "Infantile experience with suckling odors determines adult sexual behavior in male rats," *Science* 231: 729–731 (1986), and by B. D'Udine and E. Alleva, "Early experience and sexual preferences in rodents," pages 311–327 in the above-cited book by Patrick Bateson.

Finally, some other relevant papers are cited under the further readings for chapters 3, 4, 6, and 11.

CHAPTER 6. *Sexual Selection, and the Origin of Human Races*

Darwin's own classic account is still a good introduction to natural selection: Charles Darwin, *On the Origin of Species by Means of Natural Selection, or the Preservation of Favored Races in the Struggle for Life* (London: John Murray, 1859). An outstanding modern account is that of Ernst Mayr, *Animal Species and Evolution* (Cambridge, Mass.: Harvard University Press, 1963).

Three books by Carleton S. Coon describe human geographic variation, compare it to geographic variation in climate, and attempt to account for human variation in terms of natural selection: *The Origin of Races* (New York: Knopf, 1962), *The Living Races of Man* (New York: Knopf, 1965), and *Racial Adaptations* (Chicago: Nelson-Hall, 1982). Three other relevant books are by Stanley M. Garn, *Human Races,* 2nd ed. (Springfield, Ill.: Thomas, 1965), especially its chapter 5; K. F. Dyer, *The Biology of Racial*

Integration (Bristol: Scientechnica, 1974), especially its chapters 2 and 3; and A. S. Boughey, *Man and the Environment,* 2nd ed. (New York: Macmillan, 1975).

Interpretations of geographic variation in human skin color in terms of natural selection are put forward by W. F. Loomis, "Skin-pigment regulation of vitamin-D biosynthesis in man," *Science* 157: 501–506 (1967); Vernon Riley, *Pigmentation* (New York: Appleton-Century-Crofts, 1972), especially its chapter 2; R. F. Branda and J. W. Eaton, "Skin color and nutrient photolysis: an evolutionary hypothesis," *Science* 201: 625–626 (1978); P. J. Byard, "Quantitative genetics of human skin color," *Yearbook of Physical Anthropology* 24: 123–137 (1981); and W. J. Hamilton III, *Life's Color Code* (New York: McGraw-Hill, 1983). Human geographic variation in response to cold is described by G. M. Brown and J. Page, "The effect of chronic exposure to cold on temperature and blood flow of the hand," *Journal of Applied Physiology* 5: 221–227 (1952), and T. Adams and B. G. Covino, "Racial variations to a standardized cold stress," *Journal of Applied Physiology* 12: 9–12 (1958).

Just as for natural selection, Darwin's own account remains a good introduction to sexual selection: Charles Darwin, *The Descent of Man, and Selection in Relation to Sex* (London: John Murray, 1871). The further readings listed under Chapter 5 for mate selection by animals are also relevant to this chapter. Malte Andersson describes his experiments on how female widowbirds responded to males with artificially shortened or lengthened tails in an article "Female choice selects for extreme tail length in a widowbird," *Nature* 299: 818–820 (1982). Three papers describing mate choice by white, blue, and pink snow geese are by F. Cooke and C. M. McNally, "Mate selection and colour preferences in Lesser Snow Geese," *Behaviour* 53: 151–170 (1975); F. Cooke et al., "Assortative mating in Lesser Snow Geese (*Anser caerulescens*)," *Behavior Genetics* 6: 127–140 (1976); and F. Cooke and J. C. Davies, "Assortative mating, mate choice, and reproductive fitness in Snow Geese," pages 279–295 in the already-cited book *Mate Choice* by Patrick Bateson.

CHAPTER 7. *Why Do We Grow Old and Die?*

The classic paper in which George Williams presented an evolutionary theory of aging is "Pleiotropy, natural selection, and the evolution of senescence," *Evolution* 11: 398–411 (1957). Other papers that have employed evolutionary approaches are by G. Bell, "Evolutionary and nonevolutionary theories of senescence," *American Naturalist* 124: 600–603 (1984); E. Beutler, "Planned obsolescence in humans and in other biosystems," *Perspectives in Biology and Medicine* 29: 175–179 (1986); R. J. Goss, "Why mammals don't regenerate—or do they?," *News in Physiological Sciences* 2: 112–115 (1987);

L. D. Mueller, "Evolution of accelerated senescence in laboratory populations of *Drosophila,*" *Proceedings of the National Academy of Sciences* 84: 1974–1977 (1987); and T. B. Kirkwood, "The nature and causes of ageing," pages 193–206 in a book edited by D. Evered and J. Whelan, *Research and the Ageing Population* (Chichester: John Wiley, 1988).

Two books exemplifying the physiological (proximate-cause) approach to aging are by R. L. Walford, *The Immunologic Theory of Aging* (Copenhagen: Munksgaard, 1969), and MacFarlane Burnett, *Intrinsic Mutagenesis: A Genetic Approach to Ageing* (New York: John Wiley, 1974).

Some papers exemplifying the literature on biological repair and turnover are by R. W. Young, "Biological renewal: applications to the eye," *Transactions of the Ophthalmological Societies of the United Kingdom* 102: 42–75 (1982); A. Bernstein et al., "Genetic damage, mutation, and the evolution of sex," *Science* 229: 1277–1281 (1985); J. F. Dice, "Molecular determinants of protein half-lives in eukaryotic cells," *Federation of American Societies for Experimental Biology Journal* 1: 349–357 (1987); P. C. Hanawalt, "On the role of DNA damage and repair processes in aging: evidence for and against," pages 183–198 in a book edited by H. R. Warner et al., *Modern Biological Theories of Aging* (New York: Raven Press, 1987); and M. Radman and R. Wagner, "The high fidelity of DNA duplication," *Scientific American* 259, no. 2: 40–46 (August 1988).

While all readers will be aware of the changes in their own bodies with age, three papers describing the cruel facts for three different systems are: R. L. Doty et al., "Smell identification ability: changes with age," *Science* 226: 1441–1443 (1984); J. Menken et al., "Age and infertility," *Science* 233: 1389–1394 (1986); and R. Katzman, "Normal aging and the brain," *News in Physiological Sciences* 3: 197–200 (1988).

"The Adventure of the Creeping Man" will be found in Arthur Conan Doyle's *The Complete Sherlock Holmes* (New York: Doubleday, 1960). In case you think that attempts at self-rejuvenation by hormonal injections were only a fantasy of Doyle's, read about how it was actually attempted: David Hamilton, *The Monkey Gland Affair* (London: Chatto and Windus, 1986).

CHAPTER 8. *Bridges to Human Language*

How Monkeys See the World (Chicago: University of Chicago Press, 1990), by Dorothy Cheney and Robert Seyfarth, is not only a readable account of vervet vocal communications, but also a good introduction to studies of how animals in general communicate to each other and view the world.

Derek Bickerton has described his studies of creolization and his views on human language origins in two books and several papers. The books are *Roots of Language* (Ann Arbor: Karoma Press, 1981) and *Language and*

Species (Chicago: University of Chicago Press, 1990). The papers include "Creole languages," in *Scientific American* 249, no. 1: 116–122 (1983); "The language bioprogram hypothesis," in *Behavioral and Brain Sciences* 7: 173–221 (1984); and "Creole languages and the bioprogram," in *Linguistics: The Cambridge Survey,* vol. 2, pp. 267–284, edited by F. J. Newmeyer (Cambridge: Cambridge University Press, 1988). The second of those articles includes, and the third is immediately followed by, presentations by other authors whose views often diverge from Bickerton's.

Pidgin and Creole Languages, by Robert A. Hall, Jr. (Ithaca: Cornell University Press, 1966), is an older account of its subject. The best introduction to Neo-Melanesian is the book by F. Mihalic, *The Jacaranda Diary and Grammar of Melanesian Pidgin* (Milton, Queensland: Jacaranda Press, 1971). Roger Keesing's book *Melanesian Pidgin and the Oceanic Substrate* (Stanford: Stanford University Press, 1988) explores the history of Neo-Melanesian.

Among the many influential books on language by Noam Chomsky are *Language and Mind* (New York: Harcourt Brace, 1968) and *Knowledge of Language: Its Nature, Origin, and Use* (New York: Praeger, 1985).

References to some related fields that I mentioned only briefly in Chapter 8 will also be of interest. Susan Curtiss's book *Genie: a Psycholinguistic Study of a Modern-Day "Wild Child"* (New York: Academic Press, 1977) relates both a gut-wrenching human tragedy and a detailed study of a child whose parents' pathologies isolated her from normal human language and contact until age thirteen. Recent accounts of efforts to teach languagelike communication to captive apes include Carolyn Ristau and Donald Robbins's paper "Language and the great apes: a critical review," in *Advances in the Study of Behavior,* vol. 12, pp. 141–255, edited by J. S. Rosenblatt et al. (New York: Academic Press, 1982); E. S. Savage-Rumbaugh, *Ape Language: From Conditioned Response to Symbol* (New York: Columbia University Press, 1986); and "Symbols: their communicative use, comprehension, and combination by bonobos (*Pan paniscus*)," by E. S. Savage-Rumbaugh et al., in *Advances in Infancy Research,* vol. 6, pp. 221–278, edited by Carolyn Rovee-Collier and Lewis Lipsitt (Norwood, N.J.: Ablex Publishing Corporation, 1990). Some starting points in the large literature on early language learning by children include Melissa Bowerman's chapter "Language Development" in *Handbook of Cross-cultural Psychology: Developmental Psychology,* vol. 4, pp. 93–185, edited by Harvey Triandis and Alastair Heron (Boston: Allyn and Bacon, 1981); Eric Wanner and Lila Gleitman, *Language Acquisition: The State of the Art* (Cambridge, Mass.: Cambridge University Press, 1982); Dan Slobin, *The Crosslinguistic Study of Language Acquisition,* vols. 1 and 2 (Hillsdale, N.J.: Lawrence Erlbaum Associates, 1985); and Frank S. Kessel, *The Development of Language and Language Researchers: Essays in Honor of Roger Brown* (Hillsdale, N.J.: Lawrence Erlbaum Associates, 1988).

CHAPTER 9. *Animal Origins of Art*

The book that describes elephant art and illustrates it with photographs of the artist and of her drawings is by David Gucwa and James Ehmann, *To Whom It May Concern: An Investigation of the Art of Elephants* (New York: Norton, 1985). For a similar account of ape art, see Desmond Morris, *The Biology of Art* (New York: Knopf, 1962). Animal art is also treated by Thomas Sebeok, *The Play of Musement* (Bloomington: Indiana University Press, 1981).

There are two fine illustrated books on bowerbirds and birds of paradise, with pictures of their bowers: E. T. Gilliard, *Birds of Paradise and Bower Birds* (Garden City, N.Y.: Natural History Press, 1969), and W. T. Cooper and J. M. Forshaw, *The Birds of Paradise and Bower Birds* (Sydney: Collins, 1977). For a more recent technical account, see my article "Biology of birds of paradise and bowerbirds," *Annual Reviews of Ecology and Systematics* 17: 17–37 (1986). I published two accounts of the bowerbird species with the fanciest bower: "Bower building and decoration by the bowerbird *Amblyornis inornatus,*" *Ethology* 74: 177–204 (1987); and "Experimental study of bower decoration by the bowerbird *Amblyornis inornatus,* using colored poker chips," *American Naturalist* 131: 631–653 (1988). Gerald Borgia proved by experiments that female bowerbirds really do care about males' bower decorations, in his paper, "Bower quality, number of decorations and mating success of male satin bowerbirds (*Ptilonorhynchus violaceus*): an experimental analysis," *Animal Behaviour* 33: 266–271 (1985). Birds of paradise with somewhat similar habits are described by S. G. and M. A. Pruett-Jones, "The use of court objects by Lawes' Parotia," *Condor* 90: 538–545 (1988).

CHAPTER 10. *Agriculture's Mixed Blessings*

The health consequences of giving up hunting for farming receive detailed treatment in a book edited by Mark Cohen and George Armelagos, *Paleopathology at the Origins of Agriculture* (Orlando: Academic Press, 1984), and in *The Paleolithic Prescription* (New York: Harper & Row, 1988) by S. Boyd Eaton, Marjorie Shostak, and Melvin Konner. The world's hunter-gatherers are summarized in a book edited by Richard B. Lee and Irven DeVore, *Man the Hunter* (Chicago: Aldine, 1968). References describing the work schedule of hunter-gatherers, and in some cases comparing it with that of farmers, include the same book, plus the book by Richard Lee *The !Kung San* (Cambridge, Mass.: Cambridge University Press, 1979), and the following articles: K. Hawkes et al., "Aché at the settlement: contrasts between farming and foraging," *Human Ecology* 15: 133–161 (1987); K. Hawkes et al., "Hardworking Hadza grandmothers," pages 341–366 in *Comparative Socioecology of Mammals and Man,* edited by V. Standen and R.

Foley (London: Blackwell, 1987); and K. Hill and A. M. Hurtado, "Hunter-gatherers of the New World," *American Scientist* 77: 437–443 (1989). The slow spread of ancient farmers across Europe is described by Albert J. Ammerman and L. L. Cavalli-Sforza, *The Neolithic Transition and the Genetics of Populations in Europe* (Princeton: Princeton University Press, 1984).

CHAPTER 11. *Why Do We Smoke, Drink, and Use Dangerous Drugs?*

Amotz Zahavi explains his handicap theory in two papers: "Mate selection—a selection for a handicap," *Journal of Theoretical Biology* 53: 205–214 (1975), and "The cost of honesty (further remarks on the handicap principle)," *Journal of Theoretical Biology* 67: 603–605 (1977). Two other well-known models of how animals evolve to choose their mates are the runaway selection model and the truth-in-advertising model. The former was developed in a book by R. A. Fisher, *The Genetical Theory of Natural Selection* (Oxford: Clarendon Press, 1930); the latter, in a paper by A. Kodric-Brown and J. H. Brown, "Truth in advertising: the kinds of traits favored by sexual selection," *American Naturalist* 14: 309–323 (1984). These various models are evaluated by Mark Kirkpatrick and Michael Ryan, "The evolution of mating preferences and the paradox of the lek," *Nature* 350: 33–38 (1991). Melvin Konner develops another perspective on risky human behaviors in a chapter "Why the reckless survive" from his book with the same title (New York: Viking, 1990). For discussions of American Indian enemas, see Peter Furst's and Michael Coe's account of the discovery of Maya enema vases in their article "Ritual enemas," *Natural History Magazine* 86: 88–91 (March 1977); Johannes Wilbert's book *Tobacco and Shamanism in South America* (New Haven: Yale University Press, 1987); and Justin Kerr's *The Maya Vase Book*, 2 vols. (New York: Kerr Associates, 1989 and 1990), illustrating Maya vases and analyzing one enema vase in detail on pages 349–361 of volume 2. Also relevant are the many further readings on sexual selection and mate choice already listed under Chapters 5 and 6.

CHAPTER 12. *Alone in a Crowded Universe*

Pioneering calculations arguing for the existence of intelligent extraterrestrial life were carried out by I. S. Shklovskii and Carl Sagan, *Intelligent Life in the Universe* (San Francisco: Holden-Day, 1966). Arguments for and against, and what it might mean for us if we do discover extraterrestrials out there, form the subject of the book *Extraterrestrials: Science and Alien Intelligence,* edited by E. Regis, Jr. (Cambridge, Mass.: Cambridge University Press, 1985).

CHAPTER 13. *The Last First Contacts*

Bob Connolly's and Robin Anderson's book *First Contact* (New York: Viking Penguin, 1987) describes first contact in the New Guinea highlands through the eyes of both the whites and the New Guineans who met there. The quotation on page 229 is taken from their book. Other gripping accounts of first contacts and of precontact conditions include Don Richardson's *Peace Child* (Ventura: Regal Books, 1974) for the Sawi people of southwest New Guinea, and Napoleon A. Chagnon's *Yanomamo, The Fierce People,* 3rd edition (New York: Holt, Rinehart and Winston, 1983) for the Yanomamo Indians of Venezuela and Brazil. A good history of the exploration of New Guinea is by Gavin Souter, *New Guinea: The Last Unknown* (London: Angus and Robertson, 1963). The leaders of the Third Archbold Expedition describe their entrance into the Grand Valley of the Balim River in the report by Richard Archbold et al., "Results of the Archbold Expedition," *Bulletin of the American Museum of Natural History* 79: 197–288 (1942). Two accounts by earlier explorers who attempted to penetrate the mountains of New Guinea are by A. F. R. Wollaston, *Pygmies and Papuans* (London: Smith Elder, 1912), and A. S. Meek, *A Naturalist in Cannibal Land* (London: Fisher Unwin, 1913).

CHAPTER 14. *Accidental Conquerors*

Books that discuss plant as well as animal domestication in relation to the development of civilization include C. D. Darlington, *The Evolution of Man and Society* (New York: Simon and Schuster, 1969); Peter J. Ucko and G. W. Dimbleby, *The Domestication and Exploitation of Plants and Animals* (Chicago: Aldine, 1969); Erich Isaac, *Geography of Domestication* (Englewood Cliffs, N.J.: Prentice-Hall, 1970); and David R. Harris and Gordon C. Hillman, *Foraging and Farming* (London: Unwin Hyman, 1989).

References on animal domestication include S. Bokonyi, *History of Domestic Mammals in Central and Eastern Europe* (Budapest: Akademiai, 1974); S. J. M. Davis and F. R. Valla, "Evidence for domestication of the dog 12,000 years ago in the Natufian of Israel," *Nature* 276: 608–610 (1978); Juliet Clutton-Brock, "Man-made dogs," *Science* 197: 1340–1342 (1977), and *Domesticated Animals from Early Times* (London: British Museum of Natural History, 1981); Andrew Sherratt, "Plough and pastoralism: aspects of the secondary products revolution," pages 261–305 in a book edited by Ian Hodder et al., *Pattern of the Past* (Cambridge: Cambridge University Press, 1981); Stanley J. Olsen, *Origins of the Domestic Dog* (Tucson: University of Arizona Press, 1985); E. S. Wing, "Domestication of Andean mammals," pages 246–264 in a book edited by F. Vuilleumier and M. Monasterio, *High Altitude Tropical Biogeography* (New York: Oxford University Press, 1986); Simon N. J. Davis, *The Archaeology of Animals* (New Haven: Yale University Press, 1987); Dennis C. Turner and Patrick Bateson, *The Domestic Cat: The Biology of Its Behavior* (Cambridge: Cambridge University Press, 1988); and Wolf Herre and Manfred Rohrs, *Haustiere—zoologisch gesehen,* 2nd ed. (Stuttgart: Fischer, 1990).

Domestication specifically of the horse, and its importance, are the subjects of books by Frank G. Row, *The Indian and the Horse* (Norman: University of Oklahoma Press, 1955); Robin Law, *The Horse in West African History* (Oxford: Oxford University Press, 1980); and Matthew J. Kust, *Man and Horse in History* (Alexandria, Va.: Plutarch Press, 1983). The development of wheeled vehicles, including war chariots, is treated in books by M. A. Littauer and J. H. Crouwel, *Wheeled Vehicles and Ridden Animals in the Ancient Near East* (Leiden: Brill, 1979), and by Stuart Piggott, *The Earliest Wheeled Transport* (London: Thames and Hudson, 1983). Edward Shaughnessy describes the arrival of the horse and chariot in China in an article "Historical perspectives on the introduction of the chariot into China," *Harvard Journal of Asiatic Studies* 48: 189–237 (1988).

For general accounts of plant domestication, see Kent V. Flannery, "The origins of agriculture," *Annual Review of Anthropology* 2: 271–310 (1973); Charles B. Heiser, Jr., *Seed to Civilization,* new edition (Cambridge, Mass.: Harvard University Press, 1990), and *Of Plants and Peoples* (Norman: University of Oklahoma Press, 1985); David Rindos, *The Origins of Agriculture: An Evolutionary Perspective* (New York: Academic Press, 1984); and Hugh H. Iltis, "Maize evolution and agricultural origins," pages 195–213 in a book edited by T. R. Soderstrom et al., *Grass Systematics and Evolution* (Washington, D.C.: Smithsonian Institution Press, 1987). This and other papers by Iltis are a stimulating source of ideas about the differing ease of cereal domestication in the Old and New World.

Plant domestication specifically in the Old World is treated by Jane Renfrew, *Palaeoethnobotany* (New York: Columbia University Press, 1973), and by Daniel Zohary and Maria Hopf, *Domestication of Plants in the Old World* (Oxford: Clarendon Press, 1988). Corresponding accounts for the New World include Richard S. MacNeish, "The food-gathering and incipient agricultural stage of prehistoric Middle America," pages 413–426 in a book edited by Robert Wauchope and Robert C. West, *Handbook of Middle American Indians,* Volume 1: *Natural Environment and Early Cultures* (Austin: University of Texas Press, 1964); P. C. Mangelsdorf et al., "Origins of agriculture in Middle America," pages 427–445 in the just-cited book by Wauchope and West; D. Ugent, "The potato," *Science* 170: 1161–1166 (1970); C. B. Heiser, Jr., "Origins of some cultivated New World plants," *Annual Reviews of Ecology and Systematics* 10: 309–326 (1979); H. H. Iltis, "From teosinte to maize: the catastrophic sexual dismutation," *Science* 222: 886–894 (1983); William F. Keegan, *Emergent Horticultural Economies of the Eastern Woodlands* (Carbondale: Southern Illinois University, 1987); and B. D. Smith, "Origins of agriculture in eastern North America," *Science* 246: 1566–1571 (1989). Three pioneering books point out the asymmetrical intercontinental spread of diseases, pests, and weeds: William H. McNeill, *Plagues and Peoples* (Garden City, N.Y.: Anchor Press, 1976); and Alfred W. Crosby, *The Columbian Exchange: Biological and Cultural Consequences of 1492* (Westport: Greenwood Press, 1972), and *Ecological Imperialism: The Biological Expansion of Europe, 900–1900* (Cambridge: Cambridge University Press, 1986).

CHAPTER 15. *Horses, Hittites, and History*

Two stimulating, knowledgeable recent books summarizing the Indo-European problem are by Colin Renfrew, *Archaeology and Language* (Cambridge: Cambridge University Press, 1987), and J. P. Mallory, *In Search of the Indo-Europeans* (London: Thames and Hudson, 1989). For the reasons explained in my chapter 15, I agree with Mallory's conclusions, and disagree with Renfrew's, concerning the approximate time and place of proto-Indo-European origins.

An older but still useful comprehensive multiauthored book is by George Cardona et al., *Indo-European and Indo-Europeans* (Philadelphia: University of Pennsylvania Press, 1970). A journal titled (what else?) *The Journal of Indo-European Studies* is the main outlet for technical publication in this field.

The view that both Mallory and I find persuasive is supported in the writings of Marija Gimbutas, who is the author of four books in this field: *The Balts* (New York: Praeger, 1963), *The Slavs* (London: Thames and Hudson, 1971), *The Goddesses and Gods of Old Europe* (London: Thames and Hudson, 1982), and *The Language of the Goddess* (New York: Harper & Row, 1989). Gimbutas also described her work in chapters in the book by Cardona et al. cited above, in the books by Polomé and by Bernhard and Kandler-Pálsson cited below, and in the *Journal of Indo-European Studies* 1: 1–20 and 163–214 (1973), 5: 277–338 (1977), 8: 273–315 (1980), and 13: 185–201 (1985).

Books or monographs dealing with early Indo-European peoples themselves are by Emile Benveniste, *Indo-European Language and Society* (London: Faber and Faber, 1973); Edgar Polomé, *The Indo-Europeans in the Fourth and Third Millennia* (Ann Arbor: Karoma, 1982); Wolfram Bernhard and Anneliese Kandler-Pálsson, *Ethnogenese europäischer Völker* (Stuttgart: Fischer, 1986); and Wolfram Nagel, "Indogermanen und Alter Orient: Rückblick und Ausblick auf den Stand des Indogermanenproblems," *Mitteilungen der Deutschen Orient-Gesellschaft zu Berlin* 119: 157–213 (1987). Books on the languages themselves include those by Henrik Birnbaum and Jaan Puhvel, *Ancient Indo-European Dialects* (Berkeley: University of California Press, 1966); W. B. Lockwood, *Indo-European Philology* (London: Hutchinson, 1969); Norman Bird, *The Distribution of Indo-European Root Morphemes* (Wiesbaden: Harrassowitz, 1982); and Philip

Baldi, *An Introduction to the Indo-European Languages* (Carbondale: Southern Illinois University Press, 1983). Paul Friedrich's book *Proto-Indo-European Trees* (Chicago: University of Chicago Press, 1970) uses the evidence of tree names in an attempt to deduce the Indo-European homeland.

W. P. Lehmann and L. Zgusta provide and discuss a sample of reconstructed Proto-Indo-European in their chapter "Schleicher's tale after a century," pages 455–466 in *Studies in Diachronic, Synchronic, and Topological Linguistics,* edited by Bela Brogyany (Amsterdam: Benjamins, 1979). For a slightly altered version of their sample, see page 274 of this book.

The references to the domestication and importance of horses cited under Chapter 14 are also relevant to the role of horses in Indo-European expansion. Papers specifically on this subject are by David Anthony, "The 'Kurgan culture,' Indo-European origins and the domestication of the horse: a reconsideration," *Current Anthropology* 27: 291–313 (1986); and by David Anthony and Dorcas Brown, "The origins of horseback riding," *Antiquity* 65: 22–38 (1991).

CHAPTER 16. *In Black and White*

Three books providing general surveys of genocide are by Irving Horowitz, *Genocide: State Power and Mass Murder* (New Brunswick: Transaction Books, 1976); Leo Kuper, *The Pity of It All* (London: Gerald Duckworth, 1977); and Leo Kuper, *Genocide: Its Political Use in the 20th Century* (New Haven, Yale University Press, 1981). A gifted psychiatrist, Robert J. Lifton, has published studies of the psychological effects of genocide on its perpetrators and survivors, including *Death in Life: Survivors of Hiroshima* (New York: Random House, 1967) and *The Broken Connection* (New York: Simon and Schuster, 1979).

Books that describe the extermination of the Tasmanians and other native Australian groups include: N. J. B. Plomley, *Friendly Mission: The Tasmanian Journals and Papers of George Augustus Robinson 1829–1834* (Hobart: Tasmanian Historical Research Association, 1966); C. D. Rowley, *The Destruction of Aboriginal Society,* vol. 1 (Canberra: Australian National University Press, 1970); and Lyndall Ryan, *The Aboriginal Tasmanians* (St. Lucia: University of Queensland Press, 1981). Patricia Cobern's letter indignantly denying that Australian whites exterminated the Tasmanians has been reprinted as an appendix to the book by J. Peter White and James F. O'Connell, *A Prehistory of Australia, New Guinea, and Sahul* (New York: Academic Press, 1982).

Among the many books and articles detailing the extermination of American Indians by white settlers are: Wilcomb E. Washburn, "The moral and legal justification for dispossessing the Indians," pages 15–32 in a book edited by James Morton Smith, *Seventeenth Century America* (Chapel Hill:

University of North Carolina Press, 1959); Alvin M. Josephy, Jr., *The American Heritage Book of Indians* (New York: Simon and Schuster, 1961); Howard Peckham and Charles Gibson, *Attitudes of Colonial Powers Towards the American Indian* (Salt Lake City: University of Utah Press, 1969); Francis Jennings, *The Invasion of America: Indians, Colonialism, and the Cant of Conquest* (Chapel Hill: University of North Carolina Press, 1975); Wilcomb E. Washburn, *The Indian in America* (New York: Harper & Row, 1975); Arrell Morgan Gibson, *The American Indian, Prehistory to the Present* (Lexington, Mass.: Heath, 1980); and Wilbur H. Jacobs, *Dispossessing the American Indian* (Norman: University of Oklahoma Press, 1985). The extermination of the Yahi Indians, and the survival of Ishi, are the subjects of Theodora Kroeber's classic book *Ishi in Two Worlds: A Biography of the Last Wild Indian in North America* (Berkeley: University of California Press, 1961). The extermination of Brazil's Indians is treated by Sheldon Davis, *Victims of the Miracle* (Cambridge: Cambridge University Press, 1977).

Genocide under Stalin is described in books by Robert Conquest, including *The Harvest of Sorrow* (New York: Oxford University Press, 1986).

Accounts of murder and mass murder of animals by other animals of the same species are given by E. O. Wilson, *Sociobiology* (Cambridge, Mass: Harvard University Press, 1975); Cynthia Moss, *Portraits in the Wild,* 2nd ed. (Chicago: University of Chicago Press, 1982); and Jane Goodall, *The Chimpanzees of Gombe* (Cambridge, Mass.: Harvard University Press, 1986). Hans Kruuk's account of hyena murder that I quote is from his book *The Spotted Hyena: a Study of Predation and Social Behavior* (Chicago: University of Chicago Press, 1972).

CHAPTER 17. *The Golden Age That Never Was*

Extinctions of animals in the Late Pleistocene and Early Recent era are described comprehensively in a book edited by Paul Martin and Richard Klein, *Quaternary Extinctions* (Tucson: University of Arizona Press, 1984). For the history of deforestation, see John Perlin's book *A Forest Journey* (New York: Norton, 1989).

Comprehensive accounts of New Zealand's plants, animals, geology, and climate will be found in a book edited by G. Kuschel, *Biogeography and Ecology in New Zealand* (Hague: Junk, V. T., 1975). New Zealand extinctions are summarized in chapters 32–34 of the above-cited book by Martin and Klein. Atholl Anderson summarized our knowledge of moas in his book *Prodigious Birds* (Cambridge: Cambridge University Press, 1989). Moas are also the subject of a supplement to the *New Zealand Journal of Ecology,* vol. 12 (1989); see especially the articles by Richard Holdaway on pages 11–25, and by Ian Atkinson and R. M. Greenwood on pages 67–96. Other key articles relevant to moas are by G. Caughley, "The colonization of New Zealand by the Polynesians," *Journal of the Royal Society of New*

Zealand 18: 245–270 (1988), and by A. Anderson, "Mechanics of overkill in the extinction of New Zealand moas," *Journal of Archaeological Science* 16: 137–151 (1989).

Extinctions in Madagascar and Hawaii are described in Chapters 26 and 35 respectively of the above-cited book by Martin and Klein. The Henderson Island story is told by David Steadman and Storrs Olson, "Bird remains from an archaeological site on Henderson Island, South Pacific: man-caused extinctions on an 'uninhabited' island," *Proceedings of the National Academy of Sciences* 82: 6191–6195 (1985). See under suggested readings for Chapter 18 for accounts of extinctions in the Americas.

The grisly end of Easter Island civilization is recounted by Patrick V. Kirch in his book *The Evolution of the Polynesian Chiefdoms* (Cambridge: Cambridge University Press, 1984). Easter's deforestation was reconstructed by J. Flenley, "Stratigraphic evidence of environmental change on Easter Island," *Asian Perspectives* 22: 33–40 (1979), and by J. Flenley and S. King, "Late Quaternary pollen records from Easter Island," *Nature* 307: 47–50 (1984).

Some accounts of the rise and fall of Anasazi settlement at Chaco Canyon are: J. L. Betancourt and T. R. Van Devender, "Holocene vegetation in Chaco Canyon, New Mexico," *Science* 214: 656–658 (1981); M. L. Samuels and J. L. Betancourt, "Modeling the long-term effects of fuelwood harvests on pinyon-juniper woodlands," *Environmental Management* 6: 505–515 (1982); J. L. Betancourt et al., "Prehistoric long-distance transport of construction beams, Chaco Canyon, New Mexico," *American Antiquity* 51: 370–375 (1986); Kendrick Frazier, *People of Chaco: A Canyon and Its Culture* (New York: Norton, 1986); and Alden C. Hayes et al., *Archaeological Surveys of Chaco Canyon* (Albuquerque: University of New Mexico Press, 1987).

Everything that anyone would want to know about *Packrat Middens* is described in a book with that title by Julio Betancourt, Thomas Van Devender, and Paul Martin (Tucson: University of Arizona Press, 1990). In particular, Chapter 19 of that book analyzes the hyrax middens from Petra.

The possible link between environmental damage and the decline of Greek civilization is explored by K. O. Pope and T. H. van Andel, "Late Quaternary civilization and soil formation in the southern Argolid: its history, causes and archaeological implications," *Journal of Archaeological Science* 11: 281–306 (1984); T. H. van Andel et al., "Five thousand years of land use and abuse in the southern Argolid," *Hesperia* 55: 103–128 (1986); and C. Runnels and T. H. van Andel, "The evolution of settlement in the southern Argolid, Greece: an economic explanation," *Hesperia* 56: 303–334 (1987).

Books on the rise and fall of Maya civilization include those by T. Patrick Culbert, *The Classic Maya Collapse* (Albuquerque: University of New Mexico Press, 1973); Michael D. Coe, *The Maya,* 3rd. ed. (London: Thames and Hudson, 1984); Sylvanus G. Morley et al., *The Ancient Maya,* 4th ed. (Stanford: Stanford University Press, 1983); Charles Gallenkamp, *Maya: The*

Riddle and Rediscovery of a Lost Civilization, 3rd rev. ed. (New York: Viking Penguin, 1985); and Linda Schele and David Freidel, *A Forest of Kings* (New York: William Morrow, 1990).

For a comparative account of collapses of civilizations, see the book edited by Norman Yoffee and George L. Cowgill, *The Collapse of Ancient States and Civilizations* (Tucson: University of Arizona Press, 1988).

CHAPTER 18. *Blitzkrieg and Thanksgiving in the New World*

Three books provide good starting points and many references to the large, contentious literature on human settlement and extinctions of large animals in the New World. They are: the book by Paul Martin and Richard Klein cited under Chapter 17; Brian Fagan, *The Great Journey* (New York: Thames and Hudson, 1987); and Ronald C. Carlisle (editor), *Americans Before Columbus: Ice-Age Origins* (Ethnology Monograph No. 12, Department of Anthropology, University of Pittsburgh, 1988).

The blitzkrieg hypothesis was outlined by Paul Martin in his article "The discovery of America," *Science* 179: 969–974 (1973), and modeled mathematically by J. E. Mosimann and Martin in "Simulating overkill by Paleoindians," *American Scientist* 63: 304–313 (1975).

The series of articles that C. Vance Haynes, Jr., has published on Clovis culture and its origins include a chapter (pages 345–353) of the book by Martin and Klein (above), plus the following selected articles: "Fluted projectile points: their age and dispersion," *Science* 145: 1408–1413 (1961); "The Clovis culture," *Canadian Journal of Anthropology* 1: 115–121 (1980); and "Clovis origin update," *The Kiva* 52: 83–93 (1987).

For the simultaneous extinction of the Shasta ground sloth and Harrington's mountain goat, see J. I. Mead et al., "Extinction of Harrington's mountain goat," *Proceedings of the National Academy of Sciences* 83: 836–839 (1986). Critiques of pre-Clovis claims are provided by Roger Owen in a chapter "The Americas: the case against an Ice-Age human population," pages 517–563 in a book edited by Fred H. Smith and Frank Spencer, *The Origins of Modern Humans* (New York: Liss, 1984); by Dena Dincauze, "An archaeo-logical evaluation of the case for pre-Clovis occupations," in *Advances in World Archaeology* 3: 275–323 (1984); and by Thomas Lynch, "Glacial-age man in South America? A critical review," in *American Antiquity* 55: 12–36 (1990). Arguments in support of a pre-Clovis date for human occupation levels at Meadowcroft Rockshelter are summarized by James Adovasio in "Meadowcroft Rockshelter, 1973–1977: a synopsis," pages 97–131 in J. E. Ericson et al., *Peopling of the New World* (Los Altos, Calif.: Ballena Press, 1982), and in "Who are those guys?: some biased thoughts on the initial peopling of the New World," pages 45–61 in the above-cited *Americans Before Columbus: Ice-Age Origins,* edited by Ronald C. Carlisle. The first of several projected volumes with a detailed description of the Monte Verde site is by T. D. Dillehay, *Monte Verde: A Late Pleistocene Settlement in Chile,* Volume I: *Palaeoenvironment and Site Contexts* (Washington, D.C.: Smithsonian Institution Press, 1989).

Readers interested in keeping up on the story of the first Americans and the last mammoths will enjoy subscribing to a quarterly newspaper, *Mammoth Trumpet,* obtainable from the Center for the Study of the First Americans, Anthropology Department, Oregon State University, Corvallis, Ore., 97331.

CHAPTER 19. *The Second Cloud*

Species-by-species accounts of extinct and endangered species are contained in the Red Data Books published by the International Union for Conservation of Nature and Natural Resources (abbreviated IUCN). There are separate books for various groups of plants and animals, and separate books are also now appearing for different continents. Corresponding books for birds have been prepared by the International Council for Bird Preservation, abbreviated ICBP: Warren B. King, editor, *Endangered Birds of the World: The ICBP Red Data Book* (Washington, D.C.: Smithsonian Institution Press, 1981); and N. J. Collar and P. Andrew, *Birds to Watch: The ICBP World Checklist of Threatened Birds* (Cambridge: ICBP, 1988).

A summary and analysis of modern and Ice Age extinctions and their mechanisms are provided by my article "Historic extinctions: a Rosetta Stone for understanding prehistoric extinctions," pages 824–862 in the book *Quaternary Extinctions* by Martin and Klein, cited under Chapter 17. The problem of overlooked extinctions is discussed in my article "Extant unless proven extinct? Or extinct unless proven extant?" in *Conservation Biology* 1: 77–79 (1987). Terry Erwin estimates the total number of living species in a paper "Tropical forests: their richness in Coleoptera and other arthropod species," *The Coleopterists' Bulletin* 36: 74–75 (1982).

Further readings on Pleistocene and Early Recent extinctions are given under Chapters 17 and 18. In addition, Storrs Olson reviews extinctions of island birds in an article "Extinction on islands: man as a catastrophe," pages 50–53 of a book edited by David Western and Mary Pearl, *Conservation for the Twenty-first Century* (New York: Oxford University Press, 1989). Ian Atkinson's article on pages 54–75 of the same book, "Introduced animals and extinctions," summarizes the havoc wrought by rats and other pests.

EPILOGUE: *Nothing Learned, and Everything Forgotten?*

Many excellent books discuss the present and future of the extinction crisis and the other crises now facing humanity, their causes, and what to do about them. Among them are the following:

JOHN J. BERGER, *Restoring the Earth: How Americans are Working to Renew Our Damaged Environment* (New York: Knopf, 1985).

———, editor, *Environmental Restoration: Science and Strategies for Restoring the Earth* (Washington, D.C.: Island Press, 1990).

JOHN CAIRNS, JR., *Rehabilitating Damaged Ecosystems* (Boca Raton, Fl.: CRC Press, 1988).

JOHN CAIRNS, JR., K. L. Dickson, and E. E. Herricks, *Recovery and Restoration of Damaged Ecosystems* (Charlottesville: University Press of Virginia, 1977).

ANNE AND PAUL EHRLICH, *Earth* (New York: Franklin Watts, 1987).

PAUL AND ANNE EHRLICH, *Extinction* (New York: Random House, 1981).

———, *The Population Explosion* (New York: Simon and Schuster, 1990).

———, *Healing Earth* (New York: Addison Wesley, 1991).

PAUL EHRLICH ET AL., *The Cold and the Dark* (New York: Norton, 1984).

D. FURGUSON AND N. FURGUSON, *Sacred Cows at the Public Trough* (Bend, Ore.: Maverick Publications, 1983).

SUZANNE HEAD AND ROBERT HEINZMAN, editors, *Lessons of the Rainforest* (San Francisco: Sierra Club Books, 1990).

JEFFREY A. McNEELY, *Economics and Biological Diversity* (Gland: International Union for the Conservation of Nature, 1988).

JEFFREY A. McNEELEY ET AL., *Conserving the World's Biological Diversity* (Gland: International Union for the Conservation of Nature, 1990).

NORMAN MYERS, *Conversion of Tropical Moist Forests* (Washington, D.C.: National Academy of Sciences, 1980).

———, *Gaia: An Atlas of Planet Management* (New York: Doubleday, 1984).

———, *The Primary Source* (New York: Norton, 1985).

MICHAEL OPPENHEIMER AND ROBERT BOYLE, *Dead Heat: The Race against the Greenhouse Effect* (New York: Basic Books, 1990).

WALTER V. REID AND KENTON R. MILLER, *Keeping Options Alive: The Scientific Basis for Conserving Biodiversity* (Washington, D.C.: World Resources Institute, 1989).

SHARON L. ROAN, *Ozone Crisis: The Fifteen-Year Evolution of a Sudden Global Emergency* (New York: Wiley, 1989).

ROBIN RUSSELL JONES AND TOM WIGLEY, editors, *Ozone Depletion: Health and Environmental Consequences* (New York: Wiley, 1989).

STEVEN H. SCHNEIDER, *Global Warming: Are We Entering the Greenhouse Century?* 2nd ed. (San Francisco: Sierra Club Books, 1990).

MICHAEL E. SOULÉ, editor, *Conservation Biology: The Science of Scarcity and Diversity* (Sunderland, Mass.: Sinauer, 1986).

JOHN TERBORGH, *Where Have All the Birds Gone?* (Princeton: Princeton University Press, 1990).

E. O. WILSON, *Biophilia* (Cambridge, Mass.: Harvard University Press, 1984).

———, editor, *Biodiversity* (Washington, D.C.: National Academy Press, 1988).

Finally, readers interested enough to want to pursue further readings may also want suggestions about what to do to reduce the risk that our children's generation will be the one to go extinct. As I explained in the text, the average citizen can do a good deal, both by being active politically and by giving even modest amounts of money to conservation organizations. Here are the names, addresses, and telephone numbers of a few of the best-known and largest such organizations, among the many that are worthy of support:

Conservation International, 1015 Eighteenth Street NW, Suite 1000, Washington, D.C. 20036 (202-429-5660).

Defenders of Wildlife, 1244 Nineteenth Street NW, Washington, D.C. 20036 (202-659-9510).

Ducks Unlimited, 1 Waterfowl Way, Long Grove, IL 60047 (708-438-4300).

Environmental Defense Fund, 257 Park Avenue South, New York, NY 10010 (212-505-2100).

Friends of the Earth, 218 D Street SE, Washington, D.C. 20002 (202-544-2600).

Greenpeace, 436 U Street NW, Box 3720, Washington, D.C. 20007 (202-462-8817).

League of Conservation Voters, 1150 Connecticut Avenue NW, Washington, D.C. 20036 (202-785-8683).

National Audubon Society, 950 Third Avenue, New York, NY 10022 (212-546-9100).

National Resources Defense Council, 40 West Twentieth Street, New York, NY 10011 (212-727-2700).

Nature Conservancy, 1815 Lynn Street, Arlington, VA 22209 (703-841-5300).

Rainforest Action Network, 301 Broadway, Suite A, San Francisco, CA 94133 (415-398-4404).

Sierra Club, 730 Polk Street, San Francisco, CA 94109 (415-776-2211).

Trout Unlimited, 501 Church Street NE, Vienna, VA 22180 (703-281-1100).

Wilderness Society, 900 Seventeenth Street NW, Washington, D.C. 20006-2596 (202-833-2300).

World Wildlife Fund, National Headquarters, 1250 Twenty-Fourth Street NW, Suite 500, Washington, D.C. 20037 (202-223-8210).

Zero Population Growth, 1400 Sixteenth Street NW, Suite 320, Washington, D.C. 20036 (202-332-2200).

Index